Nanomaterials in Structural Engineering

Nanomaterials in Structural Engineering

Aviraj Chandrahauns Chavan

ARCLER
PRESS

www.arclerpress.com

Nanomaterials in Structural Engineering

Aviraj Chandrahauns Chavan

Arcler Press
4164 Lakeshore Road
Burlington, ON L7L 1A4
Canada
www.arclerpress.com
Email: orders@arclereducation.com

ISBN: 978-1-77956-971-4 (Hardcover)

ABOUT THE AUTHOR

Aviraj Chandrahauns Chavan is a management professional with a strong foundation in operations, complemented by a deep passion for continuous learning and creative expression. He holds a Master of Management Studies (MMS) in Operations from Welingkar Institute of Management Studies, Mumbai, and a Bachelor's degree in Mechanical Engineering from Mumbai University. He is pursuing Ph. D in Management studies from Jamnalal Bajaj Institute of Management Studies, University of Mumbai.

With hands-on experience at organizations such as MAHYCO and Radhakrishna Foodland Pvt. Ltd., Aviraj has worked extensively across demand forecasting, supply planning, material resource planning, and inventory optimization. In his role as Senior Executive – Supply Chain Planning, he was responsible for managing critical supply and inventory operations, including forecasting and planning. His work has involved employing efficient supply models, handling transport, and warehouse management projects to reduce lead times and improve serviceability. His ability to blend data-driven strategies with real-time operational insights has consistently delivered measurable improvements in supply chain performance. He is known for his analytical mindset, adaptability, and a clear communication style.

TABLE OF CONTENTS

LIST OF FIGURES

PREFACE

The construction sector has invariably been the epicenter of human progress, dictating how societies live, work, and interact with the built surroundings. With escalating needs for stronger, lighter, and more sustainable materials, scientists and engineers are searching towards new frontiers in material science. Of the new frontiers, nanoscale science has revolutionized untold possibilities by allowing for control of matter at the atomic and molecular levels. The nanoscale-specific properties of higher mechanical strength, improved thermal stability, improved chemical reactivity, and well-managed electrical behavior are being used to transform structural materials. The revolution is not only transforming the boundaries of performance but also upsetting traditional means of design, durability, and sustainability.

The use of nanomaterials in civil and structural engineering is no longer a concept of the future but a daily reality with growing practical uses. Nanoparticles, nanotubes, and nanoscale additives are being blended into cement, steel, polymers, and composites in order to make impossible feats possible. Buildings are now possible to construct using materials that resist cracking, extend service life, respond to environmental stimuli, or even self-heal. These technologies enable global goals of reducing the use of energy, lessening carbon emissions, and safer infrastructure. By bridging the domain of material science with the application of engineering, nanomaterials are pushing the envelope on how structures, bridges, and infrastructure systems will be conceived and constructed in the future.

Nanomaterials in Structural Engineering is written to provide a complete understanding of where nanoscale science intersects with the practice of structural engineering. The book combines theoretical discussions and practical implementations to provide the reader with an equally balanced view of fundamental principles, emerging technologies, and realistic applications. Chapters have been organized to move from fundamental knowledge to specific materials, manufacturing methods, performance assessments, and future developments.

Chapter 1 provides an overview of nanomaterials and the size-dependent properties, the history of their development, and their significance to engineering. It is foundational in describing why nanoscale gives rise to new behaviors over bulk materials, preparing readers for understanding the following chapters.

Chapter 2 addresses the nanoscience and nanotechnology principles. It introduces the scientific principles that rule nanoscale phenomena, including surface effects, quantum behavior, and scale-dependent interactions. The chapter also provides an overview of measurement tools and analysis techniques employed by scientists to quantify nanoscale characteristics.

Chapter 3 addresses the types of nanomaterials that can be used for structural engineering. It describes nanoparticles, nanofibers, nanotubes, nanoclays, and nanostructured coatings and their characteristics and respective fitness for different uses in engineering. Emphasis is placed on the contributions of shape transformation, composition, and surface chemistry to defining the performance.

Chapter 4 presents the nanomaterial fabrication and synthesis techniques. It outlines top-down and bottom-up approaches like chemical vapor deposition, sol-gel processing, electrospinning, and mechanical milling. Scalability, cost-effectiveness, and incorporation into construction-grade materials are of utmost interest.

Chapter 5 addresses nanomaterial application in cement and concrete, two of the most widely used construction materials. It elaborates on how hydration, strength, and porosity control are enhanced by nanosilica, carbon nanotubes, and other additives. The chapter also addresses the effects on long-term durability and sustainability due to these modifications.

Chapter 6 talks of nanomaterials in steel and alloys. It describes the way surface modifications and nanoscale reinforcements all increase fatigue life, hardness, and corrosion resistance. The lecture entails advancements in nanostructured metals intended for structural properties, alloying techniques, and coating technologies.

Chapter 7 considers nanomaterials in composites and polymers. It explains the way nanofillers and nanofibers increase stiffness, impact strength, and thermal stability of polymer-based materials. Fiber-reinforced composites are covered with special focus due to increasing usage in structural applications ranging from lightweight bridges to aeronautical structures.

Chapter 8 is devoted to durability and sustainability of nanomaterials in construction building. It discusses how nanoscale modifications enhance the life span of structural components, resist environmental degradation, and make resource efficiency easier. Environmental health and safety concerns, recycling, and disposal operations are also taken into account.

Chapter 9 covers smart nanomaterials for structural engineering with examples of self-sensing, self-healing, and energy-harvesting materials. These advanced materials allow the structure to adapt to evolving conditions, track their own performance, and enable intelligent infrastructure systems.

Chapter 10 sees in the future the trends and applications of nanomaterials. It looks into the future for advances in nanomanufacturing, digital convergence, and innovative applications in sustainable infrastructure. The chapter provides a glimpse of ongoing research directions and potential restructuring of construction works in the coming decades.

Readership for the book is postgraduate and graduate students of civil engineering, construction technology, and materials science who desire to gain knowledge in advanced materials. Practicing engineers, architects, and construction professionals will also find the book handy in order to gain an insight into the applications of nanomaterials in practice. Researchers from both academia and industry will find the book useful as a reference to study applications, and policymakers and sustainability practitioners can utilize its results in making informed decisions for infrastructure development. The book aims to serve as a bridge between laboratory research and field engineering application with both fundamental principles and applied perspectives.

CHAPTER 1

INTRODUCTION TO NANOMATERIALS

Contents

INTRODUCTION

Definition of Nanomaterials

Nanomaterials are materials whose structure is at the nanoscale, typically between 1 and 100 nanometers. Such a specific scale possesses distinct physical and chemical properties that are far from being similar to those of bulk material. For instance, nanomaterials can possess higher strength, lower weight, increased chemical reactivity, and improved electrical conductivity. Such characteristics are a result of the high surface area to volume ratio and quantum effects that are salient at the nanolevel. Researchers have leveraged these characteristics across many areas, leading to ground-breaking applications in medicine, electronics, and environmental studies. An understanding of the general definition of nanomaterials provides the ground from which deeper research into its versatile capabilities and functions is initiated, particularly as advancements in nanotechnology remain to be made and cultivated (Joudeh et al., 2022). One of the surprising aspects of nanomaterials is their broad classification, ranging from nanoparticles to nanocomposites and nanostructured materials. Each category has various applications based on its individual characteristics. For example, nanoparticles, typically spherical in nature, can be designed to target cancer cells more effectively in the field of medicine, demonstrating their bioactive properties (Patra et al., 2018).

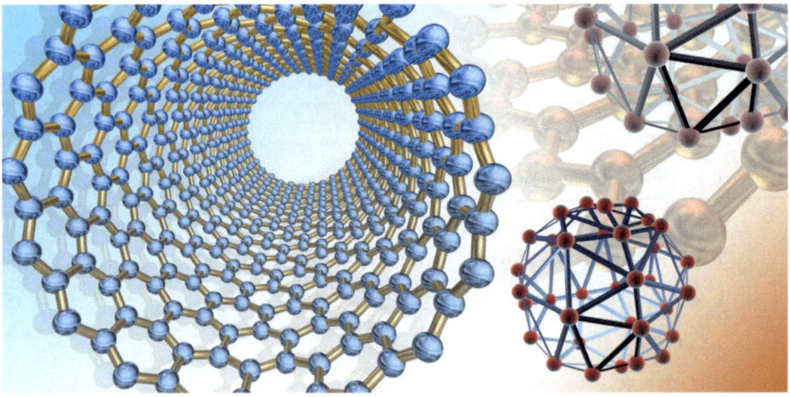

Figure 1.1. Definition of nanomaterials.

Source: https://www.niehs.nih.gov/sites/default/files/health/assets/images/nanomaterials_og.jpg.

Moreover, the physical and biological routes of synthesis also contribute to the diversity falling under this category. The biogenic route involving plant or microbial material is a green alternative compared to traditional synthetic pathways, its advantages including less cost and toxicity. As scientists strive to explore the characterization and application of nanomaterials, their definitions versus implementations prove more compatible. The scope of nanomaterials exceeds mere definitions, indicating their transformative potential in industries. For nanomedicine, the ability to target the delivery of therapeutic agents is an excellent example of how nanomaterials have the ability to revolutionize treatment platforms by making drugs more effective with lesser side effects (Patra et al., 2018). At the same time, in electronics, nanoscale-designed materials can enable the development of faster, more efficient, and smaller devices, serving as a testament to the technological upheaval brought on by nanotechnology. Generally, uses in environmental remediation and energy storage also speak to the versatility of nanomaterials. As nanotechnology grows by the day, the definition of nanomaterials and what their properties are will become the key to unlocking their potential for future inventions (Figure 1.1).

Explanation of Nanomaterials and Their Size Range

Nanomaterials are a separate group of materials that are categorized by size, typically between 1 and 100 nanometers. At this scale, intrinsic material properties have a tendency to change radically from bulk materials, making their chemical reactivity and physical properties change. This size-dependent behavior is mainly brought about by the large surface area-to-volume ratio, which enhances the performance of nanomaterials in applications. For instance, nanoparticles have been discovered to possess greater strength and lower weight, coupled with improved electrical properties, making them suitable for application in various technologies including electronics, medicine, and environmental application. It is critical to understand the intricate relationship between size and properties in order to gain maximum benefit from the potential of nanomaterials in new applications. Different types of nanomaterials possess varying properties based on their shape and size, which can be manipulated subtly during synthesis. For example, spherical ZnONPs prepared via biological methods have shown varied properties that can be used in antimicrobial activities and drug delivery systems. Spherical ZnONPs produced were reported to possess an average diameter of 7.2 nm in published reports, which is preferred to achieve maximum penetrating capacity for biological interactions. This size control is important since the specific size can have a great effect on how nanoparticles interact with biological systems and improve their site-specific delivery. The potential to control these parameters at the nanoscale acknowledges the enormous potential for advancements against antimicrobial resistance and the war against cancer and thus the importance of knowing nanomaterials and functionalities better. Aside from their structural nature, the ability to control and modify nanomaterials has opened up an universe of industrial applications. One example is metal oxide-based nanomaterials such as copper oxide (CuO) nanoparticles, which are identified for catalysis and optoelectronic uses. These recent results demonstrate that doping CuO with metals like nickel and manganese can induce dramatic optical and electronic behavior changes, demonstrating that minute changes at the nanolevel can lead to dramatic effects for application performance. As described, the introduction of dopants influences the crystallinity and particle size, thereby affecting the band-gap and the material conductivity. This illustrates not only the multiple functionality of nanomaterials but also the necessity for ongoing research to systematically maximize these materials for specific applications, thus the vital connection between composition, size, and functionality in the nanotechnology industry.

Distinction Between Nanomaterials and Bulk Materials

The distinction between nanomaterials and bulk materials is essentially driven by their size, which has a very significant influence on their physical and chemical attributes. Nanomaterials, in general, have particle sizes ranging from 1–100 nanometers. These materials exhibit certain distinctive properties that the bulk material does not possess. These properties are because of the higher surface-to-volume ratio with reducing particle size. For example, nanoparticles can have enhanced optical, electrical, and mechanical properties, which are of use in areas such as medicine, electronics, and pollution cleanup. By contrast, bulk materials have their properties largely determined by their macrostructure, which is less highly tunable than that of nanomaterials. The size at which a material is treated is thus central to determining its function and usefulness in technology and industry. In addition, the synthesis methods for nanomaterials are completely different from those employed for bulk materials, the foundation of yet another very important difference. Unlike typical quantitative methods for bulk materials that heavily rely on mechanical processes and macroscale chemistry, nanomaterials can be synthesized by quite diverse different methodology, from bottom-up (e.g., chemical vapor deposition) to top-down processing (e.g., laser ablation). In addition, recent green synthesis advances, which employ biological entities like plant extracts, have also emerged as alternative, yet effective, methods to produce metal and metal oxide nanoparticles, rendering them greener and more sustainable. These new approaches focus on the ability to engineer nanomaterials with specific properties for specific applications. In this case, the quality and reproducibility of the nanoscale material not only rely on the synthesis pathway but also on the intrinsic properties of the initial materials, which marks a shift from traditional routes towards more sustainable pathways in materials science (Shafey, 2020). Lastly, the potential applications of nanomaterials over bulk materials branch out to environmental and health safety issues. With the manufacture and use of nanomaterials increasing exponentially, their toxicity and environmental fate gain importance. In contrast to bulk materials, where due to their larger size some of the exposure hazards are reduced, the nanoscale configuration offers additional challenges, most notably because of their enhanced reactivity and penetration of biological membranes. A proper assessment of their risks, including their potential toxic effects on tissues and cells, is essential to regulatory frameworks that regulate their use and application. Furthermore, variations between nanoparticles that are engineered and naturally occurring need thorough consideration when dealing with risks, such that not just the characteristics of nanomaterials used for technological applications, but also, they are dealt with in an environmentally sustainable manner in our ecosystems (Jeevanandam et al., 2018) (Figure 1.2).

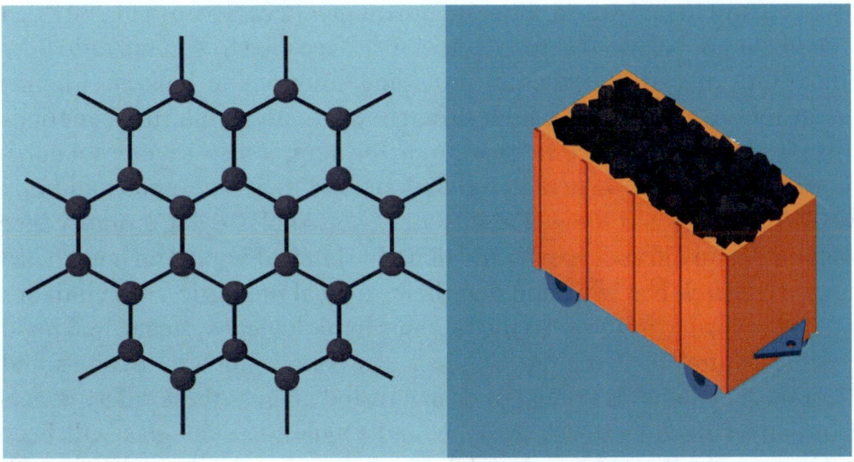

Figure 1.2. Distinction between nanomaterials and bulk materials.

Source: https://blog.polymernanocentrum.cz/content/images/2024/12/Raw-Material-Advantages-Bulk-vs-Nanoscale-PIC-A.jpg.

Importance of Surface Area to Volume Ratio in Nanomaterials

In nanomaterials, the importance of the surface area to volume ratio cannot be overstated. With decreasing particle sizes down to the nanoscale, their surface area is increased over their volume, greatly enhancing their reactivity and interaction with the environments around them. This larger surface area facilitates many chemical reactions, and hence nanomaterials are highly efficient in catalysis and drug delivery, for instance. The unusual physical and chemical behavior of such materials, due to the large surface area, allows them to adsorb more and be chemically active than their bulk forms. Hence, the understanding and manipulation of this ratio are significant to realize the full potential of nanomaterials in many technological as well as scientific applications, emphasizing the basis role of this property throughout the domain of nanotechnology (El-Kalliny et al., 2023). Moreover, the significance of a large surface area to volume ratio even extends to applications in environmental purposes, particularly in water treatment methods. Nanomaterials are becoming key drivers in the development of sophisticated filtration systems because they can efficiently adsorb contaminants and pollutants. They possess an enhanced surface area, which enables them to react with diverse toxic substances, including heavy metals and organic contaminants, making them a vital component of environmental cleanup initiatives. As placed into context in studies, synthesis and application of nano-based filtration membranes and adsorbents have been quite successful in real-world water treatment practices (Khan et al., 2019). This is a statement not only to the importance of surface area to volume ratio but also the flexibility of nanomaterials in providing solutions to bulk environmental issues at low expense and at high efficiency. Finally, while the benefits of a high surface area to volume ratio are transforming, it also becomes a concern, particularly concerning nanomaterial toxicity and their environmental impacts. Elevated reactivity and bioavailability of nanoparticles open serious concerns about their destiny in ecosystems and their detrimental impact on human health. Scientific research has indicated that their atypical properties, which their size confers, may lead to unforeseen interactions in biological systems, making their safety and control challenging to assess. As nanotechnology evolves and its applications expand, it is crucial to be aware and address these safety concerns. Preparation of integrated risk analysis and regulation strategies will become a requirement so that nanomaterials' advantages can be tapped into without undermining environmental integrity or public well-being (El-Kalliny et al., 2023). The need to take into account the surface area to volume ratio, thus, is not just an elementary factor for the use of nanomaterials but also a determining parameter in the ethical development of nanotechnology.

1.1. TYPES OF NANOMATERIALS

Nanomaterials consist of many different types; each differentiated according to their unique properties that give rise to their extensive applications. Nanoparticles, nanowires, nanorods, and nanosheets are some common building blocks of nanomaterials. Nanoparticles are either metallic, oxide, or polymeric in nature and possess a high surface area-to-volume ratio, making them more reactive and useful in drug delivery and catalysis. Nanorods and nanowires, often made of semiconductors, have exemplary electrical and optical characteristics that are of fundamental significance in electronics and photonics. Nanosheets, in contrast, mainly present in materials like graphene, have exceptional mechanical strength and elasticity. Variations in these nanomaterials indicate the tailor-made approaches scientists are using to advance functionality for specific applications, an indication of the groundbreaking prospects of nanotechnology in contemporary science (N/A, 2022). Among the many categories of nanomaterials, mesoporous silica nanoparticles (MSNs) are among the most leading contenders due to their flexibility and effectiveness in fields ranging from drug delivery to environmental remediation. The unique structural properties of MSNs, including adjustable pore diameters and broad surface area, enable prolonged and controlled delivery of therapeutic loads, with optimal drug efficacy and minimal side effects. Moreover, their ability to adsorb toxins makes MSNs potential for solving environmental pollution issues. In the agricultural industry, this is translated to providing efficient pesticide delivery systems, improve efficacy without boosting toxicity. The functional properties of MSNs thus demonstrate how certain nanomaterials can be utilized strategically as a way of solving some contemporary challenges of health and environmental sustainability (Xiang-Kong et al., 2021).

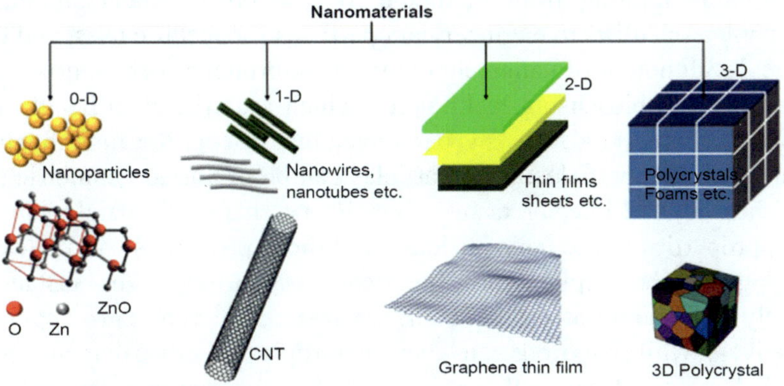

Figure 1.3. Types of nanomaterials.

Source: https://ion2.upm.edu.my/summer-uploads/20220923155810blobid0.jpg.

Exploring the varieties of nanomaterials extends beyond their structural composition to include their inherent applications in different fields, including medicine, electronics, and energy storage. For instance, carbon nanomaterials, including carbon nanotubes and graphene, have gained immense interest for the potential to enhance battery and supercapacitor performances. Their exceptional electrical and mechanical properties result in energy storage technology advances, enabling devices to function more efficiently. Nanomaterials are also significant when used to enhance the photovoltaic cell's performance, subsequently resulting in advances in renewable energy. As the research goes on to detect and fine-tune such diverse nanomaterials, nanotechnology's interdisciplinary nature allows the filling of gaps in various scientific fields, hence forming novel solutions and improved material for future applications (N/A, 2022) (Figure 1.3).

1.1.1. Nanoparticles and Their Classifications

Nanoparticles are classified based on several parameters, including their size, shape, composition, and the specific property exhibited by them. Size is likely the most fundamental categorization threshold, and nanoparticles are commonly loosely defined as materials whose dimensions are in the range 1–100 nanometers. At such dimensions, material properties are quirky and distinctly different from their bulk counterparts. Size-dependent properties include higher reactivity, altered optical properties, and engineered electrical performance, such that nanoparticles are ideal candidates for a wide array of applications across sectors ranging from medicine to electronics. Understanding these types enables scientists to engineer nanoparticles for specific uses, and this has seen the development of nanotechnology breakthroughs such as drug delivery mechanisms and biosensing technology, which are critical to healthcare and diagnostics. Shape is also a key parameter that affects the functionality and usefulness of nanoparticles. Nanoparticles can be of varied geometric shapes, from spherical, rod-like, or even cubic. Every shape of particle can impart varying properties to the nanoparticles, and those properties can be leveraged depending on the application. Spherical nanoparticles, for instance, are extremely stable and possess high surface area and therefore are applicable in drug delivery, while nanorods can have anisotropic properties that enhance their optical behavior and render them good for photothermal therapy applications. Besides, certain geometries can be used to maximize nanoparticle-biological system interaction to facilitate improved diagnostic uses (Naresh et al., 2021). With continued research in these aspects, particle shape control can result in the development of targeted diagnostics and therapy with significant advantage over conventional methods. Another critical nanoparticle categorization is based on their composition, where metals, oxides, or organic compounds are part of the composition. Metal nanoparticles such as silver and gold are studied widely due to their beneficial optical properties, stability, and biocompatibility. These are widely used in biological systems, where their properties can be exploited to enhance imaging techniques or utilized as drugs. Metal oxide nanoparticles such as titanium dioxide and zinc oxide, on the other hand, are essential in photocatalysis and as components of various electronic devices. Organic nanoparticles, including dendrimers and liposomes, are particularly significant for drug delivery and gene therapy applications and have the capability to encapsulate and deliver therapeutic entities efficiently. By employing these diverse compositions, researchers can tailor nanoparticles specifically for use, thereby driving the nanomaterials discipline (Yuan et al., 2023) (Figure 1.4).

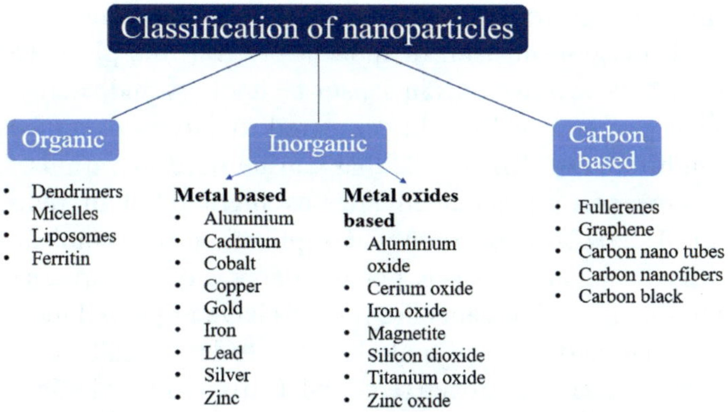

Figure 1.4. Nanoparticles and their classifications.

Source: https://www.researchgate.net/profile/Pragya-Mehta/publication/370871974/figure/fig1/AS:1143128116004383 7@1684585881374/Classification-of-nanoparticles.png.

1.1.2. Nanotubes and Their Unique Properties

The distinct characteristics of nanotubes have made them central players in the world of nanomaterials. Carbon nanotubes (CNTs) with their cylindrical nanostructures offer high mechanical strength, electrical conductivity, and thermal stability. These are because of their graphene-like structure, where carbon atoms are aligned in a hexagonal lattice. Direction and chirality of the CNTs control their electrical properties, enabling them to function as either metals or semiconductors, depending on their specific orientation. This versatility has rendered CNTs attractive for numerous applications, ranging from prospective electronics to reenforcing agents in composite materials. Furthermore, their lower weight with higher strength enables one to obtain spectacular improvements in the performance of materials employed in aerospace and automotive applications, underlining their vast versatility in revolutionizing material science and engineering fields. Besides carbon nanotubes, boron nitride nanotubes (BNNTs) represent another form of re-emerging nanomaterial endowed with distinct benefits. Although BNNTs have structural similarities with their carbon-based equivalents, they possess distinct properties that greatly enhance their functionality. Importantly, BNNTs prove to be highly thermally stable and have great electrical insulating properties, which qualify them for use in applications involving thermal management of electronic components and devices. Their optical transparency makes them even more attractive since they can be used in optoelectronic devices without interfering with the transmission of light. Moreover, the piezoelectric nature of BNNTs opens up new avenues in sensor technology and energy harvesting. This synergetic property of BNNTs compared to carbon nanotubes is crucial in creating multifunctional materials that can be used in various fields, from electronics, aerospace, to nanomedicine, as highlighted in recent research (Jakubinek et al., 2022). Utilization and application of nanotubes are rapidly evolving, with researchers intensely pursuing new ways to harness their unique properties. Ongoing developments in synthesis have improved the accessibility and purity of CNTs and BNNTs, and obstacles to large-scale applications and studies of these materials have been removed. Current trends indicate a growing interest in functionalizing the nanotubes for further enhancing their properties, resulting in the creation of composite materials with the best attributes of both CNTs and BNNTs and hence paving the way for novel applications. As described in recent literature, for example, novel use of carbon nanotubes as nanoreactors for fullerene reactions, there exists vast potential for both types of nanotubes that continues to grow (Paukov et al., 2023). By leveraging the multifunctionality and new properties of nanotubes, scientists can create better materials that will redefine performance standards across many industries.

1.1.3. Nanocomposites and Their Applications

Study of the recent field of nanocomposites includes understanding their new properties, which are due to the inclusion of nanoscale material in polymers. These composites leverage nanoscale reinforcement properties with significant enhancement in mechanical and thermal performance. The best example of this is the addition of montmorillonite (MMT) nanoclay, which has been shown to increase strength, modulus of elasticity, and stiffness of composite structures. The intrinsic property of MMT, such as high aspect ratio and hydrophilicity, provides enhanced dispersion in water-soluble polymers, leading to higher intercalation and exfoliation efficiency in the composite matrix. The result is a material class with improved properties, and MMT-based polymer nanocomposites are particularly well-suited for structural applications in industries like automotive and construction, which creates opportunities for more potential for the creation of new products (Alias et al., 2021). Aside from MMT, other renewable resources like chitin and chitosan are also of interest due to their biocompatibility and biodegradability, which are highly significant in the context of sustainable materials these days. Such polysaccharides are synergistically blended with carbon nanomaterials (CNMs) to give bio-nanocomposites with enhanced electrical and mechanical properties. The synthesis of such graphene-based nanocomposites takes advantage of organic as well as inorganic components, addressing the need for not only efficient but also environmental-friendly material.

Their applications span energy-related domains, including the fabrication of biosensors, batteries, and fuel cells, to illustrate the ways in which natural polymers of the past could be designed to produce high-performance materials for today's problems (Ikram et al., 2021). The disparate applications of nanocomposites reflect their multifunctionality and strategic significance in material science development. From automotive to biomedical engineering, hybrids of this sort can radically alter product performance through enhanced durability, efficiency, and beauty. Novel approaches to integrating nanomaterials are regularly being found, with ongoing studies into balancing performance and cost, and sustainability. As industries continue to focus more on environmental-friendly options, nanocomposites from renewable sources are poised to be at the center of developments to improve existing uses as well as produce new ones to meet future demands. Therefore, ongoing studies and reviews are required for further modification of their properties and extending their areas of applications, forming the basis of technological progress in various areas (Alias et al., 2021; Ikram et al., 2021) (Figure 1.5).

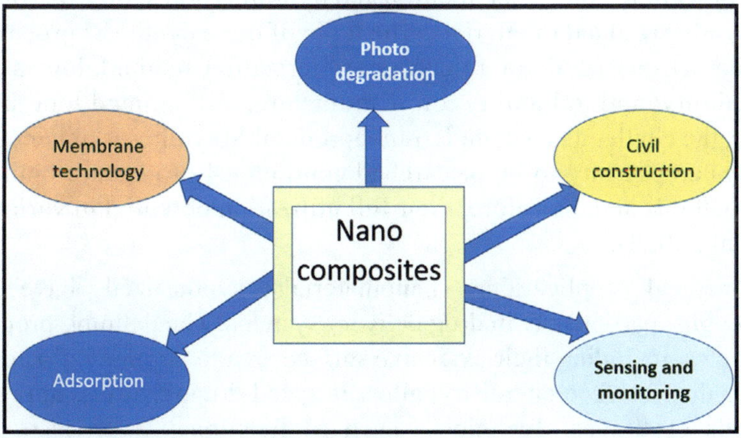

Figure 1.5. Nanocomposites and their applications.

Source: https://www.researchgate.net/publication/371282215/figure/fig2/AS:11431281168703535@1687093024083/Nanocomposites-and-their-applications-55.png.

1.2. SYNTHESIS METHODS

The synthesis methods of nanomaterials are a cornerstone foundation for the field of nanotechnology that is largely characterized by two main approaches: top-down and bottom-up. Top-down is the approach of breaking down bulk materials to particles in the nanoscale level, by methods such as milling, lithography, and etching. Top-down offers highly controlled sizes and properties of the resulting nanomaterials but is generally coupled with limitations of scalability and loss of material. In comparison, the bottom-up methodology constructs nanomaterials from molecular or atomic building blocks through synthesis strategies such as sol-gel synthesis or chemical vapor deposition. Each strategy carries its own advantages and disadvantages that influence the final product resulting from it as well as its application in many fields of research, from electronics to drug delivery and environmental remediation. Understanding these prevalent synthesis methods is important in formulating research and new applications for nanotechnology. Method choice in nanomaterial synthesis is important in determining the physicochemical attributes of the final products.

For instance, optical properties, structural integrity, and surface functionality for nanoparticles can be designed based on synthesis conditions. The final properties of the nanoparticles are influenced by precursor materials, reaction time, and temperature. Recent advances have made it possible to enhance procedures like laser ablation and hydrothermal synthesis with higher quality synthesized nanomaterials. The topic of these synthesis processes also shows the serious need for robust characterization methodologies towards reproducibility and reliability of manufacture. As pointed out in recent literature, the challenges that underpin reproducible syntheses of nanoparticles are always a necessity to be researched continuously to ensure optimization of such methods and, therefore, their full utilization potential in various fields (Patle et al., 2024).

In medical applications, nanomaterials synthesized have led to breakthroughs, particularly in drug delivery systems. The defining properties of nanoparticles, including their extensive surface area to volume ratio and ability to modify their surface chemistry, allow targeted drugs that can appropriately target cells or tissues. Techniques such as functionalization also optimize the utility of these nanocarriers, allowing appropriate drug targeting and mechanisms of drug release (Dr Andal et al., 2024). The different synthesis methods provide a versatile platform for designing nanomaterials for different complex therapeutic needs. As researchers continue to study various synthesis techniques, the developments of nanotechnology will transform traditional medical practices, producing improved patient results and more effective treatments. The integration of synthesis techniques and utilization opportunities in the pharmaceutical industry highlights the pivotal role of nanomaterials in advancing existing healthcare (Figure 1.6).

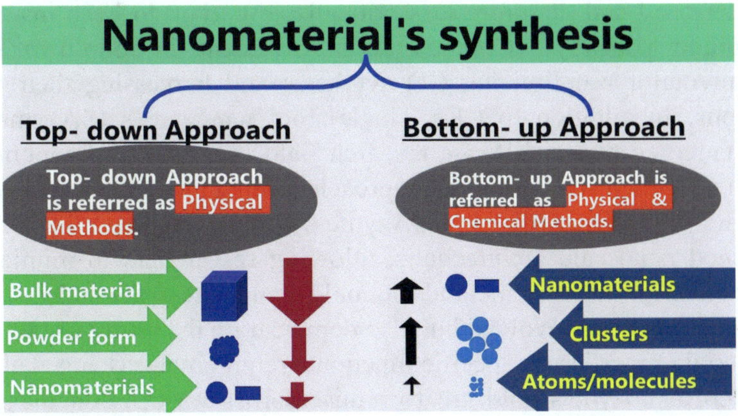

Figure 1.6. Synthesis of nanomaterials.

Source: https://i.ytimg.com/vi/YhuUFLzJSsg/maxresdefault.jpg.

1.2.1. Top-Down Approaches in Nanomaterial Synthesis

The top-down process for nanomaterial synthesis is the fabrication of nanostructures from bulk precursor materials through methods such as milling, etching, and lithography. It is characterized by the ease it gives in creating intricate geometries and confined sizes, necessary for some applications. For instance, through precision machining and photolithography, scientists can create nanoscale devices with high precision, useful for electronics and photonics. This method is particularly valuable as it is possible to scale it up and apply with existing manufacturing processes. The disadvantage of nanoscale uniformity and potential for defecting materials ensures that refinement of these methods has to be continuous. As the demand for designed nanomaterials in a wide variety of applications continues to grow, the top-down method remains an underpinning research and industrial application technique in nanotechnology. Carbon dots (CDs) are the best example of the promise offered by top-down synthesis procedures when it comes to working with nanomaterials. Researchers have been working with the development of CDs through processes such as laser ablation and electrochemical oxidation, which facilitate effective disintegration of carbon sources into materials at the nanoscale. CDs have been prominently highlighted due to their exceptional optical properties and are consequently of use in bioimaging, sensing, and optoelectronic applications (Alafeef et al., 2023). The advantages of using top-down approaches to CDs are that they are capable of yielding high-quality material with tunable luminescent properties. Nevertheless, fluorescence mechanisms underlying CDs are not yet fully understood and therefore it is critically important to keep investigating the synthetic pathways. Continuous improvement in top-down methods will be pivotal for optimizing CD synthesis and harnessing their versatile applications. In addition to CDs, single-atom nanozymes (SAzymes) have recently emerged as a significant research field in the field of nanomaterials, with both top-down and bottom-up approaches being implemented in synthesis. Top-down approaches utilized in SAzyme synthesis facilitate the creation of well-defined nanoscale architectures, allowing researchers to mimic natural enzymes effectively. These methods not only contribute to the understanding of the catalytic reactions involved but also demonstrate the utility of tailoring the nanomaterial properties for specific functional requirements (Peng et al., 2023). Single-atom nanozymes hold great promise across various domains, ranging from biosensing to therapy, as their atomic-level precision offers advantages over conventional enzymes. The continued refinement of top-down synthesis techniques will be the foundation for continued advancements in designing and implementing SAzymes, illustrating the versatility and revolutionary nature of nanomaterial innovations.

1.2.2. Bottom-up Approaches in Nanomaterial Synthesis

Bottom-up approaches are a central element in nanomaterial synthesis due to their ability to build materials from atomic or molecular building blocks. This strategy is completely different from the top-down strategies that most likely consist of subtractive techniques creating defects and limiting control over the resulting structure. The bottom-up process allows the nanomaterials to be controlled precisely, leading to developing structures with precise properties tailored for specific applications. Nanostructures produced with the help of chemical vapor deposition or sol-gel techniques can, for instance, provide materials with enhanced electrical, thermal, and optical properties. Consequently, the above approaches hold great promises for the design of new nanomaterials with enhanced functionalities towards electronics, catalysis, and medicine. One of the most striking applications of bottom-up synthesized nanomaterials pertains to the area of medical therapeutics in general, and that using the case of silver nanoparticles (AgNPs). As indicated in recent research work, AgNPs have been found to exhibit fascinating antibacterial action, effective against a wide range of pathogens, including drug-resistant microorganisms. Their efficacy is founded on multiple mechanisms of action, which not only render them powerful antimicrobial agents but also perfect candidates for application as medicament ingredients with infection preventative purposes. Bottom-up synthesis of AgNPs enables the manipulation of their shape and size, factors which are crucial determinants of their biological activity. Thus, the ability to control the synthesis process in a precise manner is directly proportional to the effectiveness of AgNPs in pharmaceutical uses, addressing the urgent demand for new antibacterial drugs in medicine. Moreover, the merging of nanomaterials with biosensing technologies represents a revolutionary step ahead for diagnosis and monitoring. By using bottom-up methods, one can plan various nanostructures such as nanoparticles, nanowires, and carbon nanotubes for the optimization of biosensors. These nanomaterials possess good surface-to-volume ratios and high-quality electrical properties that allow for high sensitivity and rapid response in biological signal sensing. Proper application of such materials allows for the construction of biosensors with extremely low detection levels, which are found to be highly valuable in health care, environmental monitoring, and quality control in the food industry. By utilizing bottom-up approaches to the synthesis of nanomaterials, researchers are capable of creating innovative biosensors that can meet contemporary needs in biological detection and diagnostics, showcasing the ability of nanotechnology to propel other scientific disciplines.

1.2.3. Green Synthesis Methods and Their Benefits

Green synthesis methods for nanomaterials are a revolution in material science, where eco-friendly processes receive more emphasis compared to traditional chemical approaches. Such methods utilize natural resources, i.e., plant extracts, microbes, or fungi, to facilitate the production of nanomaterials with low environmental impact. An excellent example is the production of zinc oxide (ZnO) nanoparticles, which have gained prominence due to their excellent semiconducting, piezoelectric, and pyroelectric nature. Not only do these features improve the function of the nanomaterials but also contribute to them being produced in an eco-friendly manner. The development towards green synthesis approaches in nanotechnology is a great indication of how scientific innovation can be aligned with environmental preservation, thus opening the way for broader application of nanomaterials in numerous industries, from electronics to medicine, thus acting as a precedent for the formation of material production in the future (Sharma et al., 2022). The benefits of green synthesis approaches extend beyond their environmental advantages to encompass enhanced properties of the product nanomaterials. For instance, green synthesis approaches have the capability to yield nanoparticles that are highly pure, low in toxicity, and functionalized in a controlled fashion. In particular, nanocellulose (NC) derived from renewable cellulose resources has emerged as a paradigm case of a high-performance green material. NC exhibits excellent mechanical performance and biodegradability and can be applied in diverse applications like composites, adhesives, and biomedical devices. Its chemical adaptability inherent in its dense hydroxyl functional groups also facilitates its being subjected to varied modifications that enhance its functionality. While scientists are still exploring the possibility of NC, they are highlighting the many advantages that green synthesis can offer, making it a leading method in creating green nanomaterials (Trache et al., 2020). In addition, scaling up green synthesis protocols can radically change industrial nanotechnology protocols and drive the transition towards greener manufacturing models. Employing environmentally benign materials and biological-process–oriented protocols not only produces less waste but also mitigates the use of poisonous chemicals that traditional synthesis methods typically rely on. Moreover, green synthesis protocols are cost-effective, which becomes an attractive option for manufacturers seeking cheaper protocols without compromising product quality. As the need for ecologically sustainable materials will only grow, so will the spread of such processes, paving the way for more development and research. The future of nanomaterials rests in innovative practice like green synthesis, balancing the drive for innovative material characteristics with the requirements for environmental sustainability and bringing material science into a new era (Sharma et al., 2022; Trache et al., 2020) (Figure 1.7).

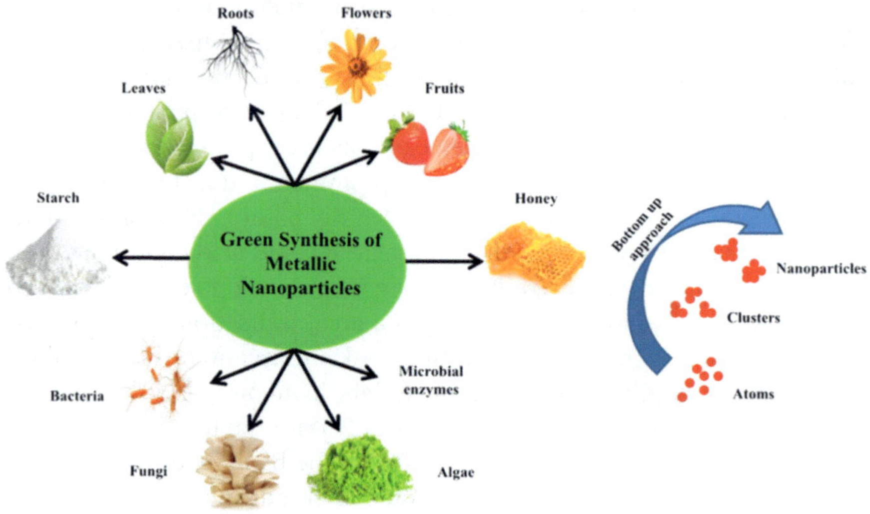

Figure 1.7. Green synthesis methods.

Source: https://www.researchgate.net/publication/340667708/figure/fig1/AS:11431281213352474@1702989563584/Different-types-of-green-synthesis-used-for-the-preparation-of-metal-nanoparticles.tif.

1.3. CHARACTERIZATION TECHNIQUES

Characterization of nanomaterials is the foundation of how to comprehend their unique properties and applications. Over time, various different techniques have been developed, each addressing specific aspects of materials at the nanoscale. Thermal lens spectrometry (TLS) is one of the methods that find significant application in the detection and analysis of compounds in liquid samples, particularly in nanomaterials research. TLS allows scientists to quantify the thermal diffusivity of materials, an important property that reflects their performance in applications such as catalysis and environmental remediation. By the potential to examine molecular dynamics, TLS supports the development of more efficient photocatalytic materials, thus fostering efficiency in nanobiophotonics. As a multifunctional tool, it becomes invaluable both in the lab and field applications, giving information on the intricate nature of various nanostructured materials.

Photoacoustic spectroscopy (PAS) contributes to the set of characterization techniques as well, providing beneficial information regarding the elastic and thermal properties of nanomaterials. PAS can be used to measure materials' reflection coefficient and thermal diffusivity, for example, that of Fe_2O_3 polymorphs, detailing the potency of techniques in the solution of material characterization challenges. PAS has also been applied to unveil ZnO nanoparticles' intense absorption properties upon deposition on diatomite substrates. The ability of this technique to overcome working frequency limitations is another aspect that contributes to its flexibility, particularly in silicon-based nanomaterial studies. The inclusion of PAS in the characterization setup gives emphasis to its significance, bridging gaps in insights into complex material interactions and setting the stage for nanotechnology breakthroughs. Beyond spectroscopic methods, experiential experiments enrich the learning process of undergraduate materials science students. For instance, a lab experiment in magnetic material synthesis and characterization allows students to directly interact with key techniques.

With the use of sol-gel combustion synthesis, students prepare metal ferrite spinels and investigate their magnetic characteristics using a magnetic susceptibility balance. Students gain key skills in interpreting results and applying characterization methods such as X-ray diffraction (XRD) and transmission electron microscopy (TEM) during this process. This approach not only emphasizes the interdisciplinarity of the field of materials science and chemistry but also provides the students with a hands-on experience of the complexity of research in nanomaterials. Through experiential learning, the students are thus able to appreciate the depth of knowledge required in the field of nanotechnology (Figure 1.8).

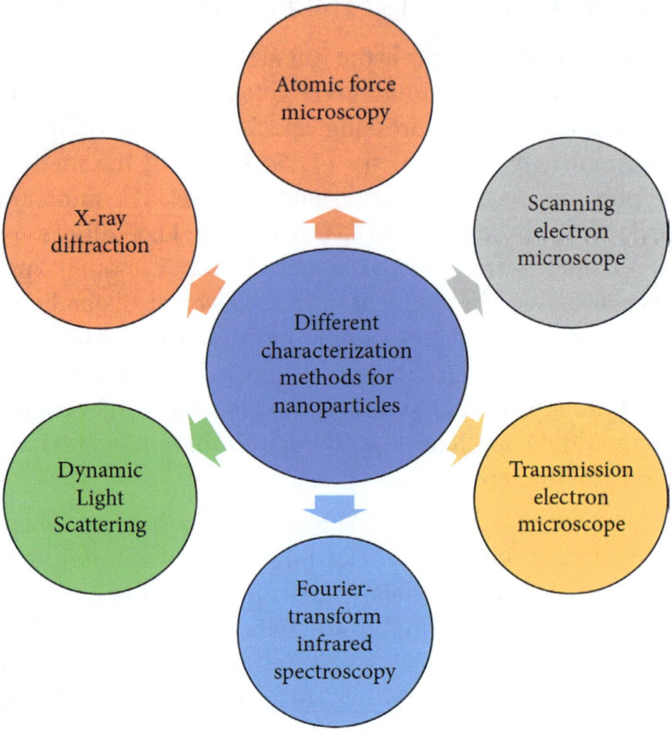

Figure 1.8. The characterization of nanomaterials.

Source: https://www.researchgate.net/publication/357385423/figure/fig3/AS:1108525770117180@1641304547697/Different-characterization-methods-for-nanoparticles.jpg.

1.3.1. Scanning Electron Microscopy (SEM)

SEM plays a focal role in the characterization of nanomaterials, providing unparalleled information about the microstructure and surface morphology of nanomaterials. SEM is distinct from standard optical microscopy in that it has the capability to achieve much higher resolution with focused electron beams. With this, researchers are able to observe materials at the nanolevel and gain valuable information on the structural features of different nanomaterials. For instance, SEM is at the forefront of making measurements of the uniformity and quality of nanostructures that are critical to the optimization of material function and synthesis in applications ranging from nanoelectronics to high-performance sensors. As such, the flexibility of SEM makes it an indispensable tool in the fast-evolving discipline of nanotechnology allowing for the accuracy of analysis needed for the expansion of both scientific understanding and utilitarian applications of nanostructured materials. The integration of SEM with advanced techniques makes it possible to conduct more through studies of nanomaterials. For example, a dedicated nanomanipulation system within an SEM system has facilitated multi-physical testing of nanostructures by combining mechanical, electrical, and photonic testing functions. This technology makes real-time testing of how various stimuli affect individual properties of nanostructures possible. In particular, novel studies of InGaN/GaN nanowires have revealed critical interactions between mechanical compression and optoelectronic properties, demonstrating the new insights that an integrated SEM approach can offer. Such integrations pave the way to stronger nanoparticle characterization methods, which are critical for guiding synthesis and enhancing the performance of nanomaterials in functional devices, including those in optoelectronics and nanophotonics (Qu et al., 2023). For nanomedical uses, SEM has also been instrumental in material characterization for biomedical uses, such as wound healing medications. Recently, utilizing SEM, scientists have been able to prove the efficacy of fish scale-derived collagen nanoparticles, which were structurally intact and biocompatible. The morphological data derived using SEM not only confirms the surface rough features of such nanoparticles, which have applications in drug delivery, but also confirms the successful formulation of collagen peptide structures comparable to naturally occurring kinds. The ability to analyze such nanoscale features provides a deeper understanding of the role played by the nanoparticles in speeding up wound healing processes, compared to traditional treatment. This capacity to create detailed visualizations underscores the significance of SEM in driving the application of nanomaterials within healthcare settings, ensuring innovations that have the potential to enhance patient outcomes and reduce healthcare costs (Shalaby et al., 2023).

1.3.2. Transmission Electron Microscopy (TEM)

TEM is a highly significant tool for characterizing nanomaterials, revealing intricate details regarding their structural and compositional features. Compared to conventional light microscopy, TEM utilizes high-energy electrons that allow much higher resolution imaging, often at the atomic scale. Such capability renders TEM an important instrument in nanotechnology in which the morphology and crystallography of materials must be characterized. Through TEM, researchers can observe features like grain boundary, defects, and size of crystallites that are associated with the functionality and performance of nanomaterials. The huge volume of information obtained by TEM research directs the synthesis processes and potential applications of such materials, particularly in those areas where material properties are intrinsically linked with their nanoscale architecture. Additionally, TEM characterization of nanostructured materials provides information on how these materials interface with biological systems, a point that is becoming increasingly significant in nanomedicine. Techniques such as electron diffraction and energy-dispersive X-ray spectroscopy (EDS) complement imaging by providing compositional and structural information that assists in distinguishing how nanomaterials can be designed for specific biomedical applications.

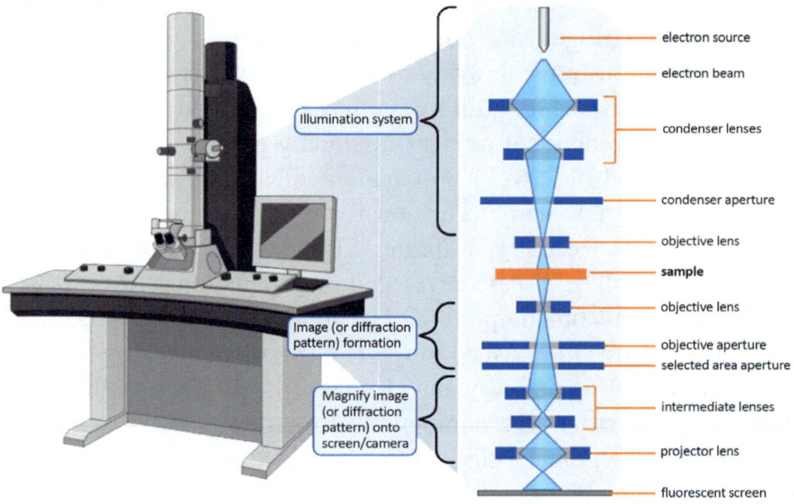

Figure 1.9. Transmission electron microscopy (TEM).

Source: https://www.nanoscience.com/wp-content/uploads/2023/01/TEM-diagram-transparent.png.

For instance, as emphasized by recent research, characterization of copper oxide nanorods synthesized via green pathways revealed their potential as future therapeutics in combating multi-drug-resistant pathogens. TEM analysis revealed their distinct morphology and allowed scientists to correlate the structural attributes with biologically relevant properties, thus emphasizing the role of advanced characterization equipment in the evaluation of nanomaterials for biomedical applications (Qamar et al., 2020). Additionally, the inherent problems of nanomaterial characterization, such as sample preparation and analysis, cannot be overlooked. It is essential to prepare samples properly for TEM since nanomaterials are often brittle and can alter when subjected to the electron beam. Contrast and resolution issues are greatly dependent on sample thickness and the material's natural traits. As explained in the literature, TEM cutting-edge methods are needed in an effort to overcome such challenges, thus enabling proper characterization of novel nanostructures like those used in biomedical applications (Jagadeesh et al., 2023). Integration of TEM with other characterization methods enhances our understanding of nanomaterials towards laying the foundation for new applications in various industries, which affirms the importance of TEM in advancing the science of nanotechnology (Figure 1.9).

1.3.3. Atomic Force Microscopy (AFM)

In the world of nanotechnology, AFM is an essential instrument for nanoscale material characterization. In contrast to traditional microscopy methods that are light-dependent, AFM utilizes a mechanical probe to scan the surface of an example and provide topographical information with tremendous resolution. This technique allows researchers to visualize and manipulate surfaces at the atomic scale and achieve deeper insights into the physical and chemical behavior of nanomaterials. As the AFM can quantify forces in various modes such as contact and tapping, it is particularly useful in the study of the mechanical properties of nanoparticles and their behavior under various environments. AFM's flexibility and precision have made it a key instrument in examining and adjusting nanomaterials in most scientific disciplines, ranging from materials science to biology. The ability of AFM to take measurements *in situ* and under ambient conditions is one of its key advantages and significant in conducting research on nanomaterials. This feature allows investigators to analyze dynamic processes and phenomena in materials under the influences of temperature, humidity, or chemical environments. AFM can acquire *in situ* information on surface morphology and changes, essential in the characterization of material behaviors at the nanoscale. Moreover, combining AFM with other experimental techniques enhances its capability and provides a vast amount of information on nanoparticle properties. Such interdiscipline approaches are essential for material optimization for certain applications because they provide knowledge about how material properties can influence performance. AFM is thus a significant step in material characterization because it opens doors to groundbreaking nanomaterial applications in technology and medicine (Mourdikoudis et al., 2018). In addition, AFM is used actively to investigate electrocatalytic behaviors of nanomaterials, particularly for energy conversion reactions such as the oxygen-reduction reaction (ORR) and oxygen-evolution reaction (OER). Surface properties and defect structures of electrocatalysts are crucial to their design and improvement in devices such as metal–air batteries. Research has further indicated that properties such as surface area, well-defined facet structures, and in-built defects perform fundamental roles in electrocatalytic performance (Yan et al., 2017). By using AFM, researchers can effectively map and quantify these surface properties and, therefore, provide evidence of the impact of defect engineering on electrocatalytic performance. This appreciation can inform the logical synthesis of more effective electrocatalysts, emphasizing AFMs important role in advancing the art of nanomaterials and enabling progress toward sustainable energy technology (Figure 1.10).

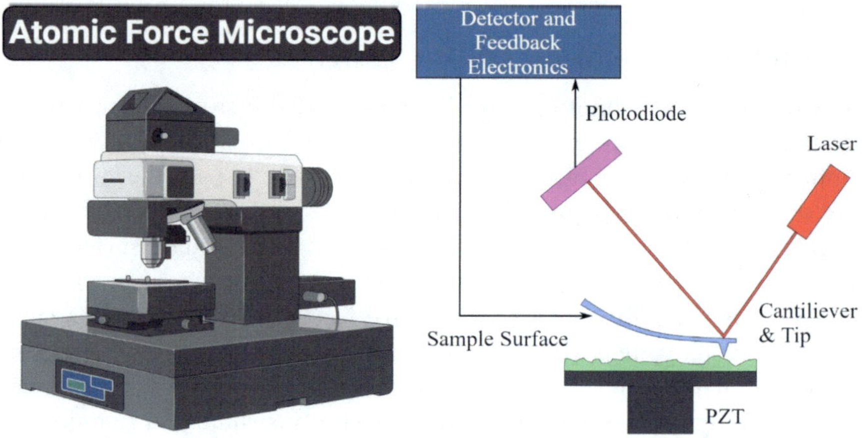

Figure 1.10. Atomic force microscopy (AFM).

Source: https://microbenotes.com/wp-content/uploads/2020/03/Atomic-Force-Microscope-AFM.jpeg.

1.4. PROPERTIES OF NANOMATERIALS

Nanomaterials properties distinguish them from their bulk counterparts primarily due to their unique size and surface-to-volume ratio. Materials at the nanoscale exhibit new physical and chemical behaviors, with enhanced reactivity, strength, and conductivity. These are due to quantum mechanical effects, which occur when materials are reduced in size to levels comparable to the atom and molecule scale. Nanoparticles, say, may have increased optical capabilities that would find use in anything from electronic systems to medicine. While researchers chase such opportunities, it is essential to develop a clearer picture of how changing the nanoscale affects the performance of materials. The ongoing advancement of nanotechnology reveals how such transformed properties can lead to new solutions for energy storage, drug delivery, and pollution clean-up, thus making the exploration of such materials timely and important to technology and scientific progression. Functionally, nanomaterials also enhance anti-corrosion properties in application, thus extending the life of metal components. Polymer nanocomposite coating technologies are gaining momentum with their excellent barrier properties and ease of modification. In addition, the addition of nanomaterials to the polymer matrices has been shown to greatly improve the anticorrosive performance of these coatings as a cost-effective solution to the eternal problem of metal corrosion. Emerging advances in the technology underscore the fact that nanomaterials not only provide physical protection but also chemical resistance through their unique interactions with corrosives.

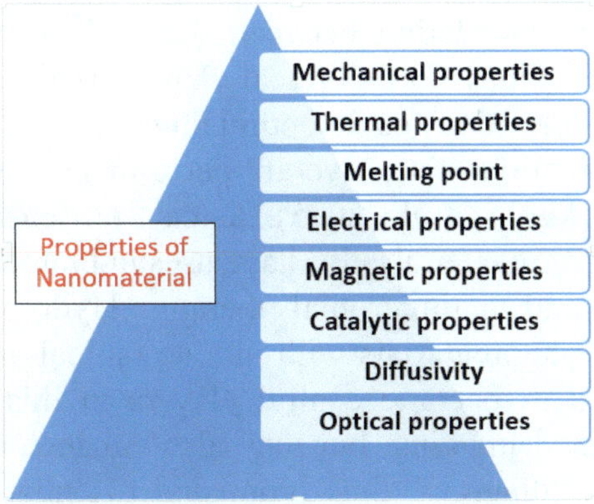

Figure 1.11. Properties of nanomaterials.

Source: https://chemistnotes.com/wp-content/uploads/2023/01/properties-of-nano-materials.png.

The advances highlight the groundbreaking use of nanotechnology in material sciences with nanomaterials playing central roles as essential ingredients in new solutions for traditional durability issues (Peng et al., 2020). The use of nanomaterials also extends to biomedical applications, as piezoelectric nano-biomaterials promise much. They are materials that generate an electric charge from mechanical stress, and they increasingly have more significant roles to play in drug delivery systems and tissue engineering. Their multifaceted interaction with biological systems is due to both their nanoscale nature and inherent properties, which allow for targeted therapeutic delivery and enhanced cellular responses. This versatility enables scientists to explore numerous avenues in theranostics and regenerative medicine, where biomaterial response control and drug release are critical. Ongoing research and adjustment of piezoelectric nanomaterials can enable groundbreaking advances in healthcare technologies, bringing together engineering and medicine to enhance patient care (Kapat et al., 2020) (Figure 1.11).

1.4.1. Optical Properties and Their Implications

The optical features of nanomaterials deeply influence their applicability in an enormously wide range of applications, touching from electronics to medicine. For instance, materials such as zinc oxide (ZnO) have been of interest for their superior electron-transfer capability, an ability to be enhanced by doping with rare-earth metal ions. This enhancement is particularly relevant in perovskite solar cell (PSC) development, whose power conversion efficiency is directly influenced by photogenerated electrons' mobility. According to recent studies, doped ZnO was shown to achieve outstanding efficiencies of up to 22.9%, demonstrating a clear correlation between the optical property and energy conversion device efficiency. Such optical characteristics are manipulated to ease the creation of tailored designs for optimizing device stability and inhibiting photocarrier recombination, thus overall performance in solar technology and applications (Mokgolo et al., 2025). Aside from energy application, the optical properties of nanomaterials also have uses in biomedicine, for example, diagnostic imaging and antimicrobial therapy. Hydrophobic carbon dots (HCDs), a family of nanomaterials with unique optical and physicochemical properties, are poised to become central players in this area. Not only do they facilitate biological imaging, but they also enhance the identification of pathogens in various situations, thus promising to enhance food safety and medicine. Synthesis pathways of HCDs, including hydrothermal processes and microwave irradiation, have been investigated at a critical scale with regard to their influence on such optical properties. Antimicrobial ability of HCDs has brought forth their potential in designing novel antibacterial coatings with tremendous potential to enhance sterilization process in the clinical and industrial settings (Molaei et al., 2025). Further, the significance of the optical property of nanomaterials is also reflected in the application of nanomaterials in optoelectronic devices, which leverage the property to advance the capabilities of technology in communications and sensors. The variability of the optical properties in nanomaterials makes it possible for scientists to develop components that react to light in some way to make devices such as photodetectors and light-emitting diodes more efficient. By tailoring these properties, researchers can fabricate devices that not only perform more efficiently but also facilitate new applications in the smart technology and Internet of Things (IoT) domains. As nanomaterials are further developed, the potential for emerging applications—ranging from chemosensors to bioimaging—is vast, placing focus on the convergence of their optical properties with technological advancement. Highlighting the significance of optical properties is pivotal in building a better comprehension of how they drive innovation in different fields, restoring their status as building blocks in nanotechnology (Molaei et al., 2025) (Figure 1.12).

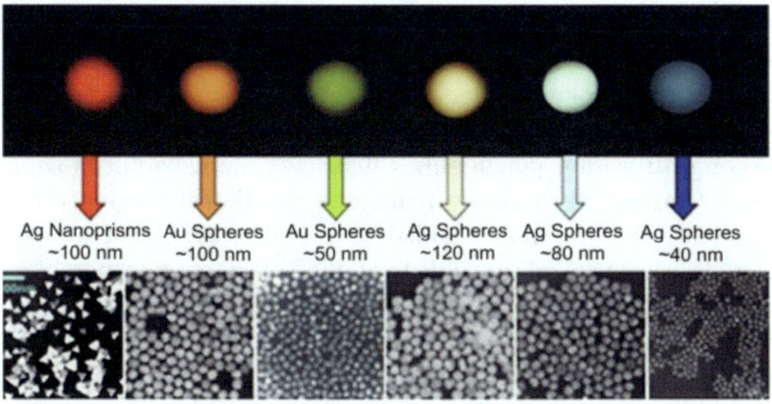

Figure 1.12. Optical property of nanomaterials.

Source: https://ars.els-cdn.com/content/image/3-s2.0-B9780128168066000157-f15-08-9780128168066.jpg.

1.4.2. Electrical Properties and Applications in Electronics

The electrical properties of nanomaterials further enhance their functionality in the electronics sector, making them part and parcel of modern technology. Nanoscale materials exhibit unique behavior as a result of quantum effects, drastically altering their resistivity, conductivity, and general electrical nature. Carbon nanotubes and graphene, for instance, exhibit exemplary electrical conductivity as a result of their immense surface area and mobility of charge carriers. These not only improve the rate of signal transmission but also enable the miniaturization of devices. Nanomaterials are therefore utilized in advanced electronics, such as transistors and sensors, where their efficiency is better than the usual materials. As technology advances, research on the electrical behavior of nanomaterials continues to be essential in developing even faster and more efficient devices to address the increasing high-performance needs (Mishra et al., 2024). Besides, their use in electronic applications is not just about enhancing performance. They also offer opportunities for the creation of cleaner and greener technologies. Nanoscale devices have the ability to consume less power, thereby reducing power usage in electronic devices. The employment of nanotechnology in energy-efficient electronic device development shows the important role that these materials play in enabling sustainable innovations. For example, use of nanomaterials in organic light-emitting diodes (OLEDs) and flexible electronics not only supports improved performance but also makes it easier to design lighter, mobile devices with reduced environmental impacts. Through lower uses of resources through greater efficiency, engineers are able to address growing issues of electronic waste and energy sustainability (Kumar et al., 2023). In line with this, the combination of nanomaterials and electronics opens up new paths for green technology. With nanomaterial advancements, study goes further to comprehend electric properties of nanomaterials and open up underlying utilization potentialities that have the capability of redefining electronics. Current research now focuses on defect engineering, where defects are used as a design mechanism to modulate the electrical behavior of nanomaterials. These defects are able to maximize charge transport phenomena and render nanotechnology devices more effective. This review of defect interactions, in particular, in carbon nanotubes and graphene, brings out the extensive implications of nanomaterials in various applications, ranging from electronics to environmental sustainability. As nanotechnology advances, it is more important than ever to keep discovering and improving these properties so that they may be tailored to meet specific requirements within the electronics industry for innovative innovations and enhancements in device functions (Kumar et al., 2023; Mishra et al., 2024).

1.4.3. Mechanical Properties and Enhancements Over Bulk Materials

Nanomaterials have drawn extensive attention in the field of materials science owing to their superior mechanical properties and advantages over bulk material. As the materials get reduced to the nanoscale, the mechanical properties get altered enormously due to the increased surface area to volume ratio and quantum confinement effects. This alteration comes in several different ways in the form of improved strength, hardness, and elasticity. For example, some research has shown that nanoscale material will possess tensile strengths several times greater than their bulk counterpart, and thus they are particularly worth utilizing in applications involving a need for lightweight yet strong material. Thus, nanomaterial synthesis can bring forth innovations in industries ranging from aerospace to biomedical engineering, calling attention to nanomaterials' power in revolutionizing conventional material applications (Podagatlapalli, 2025). In addition to their enhanced mechanical properties, nanomaterials also exhibit enhancements in toughness and ductility, which are critical in the majority of structural applications. The nanomaterial structure facilitates dislocation movements at the nanoscale, which creates an enhanced energy absorption ability during deformation. Consequently, the materials withstand more stress and strain without failing catastrophically. A wonderful example of this is that of silk fibroin nanomaterials, which possess greater mechanical strength than bulk silk fibroin. These are achieved through the nanostructure's ability to support more effective load distribution and stress transfer processes, making them promising candidates for a wide range of applications in biomedicine. These advancements indicate the significant role nanomaterials play in the enhancement of traditional materials, thus bridging the gap towards better engineering solutions (Han et al., 2025). Besides, the synthesis methods of nanomaterials also play a significant role in the definition of mechanical properties. Lasers ablation and chemical vapor deposition methods provide precise control of nanoparticle size, shape, and particle distribution, which further dictate the mechanical performance. For instance, laser ablation methods allow the production of nanoparticles with homogeneous properties, so that reproducible strengthening and ductility improvements follow. However, it is worthwhile mentioning that each synthesis pathway also possesses specific strengths and weaknesses, which influence reproducibility and scalability. Lastly, constant research on new synthesis technologies will continue to enhance the mechanical properties of nanomaterials, propelling them into widespread applications in many areas and offering potential solutions to existing limitations in bulk materials technology (Podagatlapalli, 2025).

1.5. APPLICATIONS OF NANOMATERIALS

Applications for nanomaterials cross a very broad array of industries significantly enhancing performance and sustainability. In architecture and interior design, for instance, nanostructured materials are revolutionizing conventional construction practices. Such products, such as high-strength composites and self-cleaning coatings, address contemporary issues of energy efficiency and pollution reduction. Advances, such as nano-engineered concrete, possess features such as increased strength and thermal insulation, as well as pollutant emission reduction. Green nanomaterials are part of the universal movement towards sustainability, considering that such materials can minimize the consumption of resources and trigger eco-friendly designs. The use of advanced nanotechnology not only meets architectural aesthetic demands but also enhances the functionality of a building as a whole, paving the way for energy-saving, responsive, and sustainable buildings in the future (Ar. Sultana et al., 2025). Nanomaterials are also a focus in another sector that is agriculture, where it has potential uses to combat severe issues in food production. The use of nanotechnology in this field enhances pest control, soil fertility, and yield through precise applications. For instance, nano-pesticides deliver specific pest control with minimal environmental impact while ensuring maximum efficacy. Similarly, slow-release nano-fertilizers supply nutrients effectively, cutting on wastage while promoting sustainable application. In addition, the use of nano-biosensors facilitates early detection of the pest, which facilitates interventions at the right time and thus guarantees productivity while avoiding significant use of chemical inputs. With continued efforts in agricultural practices to ensure output needs are met while maintaining environmental stewardship, the role of nanomaterials as an ingredient in the implementation of sustainable agricultural practices required to provide future food security is paramount (Revathi et al., 2025). Outside of agriculture and architecture, nanomaterials are making significant contributions in other applications of technology such as electronics and medicine. In electronics, devices used to measure nanoscale enhance performance metrics to allow for devices to be minimized in space while their efficiency and speed are optimized. This is seen in the design of nano-transistors and conductive nanomaterials to use as precursors in next-generation computing technology. Parallel to this in the medical field, nanotechnology is enabling advanced drug delivery systems, allowing targeted therapies to improve patient care while minimizing side effects. Nanocarriers, designed to deliver drugs to target cells, are a case in point. The versatility and power of nanomaterials to solve complex issues in many industries prove their revolutionary nature, and they are a crucial element in driving scientific and technological progress (Ar. Sultana et al., 2025; Revathi et al., 2025) (Figure 1.13).

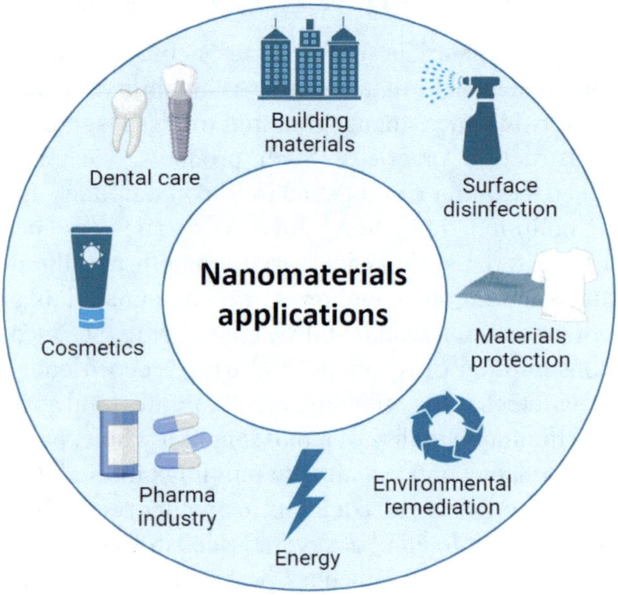

Figure 1.13. Applications of nanomaterials.

Source: https://pub.mdpi-res.com/nanomaterials/nanomaterials-13-02406/article_deploy/html/images/nanomaterials-13-02406-g001.png?1692933083.

1.5.1. Use in Medicine and Drug Delivery Systems

Nanomaterials have revolutionized the field of medicine, particularly in diagnostics and therapeutic treatment. Nanoparticles, with their unique physical and chemical properties, are new tools for the diagnosis and treatment of disease. Owing to their small size, they are able to penetrate biological barriers, reach target sites with high precision that conventional medications are unable to achieve. For instance, the ability of nanoparticles to act as contrast agents for imaging has revolutionized diagnostics, enabling earlier and more accurate detection of conditions such as cancer. Such a shift from traditional approaches to nanomaterial-based ones brings attention to improved patient outcomes and growing recognition of the role of nanotechnology's contribution in modern medicine (Joseph et al., 2023). One of the numerous applications of nanomaterials is drug delivery systems, which have gained more and more attention because they show promise in increasing the efficacy of treatment with decreased side effects. Regular drug delivery can be plagued by problems of poor solubility and non-specific delivery, which undermines the therapeutic effect. Nanomaterials, particularly those intended for targeted delivery, address these issues by covering drugs in nanoparticles that release therapeutics upon being exposed to specific stimuli, such as pH and temperature variations. DNA nanotechnology has also emerged as a pioneering discipline among nanomaterials, comprising DNA-based systems demonstrating high specificity towards tumor cells. The fabrication of multifunctional DNA nanomaterials foretells a new age of personalized medicine and therapeutic strategies, but safety and environmental issues of the usage of nanoparticles continue to be controversial (Ma et al., 2021). Despite the great potential of nanomaterials in medicine, issues regarding safety and toxicity should be closely examined. The unique interaction of nanoparticles at the cellular and molecular levels can generate unforeseen biological effects that demand meticulous investigation. Toxicological evaluations play a critical role in understanding how nanoparticles interact with different biological systems as their small size enables them to penetrate cellular membranes and concentrate in organs. Furthermore, regulatory frameworks continue to evolve to address issues surrounding the commercialization of nanomedicine. While studies continue to unveil the many opportunities with which nanomaterials can support drug delivery and therapeutic response, a balanced strategy considering both strengths and limitations will be instrumental towards making their safe integration into practice (Joseph et al., 2023).

1.5.2. Role in Environmental Remediation

Environmental remediation with nanomaterials is rapidly emerging as a solution to combating pollution. Environmental pollution is a general and emergent issue in today's world, mainly due to activities of humans that release poisonous materials into the air, water, and soil (Cerra et al., 2024). The contaminants elevate human health concerns and environmental risks, thereby necessitating the establishment of new technologies for remediation. Nanomaterials, particularly carbon-based nanomaterials (CBNs), have demonstrated promising efficacy towards addressing such environmental issues owing to their unique properties. Interestingly, their high adsorption capacities and large surface areas make it possible for them to clean and adsorb an extensive range of pollutants, including heavy metals and organic compounds. Such versatility puts CBNs in the limelight in the development of sustainable activities tailored to effectively neutralize pollution impacts.

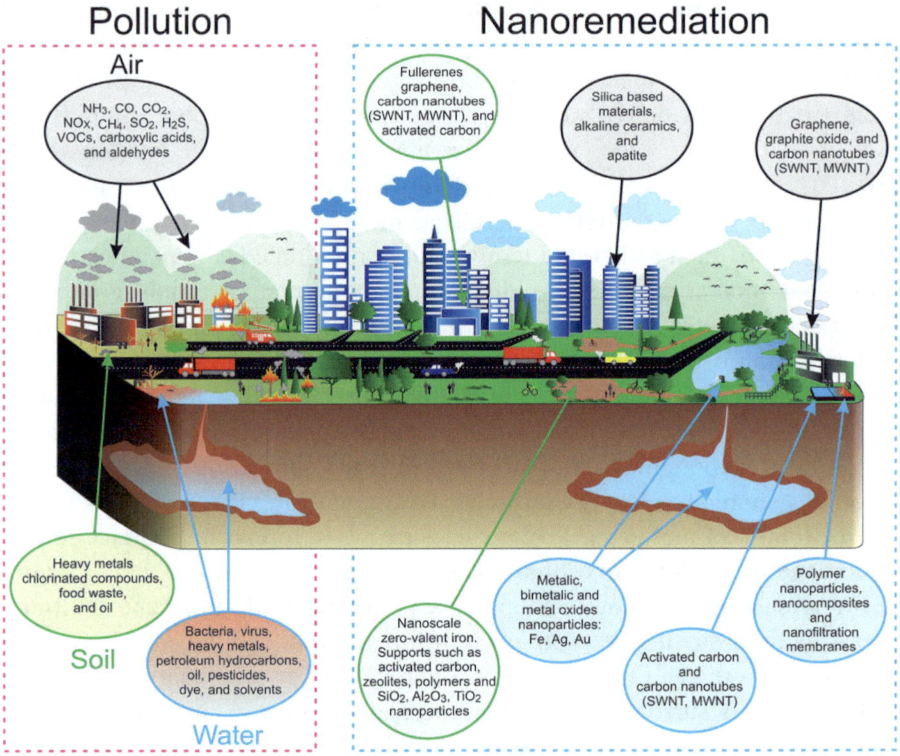

Figure 1.14. Role in environmental remediation.

Source: https://images-provider.frontiersin.org/api/ipx/w=1200&f=png/https://www.frontiersin.org/files/Articles/793765/fenvs-09-793765-HTML/image_m/fenvs-09-793765-g001.jpg.

The efficacy of CBNs is extended to various environmental processes, including soil cleanup and treatment of water. New technologies focus on their capacity to recover novel emerging pollutants that traditional technologies cannot catch. For instance, carbon nanotubes and graphene have been used for the adsorption of harmful substances in a way that facilitates their easy recovery and destruction or disposal. This capability makes it worth pursuing more research into the development of nanomaterials and their applications in environmental rehabilitation. Researchers and engineers are currently focusing their efforts on enhancing the properties of such materials to make them more applicable. Given the new environmental context, the applicability of CBNs is a beacon of hope in addressing the growing threat of pollutants in various ecosystems. The contribution of CBNs towards bringing about a world that is sustainable cannot be overstated as well. Their integration into the remediation process mirrors the transition to greener solutions not just addressing pollution but also opening floodgates towards future technological advancements. Research indicates that the prospects of CBNs in guaranteeing innovation within the area are very high, considering that they offer potential for enhancing the effectiveness of current remediation technologies and introducing new methods (Rani et al., 2025). Through the exploitation of the distinctive features of nanomaterials, researchers are discovering windows that can revolutionize the environmental management landscape. As the globe moves towards sustainability, the ongoing research on the diversity and effectiveness of CBNs is crucial in the development of long-lasting solutions that ensure cleaner and safer environments for future generations (Figure 1.14).

1.5.3. Applications in Energy Storage and Conversion

The emerging area of nanomaterials contains revolutionary potential for energy storage and conversion applications, primarily through their unprecedented properties at the nanoscale. Nanomaterials, such as transition metal dichalcogenides (TMDs), possess dominant phase-dependent physicochemical properties that can be engineered for specific applications. For instance, TMDs' stacking structures allow for efficient charge transport and greater surface area, which are essential factors in improving energy storage devices. Researchers have been employing phase engineering more to leverage such properties, leading to rising and metastable phases that significantly improve electronic and electrochemical performance. As demonstrated by relatively newer developments, TMDs can be employed as base materials for batteries and supercapacitors, thus illuminating a pathway for future energy storage technology (Zhai et al., 2024). Besides TMDs, multicomponent metallic nanomaterials have tremendous potential for applications in electrochemical energy storage and conversion due to their unique structures. The engineering of such materials allows for the modification of their surface characteristics and crystallinity, thus directly influencing their performance characteristics. For instance, it has been shown that strain engineering can modify the electronic nature of such materials, influencing their catalytic behavior in energy conversion processes. Specifically, purposeful control over lattice strains can lead to improved electrocatalytic performances, tuning reactions such as hydrogen evolution or oxygen reduction. As such, work on new methods of engineering multicomponent nanomaterials goes to even further enhance their application in energy technologies as well as provide new possibilities for efficiency (Yao et al., 2023). Moreover, the potential of nanomaterials extends far beyond energy storage alone, encompassing numerous applications in energy conversion processes too. Their incorporation into fuel cells and solar cells is a prime example of this dynamic potential, wherein they enhance efficiency as well as stability. Synergism among varied nanostructures can improve overall performance parameters, such as charge transfer and light absorption. With the growing demand for sustainable energy solutions, the role of nanomaterials in enabling this shift becomes more significant. Therefore, continued research into nanomaterials' structural interactions, charge behavior, and surface changes is the primary means of achieving even greater advancements in energy conversion and storage technology and eventually more sustainable energy systems (Yao et al., 2023).

1.6. CHALLENGES AND RISKS

The integration of nanomaterials into industries, while presenting huge strides ahead, is fraught with huge challenges and risks to be diligently evaluated. With the increasing use of nanotechnology in fields such as electronics, medicine, and agriculture, the potential health and environmental effects of exposure to engineered nanoparticles increase and become more and more significant. The challenge in assessing the safety of nanomaterials is their unique properties, which can differ enormously from their bulk counterparts. Such include greater reactivity and greater capacity for penetration into biological systems and inducing unforeseen toxicity. The growing public concern regarding nanotechnology necessitates thorough appreciation of its impacts, hence necessitating strict health, safety, and environmental regulations to counter the impacts of their application in everyday procedures (Martin et al., 2023). Apart from health concerns, the environmental impact of nanomaterials raises warning challenges that have to be tackled. Persistence of nanomaterial in the environment leads to unforeseen issues for wildlife and human populations. Moreover, their tendency to bioaccumulate leads to additional hazards where they can eventually enter the food chain and enhance their negative effects. The need for due risk assessment procedures becomes eminent as the environmental fate of nanoparticles is yet to be made clear. In particular, studies must provide answers to the life cycle of these materials, from manufacturing through utilization and final waste disposal, to offset contamination of the environment and preserve biodiversity (Prado-Audelo et al., 2021).

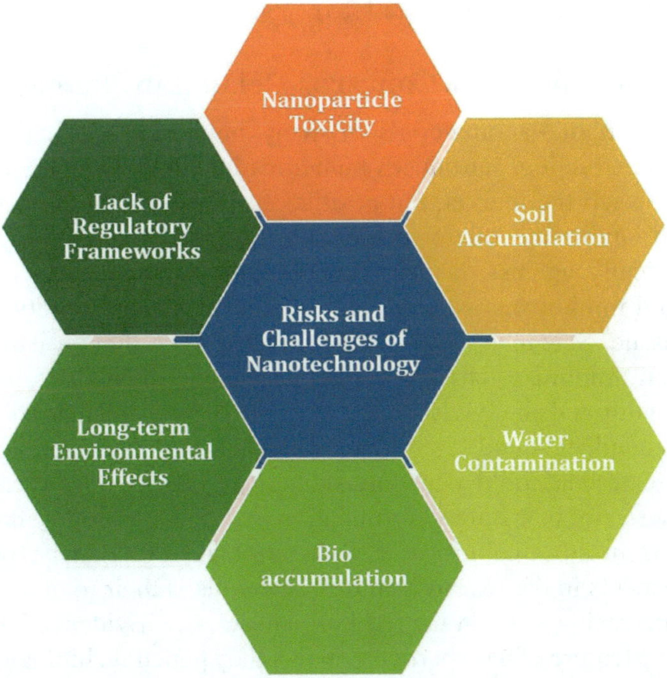

Figure 1.15. Challenges and environmental implications of nanomaterials.

Source: https://www.researchgate.net/publication/392866530/figure/fig3/AS:11431281509182655@1750440045978/Challenges-and-environmental-implications-of-nanomaterials-in-agriculture-42.jpg.

Applications of nanotechnology's relationship with environmental preservation necessitates cooperative action among researchers, regulatory bodies, and industries to establish responsible protocols. Nanotoxicity remains a focal point in an appreciation of the conundrum posed by the utilization of nanomaterials. As discussed earlier, nanoparticles' unique properties complicate their toxicity evaluation and increase the opportunity for negative effects on human health and environments. Research on toxicological processes, exposure routes, and countermeasures is urgently needed to neutralize concerns regarding the interaction between nanomaterials and biological systems. Furthermore, the application of nanoremediation technologies has emerged as a favorable choice to mitigate environmental pollution. However, these strategies will have to be implemented and tried carefully so that they do not create new hazards or enhance current issues (Martin et al., 2023). It is essential to weigh the enormous potential of nanotechnology against rigorous safety assessments so that one can meet the present challenges of its creation in an efficient and sustainable manner (Prado-Audelo et al., 2021) (Figure 1.15).

1.6.1. Health and Safety Concerns Related to Nanomaterials

The introduction of nanomaterials in many industries, particularly interior design and construction, introduces numerous health and safety concerns. As innovations sweep the market, the positive attributes of nanomaterials—such as decreased environmental footprints and increased functionality—must be weighed critically against danger. Though these materials can raise energy efficiency and minimize waste, the lack of comprehensive safety testing is a quite grim issue. Scientists have discovered that even though nanostructured materials can minimize environmental pollutants, their long-term effects on human health and the environment are still poorly understood. Thus, it is important to conduct sound research to evaluate the probable avenues through which nanomaterials could penetrate biological systems and create harmful health impacts, so that safety formulation never undermines the probable advancements in sustainability (Ar. Sultana et al., 2025). One of the most significant aspects in the research on nanomaterials is their toxicity, which is a constantly increasing issue in the field of nanoscience. Evidence indicates that the one unique feature of nanoparticles, in that they penetrate biological barriers, is of concern when safety is considered. The generation of ROS is an important mechanism through which toxicity of these substances occurs. Excessive ROS leads to oxidative stress, which will compromise cell integrity, induce damage to DNA, and even trigger oncogenic processes. The consequences of such toxicological effects require careful management of synthesis, use, and disposal of nanomaterials, reinforcing the necessity for regulative policies that couple human health and ecological safety with technology (Jyoti et al., 2022). Lastly, in order to avoid nanomaterial health and safety risks, scientists, industry experts, and regulatory agencies must collaborate. Developing guidelines encompassing safe production, handling, and disposal processes will be essential in the promotion of ethical innovation. It is equally essential that public health aspects be incorporated during the research and application stages of nanotechnology so that innovations in material science do not jeopardize workers' safety and the environmental integrity unintentionally. As technology advances, there is a requirement for ongoing commitment to meet the double challenge of reaping nanomaterials' benefits without jeopardizing health and safety. Persistent investment in the research on the implications of nanotoxicity and careful assessment of nanomaterials will facilitate the creation of a responsible and innovative future for technology and sustainability (Ar. Sultana et al., 2025; Jyoti et al., 2022).

1.6.2. Environmental Impact and Sustainability Issues

The exploitation of nanomaterials in various industries has brought unprecedented innovations; however, this also raises significant environmental effects and sustainability issues. As researchers continue to harness the possibilities of nanomaterials in applications ranging from drug delivery to agriculture, it becomes necessary to take into account the life cycle of nanomaterials. The production process of these materials typically entails poisonous chemicals and high energy consumption, contributing to environmental degradation. For example, conventional synthesis techniques of nanomaterials tend to use toxic precursors, which present environmental hazards in the event of disposal. Therefore, green synthesis alternatives are critical to alleviate these negative impacts, and hence the importance to focus on sustainability in nanomaterial research and application. Improving environmental management practice and taking a life-cycle approach towards nanomaterials can encourage the transition towards sustainable approaches, which ultimately will be beneficial to both industry and the environment. As part of their pursuit for eco-friendly substitutes, green synthesis of nanomaterials is an encouraging method, particularly that of Carbon Quantum Dots (CQDs). The nanomaterials are synthesized from renewable materials, such as plant extracts and biomass, which have been beneficial in reducing negative environmental impacts over long periods. The process points out how sustainable processes not only offer an improved ecological footprint but also the functional properties of the obtained nanomaterials. For example, integration of natural chemicals can confer bioactive functionality to CQDs and improve their application in biotech. Hydrothermal and solvothermal methods common in green synthesis add a double advantage: minimizing environmental impacts while delivering high-quality nanomaterials with beneficial applications, e.g., biomedical and environmental monitoring applications. However, it is necessary to address optimization problems in such synthesis methods to fully exploit CQDs in eco-friendly applications, and as such, further research and development are required. Its application in agriculture also reflects its potential along with related sustainability challenges. Nano-pesticides, slow-release nano-fertilizers, and nano-biosensors are some of the developments that illustrate the potential for nanomaterials to enhance crop yields and management. These approaches not only aim to improve agricultural yields but also have the capability to control chemical runoff and soil erosion. Nevertheless, environmental and health hazards of nanomaterials need to be scientifically evaluated, as random applications can potentially stimulate unforeseen ecological responses. Consequently, collaboration and interdisciplinary research are critical in formulating guidelines for their safe application balancing the benefits of nanotechnology in agriculture against environmental sustainability. Moreover, addressing these sustainability issues is essential to advancing global food security and minimizing the negative impact of agriculture on the environment, as has been emphasized in current reviews of nanotechnology on this issue (Revathi et al., 2025).

1.6.3. Regulatory Challenges in Nanomaterial Usage

The rapid evolution of nanotechnology has generated huge promise in many fields, mostly medicine, electronics, and environmental applications. However, the inherent complexity of nanomaterials calls for comprehensive regulatory frameworks that are capable of responding to their unique properties and potential hazards. Present regulations have a tendency to employ traditional substance testing that is prone to not fully capture the activities and interplays of the nanoscale substances. This lack of having special guidelines can lead to inconsistencies in risk assessments, which leaves loopholes in protecting public health and the environment. Thus, regulatory bodies must develop more focused strategies, like comprehensive analyses of nanomaterial toxicity, environmental behavior, and performance, to ensure public safety is not threatened as development progresses. Furthermore, the diversity of regulatory climates worldwide presents additional challenges for nanomaterial regulation. Different jurisdictions have varying levels of strictness in their regulatory processes, which lead to disparity in safety standards as well as approval procedures. For instance, some regions prioritize speedy innovation, while others are committed to rigorous risk assessment, resulting in a fragmented market. Such disparity may impede global cooperation and discourage investment in nanomaterial research and development. There needs to be a harmonized regulation system in order to allow the general and responsible use of nanotechnology. Such action could foster innovation while ensuring security provisions accommodate the unique complexity with nanomaterials, creating an environment conducive to healthy progress within the sector. In the health sector, nanomaterials' capacity to revolutionize treatment possibilities, particularly in oncology, evokes the need for good regulation. Though others have reached the human trial stage, safety and efficacy issues commonly impede clinical application. That it is hard to take lab data and make it into effective therapy constitutes a major regulatory problem. These problems encompass such challenges as having no universal test protocols and varying definitions of what is considered a nanomaterial adding to the approval process. By addressing these challenges, regulatory agencies are able to open the door for the integration of new therapies that utilize the unique advantages of nanotechnology in order to eventually improve patient outcomes in the fight against disease like cancer. This two-pronged strategy of risk management and adaptation is crucial to developing a robust regulatory infrastructure able to effectively control the new technology of nanotechnology (Chehelgerdi et al., 2023; Muteeb et al., 2023).

SUMMARY

Research of nanomaterials, to mention their great promise in numerous areas of science and technology, particularly in biomedical applications. The unique properties of nanoparticles (NPs) open up prospects for advancements in drug delivery, imaging, and regenerative medicine. With their controlled size, shape, and chemical composition, researchers can tailor NPs to react with desired biological activity, critical for better therapeutic effectiveness. However, the intricate interaction between NPs and living systems calls for better understanding of the resulting biological impacts, including the generation of reactive oxygen species (ROS) related to metallic NPs. The relationship between NP characteristics and their consequent biological impacts necessitates increased study of NP-induced toxicity and their roles in cellular processes (Dayem et al., 2017). Besides, the implications of nanomaterials extend beyond biomedical fields and contribute to environmental sustainability as well as policy-making. The planetary boundary (PB) concept stresses the most important boundaries that humans must not exceed in order to maintain an equilibrium world. With the advancement of nanotechnology, its incorporation into sustainable processes offers worthwhile pathways for driving pertinent global concerns like the depletion of resources and environmental degradation. Nanomaterial innovation has the potential to bridge technology and sustainability via solutions that reduce environmental footprints but enhance human health (Steffen et al., 2015).

This intersection of nanotechnology and sustainability is a forward-looking step in reducing risk in relation to environmental limits. In the end, the innovation and application of nanomaterials are a great leap in science and technology with potential opportunities and risks. As researchers delve more into the potential of NPs, so too does it become increasingly important to have some sort of ethical code governing their use. There will need to be extensive safety evaluations and regulatory measures to ensure the benefits of nanomaterials are not being paid for in too high a cost in health or environment. The ongoing dialogue between scientific progress and social implications can ensure a better ethical application of nanotechnology, paving the way for its integration into other disciplines while guaranteeing public interest. Emphasis on the balance dimension of nanomaterials research will enhance our awareness and innovations that foster sustainable development and human well-being.

In exploring the key advancements in the field of nanomaterials, the talks have highlighted a series of crucial aspects pertinent to their application and marketability. Among these, one has been the remarkable characteristics of nanomaterials like increased surface area, unique optical properties, and enhanced electrical conductivity. These aspects enable their application in numerous fields from electronics, energy, to medicine. The perovskite solar cell (PSC) review is a case in point, referencing that even at very high power conversion efficiencies, stability and potential lead leakage must be taken very seriously. The integration of self-healing mechanisms in PSCs brings innovation in device lifespan extension while maintaining efficiency. Such integration of findings focuses on the need for ongoing research to address issues of stability and enhance useful applications of nanomaterials in green technology (Liu et al., 2025). Additionally, the comparison between vaccinated and unvaccinated groups in the COVID-19 case holds interesting insights that connect with developments in nanotechnology. The vaccination profile of elderly patients detailing their vulnerability and comorbidities offset against the reality that they reduce the risk of death categorically justifies the adoption of a holistic approach to technological advancements. Just as the COVID-19 vaccine has revolutionized patient outcomes, the evolution of nanomaterials can revolutionize the performance and reliability of technology in hostile environments. The interaction between various factors that affect outcomes is the same as the intricate considerations involved in designing and utilizing nanomaterials, and this means that additional research is vital in understanding how these materials can be optimized in efficiency and safety (Smits et al., 2024). The conclusion and analysis of the main points presented bring out the great potential of nanomaterials in various applications but also acknowledging the issues that must be confronted. The discussions surrounding PSCs, the purported self-healing properties specifically, emphasize the need for solutions that reconcile present stability and toxicity problems. As nanotechnology is increasingly researched by scientists, utmost importance must be given not only to the performance metrics but also to the long-term impacts the materials have on safety and health. Through extended investigation of these fronts, the field can strive to realize the complete potential of nanomaterials and push a responsible approach in utilizing them in various industries. Comprehensive understanding in this regard is essential in facilitating innovation and effective integration of nanotechnology into traditional uses (Liu et al., 2025; Smits et al., 2024).

REFERENCES

1. Abdelrahman, S. E. A. H. A., El Hawary, S. E., Mohsen, E., El Raey, M. A., Selim, H. M. R. M., Hamdan, A. M., Ghareeb, M., et al. (2024). Bio-fabricated zinc oxide nanoparticles mediated by endophytic fungus Aspergillus sp. SA17 with antimicrobial and anticancer activities: *In vitro* supported by *in silico* studies. Frontiers in Microbiology, 15. https://www.semanticscholar.org/paper/497a6e73bc5884f0513045b5fe44c2207eb365b4 (accessed on 25th September 2025).
2. Alafeef, M., Srivastava, I., Aditya, T., & Pan, D. (2023). Carbon dots: From synthesis to unraveling the fluorescence mechanism. Small, 20. https://doi.org/10.1002/smll.202303937 (accessed on 25th September 2025).
3. Alias, A. H. H., Norizan, M. N., Sabaruddin, F. A., Asyraf, M. R. M., Norrahim, M. N. F., Ilyas, R. A., Kuzmin, A. M., et al. (2021). Hybridization of MMT/lignocellulosic fiber reinforced polymer nanocomposites for structural applications: A review. Coatings, 11, 1355–1355. https://doi.org/10.3390/coatings11111355 (accessed on 25th September 2025).
4. Andal, C., Kavitha, E., Bharathi Devi, V., & Suchithra, J. P. (2024). A Textbook on Nanomaterials and Nanotechnology. https://www.semanticscholar.org/paper/9752c60376938f531140e5639cdf1749e881f695 (accessed on 25th September 2025).
5. Bruna, T., Maldonado-Bravo, F., Jara, P., & Caro, N. (2021). Silver nanoparticles and their antibacterial applications. International Journal of Molecular Sciences, 22, 7202–7202. https://doi.org/10.3390/ijms22137202 (accessed on 25th September 2025).
6. Cerra, S., & Fratoddi, I. (2024). Nanoscale solutions: The transformative applications of functionalized nanomaterials in environmental remediation. Applied Nano. https://www.semanticscholar.org/paper/7bfb8df1adbb3f267ed75a18641c04d505cf041d (accessed on 25th September 2025).
7. Chehelgerdi, M., Chehelgerdi, M., Allela, O. Q. B., Cosme Pecho, R. D., Narayanan, J., Rao, D. P., Thamaraikani, T., et al. (2023). Progressing nanotechnology to improve targeted cancer treatment: Overcoming hurdles in its clinical implementation. Molecular Cancer, 22. https://doi.org/10.1186/s12943-023-01865-0 (accessed on 25th September 2025).
8. Dayem, A. A., Hossain, M., Lee, S., Kim, K., Saha, S. K., Yang, G. M., Choi, H. J., et al. (2017). The role of reactive oxygen species (ROS) in the biological activities of metallic nanoparticles. International Journal of Molecular Sciences, 18, 120–120. https://doi.org/10.3390/ijms18010120 (accessed on 25th September 2025).

9. Del Prado-Audelo, M. L., García Kerdan, I., Escutia-Guadarrama, L., Reyna-González, J. M., Magaña, J. J., & Leyva-Gómez, G. (2021). Nanoremediation: Nanomaterials and nanotechnologies for environmental cleanup. Frontiers in Environmental Science, 9. https://doi.org/10.3389/fenvs.2021.793765 (accessed on 25th September 2025).
10. El Shafey, A. M. (2020). Green synthesis of metal and metal oxide nanoparticles from plant leaf extracts and their applications: A review. Green Processing and Synthesis, 9, 304–339. https://doi.org/10.1515/gps-2020-0031 (accessed on 25th September 2025).
11. El-Kalliny, A. S., Abdel-Wahed, M. S., El-Zahhar, A. A., Hamza, I. A., & Gad-Allah, T. A. (2023). Nanomaterials: A review of emerging contaminants with potential health or environmental impact. Discover Nano, 18. https://doi.org/10.1186/s11671-023-03787-8 (accessed on 25th September 2025).
12. Han, C., Chen, S., Shi, Z., Guan, X., Zou, W., & Liu, J. (2025). An innovative treatment for hypoxic-ischemic encephalopathy: Silk fibroin nanomaterials improve neural stem cell axon formation and facilitate cognitive improvement. Neural Regeneration Research. https://www.semanticscholar.org/paper/8d7cc0459bac6fabdd5fc182ba3abb6e1d3e7b2e (accessed on 25th September 2025).
13. Ikram, R., Jan, B. M., Qadir, M. A., Sidek, A., Stylianakis, M. M., & Kenanakis, G. (2021). Recent advances in chitin and chitosan/graphene-based bio-nanocomposites for energetic applications. Polymers, 13, 3266–3266. https://doi.org/10.3390/polym13193266 (accessed on 25th September 2025).
14. Iqbal, A., Zafar, N., Bashir, R., Sultan, M., Naz, A., Bashir, A., & Akhter, Z. (2022). Influence of manganese and nickel doping on optical and electric properties of CuO nanostructures for optoelectronic applications. International Journal of Nanoparticles. https://www.semanticscholar.org/paper/ff57239ef57b62a3fddee1e36d232b3ee04275fc (accessed on 25th September 2025).
15. Jagadeesh, P., Rangappa, S. M., & Siengchin, S. (2023). Advanced characterization techniques for nanostructured materials in biomedical applications. Advanced Industrial and Engineering Polymer Research. https://doi.org/10.1016/j.aiepr.2023.03.002 (accessed on 25th September 2025).
16. Jakubinek, M. B., Kim, K. S., Kim, M. J., Martí, Á. A., & Pasquali, M. (2022). Recent advances and perspective on boron nitride nanotubes: From synthesis to applications. Journal of Materials Research, 37, 4403–

4418. https://doi.org/10.1557/s43578-022-00841-6 (accessed on 25th September 2025).

17. Jeevanandam, J., Barhoum, A., Chan, Y. S., Dufresne, A., & Danquah, M. K. (2018). Review on nanoparticles and nanostructured materials: History, sources, toxicity and regulations. Beilstein Journal of Nanotechnology, 9, 1050–1074. https://doi.org/10.3762/bjnano.9.98 (accessed on 25th September 2025).
18. Joseph, T. M., Mahapatra, D. K., Esmaeili, A., Piszczyk, Ł., Hasanin, M. S., Kattali, M., Haponiuk, J. T., et al. (2023). Nanoparticles: Taking a unique position in medicine. Nanomaterials, 13, 574–574. https://doi.org/10.3390/nano13030574 (accessed on 25th September 2025).
19. Joudeh, N., & Linke, D. (2022). Nanoparticle classification, physicochemical properties, characterization, and applications: A comprehensive review for biologists. Journal of Nanobiotechnology, 20. https://doi.org/10.1186/s12951-022-01477-8 (accessed on 25th September 2025).
20. Kanase, J. A., Gosavi, A. A., Kanse, A. S., Chaugule, A. M., & Adhalrao, S. B. (2022). Nanotechnology to nanotoxicity, showing its injuries concerns on human health. International Journal of Advanced Research in Science, Communication and Technology. https://www.semanticscholar.org/paper/5ba492c6c3b3fdf40c911a476c99be7c0db10521 (accessed on 25th September 2025).
21. Kapat, K., Shubhra, Q. T. H., Zhou, M., & Leeuwenburgh, S. C. G. (2020). Piezoelectric nano-biomaterials for biomedicine and tissue regeneration. Advanced Functional Materials, 30. https://doi.org/10.1002/adfm.201909045 (accessed on 25th September 2025).
22. Khan, N. A., Khan, S. U., Ahmed, S., Farooqi, I. H., Dhingra, A., Hussain, A., & Changani, F. (2019). Applications of nanotechnology in water and wastewater treatment: A review. Asian Journal of Water Environment and Pollution, 16, 81–86. https://doi.org/10.3233/ajw190051 (accessed on 25th September 2025).
23. Kong, X. P., Zhang, B., & Wang, J. (2021). Multiple roles of mesoporous silica in safe pesticide application by nanotechnology: A review. Journal of Agricultural and Food Chemistry, 69, 6735–6754. https://doi.org/10.1021/acs.jafc.1c01091 (accessed on 25th September 2025).
24. Kumar, S. M., Yadav, R., & Prasad, T. S. (2023). Recent advances in nanotechnology. International Journal of Nanomaterials Nanotechnology and Nanomedicine, 9, 015–023. https://doi.org/10.17352/2455-3492.000053 (accessed on 25th September 2025).
25. Liu, T., & Yang, Z. Y. (2025). Self-healing perovskite solar cells: Introduction,

recent progresses and perspective. SmartMat. https://www.semanticscholar.org/paper/e88445f8e557c06e4dbcbb49686fa8ad0561dd8e (accessed on 25th September 2025).

26. Ma, W., Zhan, Y., Zhang, Y., Mao, C., Xie, X., & Lin, Y. (2021). The biological applications of DNA nanomaterials: Current challenges and future directions. Signal Transduction and Targeted Therapy, 6. https://doi.org/10.1038/s41392-021-00727-9 (accessed on 25th September 2025).

27. Maheshwari, S., Verma, A., & Singh, A. (2025). Green synthesis of multi-functional carbon dots and their applications. Current Topics in Chemistry. https://www.semanticscholar.org/paper/6c35accb65ebc66fde5817a5ba9f63a351b2701d (accessed on 25th September 2025).

28. Martin, C., Nourian, A., Babaie, M., & Nasr, G. G. (2023). Environmental, health and safety assessment of nanoparticle application in drilling mud–Review. Geoenergy Science and Engineering, 226, 211767–211767. https://doi.org/10.1016/j.geoen.2023.211767 (accessed on 25th September 2025).

29. Mishra, R. K., Verma, K., & Sethi Singh, D. (2024). Defect engineering in nanomaterials: Impact, challenges, and applications. Smart Materials in Manufacturing, 2, 100052–100052. https://doi.org/10.1016/j.smmf.2024.100052 (accessed on 25th September 2025).

30. Mokgolo, P. J., Gumede, T., Ocaya, R., & Malevu, T. (2025). Enhancing perovskite solar cells with rare-earth metal doped zinc oxide: A review of electron mobility, stability, and photocarrier recombination. International Journal of Energy Research. https://www.semanticscholar.org/paper/5a4f09e4fbba3e1b040a014678b41ca646a23b00 (accessed on 25th September 2025).

31. Molaei, R., Razavi, R., Omer, A. K., Lotfi Javid, A., Ezati, P., Nikfarjam, N., Lim, L. T., et al. (2025). Hydrophobic carbon dots: An overview of the synthesis, purification, cytotoxicity, and potential applications in food safety and analytical chemistry. Quality Assurance and Safety of Crops & Foods. https://www.semanticscholar.org/paper/6d1688d89d66065d6141a457940e8a19faa0a77c (accessed on 25th September 2025).

32. Mourdikoudis, S., Pallares, R. M., & Nguyễn, T. K. T. (2018). Characterization techniques for nanoparticles: Comparison and complementarity upon studying nanoparticle properties. Nanoscale, 10, 12871–12934. https://doi.org/10.1039/c8nr02278j (accessed on 25th September 2025).

33. Muteeb, G., Rehman, M. T., Shahwan, M., & Aatif, M. (2023). Origin of antibiotics and antibiotic resistance, and their impacts on drug

development: A narrative review. Pharmaceuticals, 16, 1615–1615. https://doi.org/10.3390/ph16111615 (accessed on 25th September 2025).

34. N/A. (2022). Synthesis and Applications of Nanoparticles. https://doi.org/10.1007/978-981-16-6819-7 (accessed on 25th September 2025).

35. Naresh, V., & Lee, N. (2021). A review on biosensors and recent development of nanostructured materials-enabled biosensors. Sensors, 21, 1109–1109. https://doi.org/10.3390/s21041109 (accessed on 25th September 2025).

36. Patle, B. C., & Dhapake, P. (2024). Introduction to nanoparticles in pharmacy: A review. Asian Journal of Research in Pharmaceutical Sciences. https://www.semanticscholar.org/paper/42891310d1d33d1538613dc170e13856bb2eedfd (accessed on 25th September 2025).

37. Patra, J. K., Das, G., Fraceto, L. F., Campos, E. V. R., Rodriguez-Torres, M. P., Acosta-Torres, L. S., Díaz-Torres, L. A., et al. (2018). Nano based drug delivery systems: Recent developments and future prospects. Journal of Nanobiotechnology, 16. https://doi.org/10.1186/s12951-018-0392-8 (accessed on 25th September 2025).

38. Paukov, M. I., Kramberger, C., Begichev, I. V., Kharlamova, M. V., & Burdanova, M. G. (2023). Functionalized fullerenes and their applications in electrochemistry, solar cells, and nanoelectronics. Materials, 16, 1276–1276. https://doi.org/10.3390/ma16031276 (accessed on 25th September 2025).

39. Peng, C., Pang, R., Li, J., & Wang, E. (2023). Current advances on the single-atom nanozyme and its bio applications. Advanced Materials, 36. https://doi.org/10.1002/adma.202211724 (accessed on 25th September 2025).

40. Peng, T., Xiao, R., Rong, Z., Liu, H., Hu, Q., Wang, S., Li, X., et al. (2020). Polymer nanocomposite-based coatings for corrosion protection. Chemistry–An Asian Journal, 15, 3915–3941. https://doi.org/10.1002/asia.202000943 (accessed on 25th September 2025).

41. Podagatlapalli, G. K. (2025). The fundamentals of synthesis of the nanomaterials, properties, and emphasis on laser ablation in liquids: A brief review. Discover Nano, 20. https://www.semanticscholar.org/paper/aea46d863a9372460aace7d4e375028f9231ed2d (accessed on 25th September 2025).

42. Qamar, H., Rehman, S., Chauhan, D. K., Tiwari, A. K., & Upmanyu, V. (2020). Green synthesis, characterization and antimicrobial activity of copper oxide nanomaterial derived from Momordica charantia. International Journal of Nanomedicine, 15, 2541–2553. https://doi.org/10.2147/ijn.s240232 (accessed on 25th September 2025).

43. Qu, J., Wang, R., Pan, P., Du, L., Mi, Z., Sun, Y., & Liu, X. (2023). An SEM-based nanomanipulation system for metaphysical characterization of single InGaN/GaN nanowires. IEEE Transactions on Automation Science and Engineering, 20, 233–243. https://www.semanticscholar.org/paper/25e143cf8ec322add4016eef2210a9dea399ce8f (accessed on 25th September 2025).

44. Rani, D., Kumar, R., Kumar, S., & Rani, J. (2025). Nanomaterials in environmental remediation: Harnessing the versatility of carbon-based nanomaterials for sustainable solutions. Ecology, Environment and Conservation. https://www.semanticscholar.org/paper/1129abca7c3318eb6a720fef1af86eb331b0d13b (accessed on 25th September 2025).

45. Regan, A., O'Donoghue, J., Poree, C., & Dunne, P. (2023). Introducing materials science: Experimenting with magnetic nanomaterials in the undergraduate chemistry laboratory. Journal of Chemical Education, 100, 2387–2393. https://www.semanticscholar.org/paper/4380cb42736467d0f86eae7b6e43db0bf54493f9 (accessed on 25th September 2025).

46. Revathi, J., Chougule, P. S., Hm, A., Shenoy, H., Singh, D. B., Neeraj, & Reddypriya, P. (2025). Applications of nanotechnology in agriculture: A review. Journal of Advances in Biology & Biotechnology. https://www.semanticscholar.org/paper/388476l0aada661edd4cfb3ea4a70e81cc8dc142 (accessed on 25th September 2025).

47. Shalaby, M., Ghareeb, A. Z., Khedr, S. M., Mostafa, H. M., Saeed, H., & Hamouda, D. (2023). Nanoparticles of bioactive natural collagen for wound healing: Experimental approach. https://www.semanticscholar.org/paper/6a1eed3c35748c9b0191470c2e71323e74235eac (accessed on 25th September 2025).

48. Sharma, P., Hasan, M. R., Mehto, N. K., Deepak, D., & Narang, J. (2022). 92 years of zinc oxide: Has been studied by the scientific community since the 1930s–An overview. Sensors International, 3, 100182–100182. https://doi.org/10.1016/j.sintl.2022.100182 (accessed on 25th September 2025).

49. Smits, R. A. L., van Raaij, B. V., Trompet, S., van der Linden, C. V. D., van der Bol, J. M., Jansen, S. W. M., Polinder-Bos, H., et al. (2024). Differences in characteristics and outcomes of older patients hospitalized for COVID-19 after introduction of vaccination. European Geriatric Medicine, 15, 941–949. https://www.semanticscholar.org/paper/eebbb550b5d13efde2a707156bf408a621231237 (accessed on 25th September 2025).

50. Steffen, W., Richardson, K., Rockström, J., Cornell, S., Fetzer, I., Bennett, E. M., Biggs, R., et al. (2015). Planetary boundaries: Guiding human

development on a changing planet. Science, 347. https://doi.org/10.1126/science.1259855 (accessed on 25th September 2025).

51. Sultana, A., & Kiran, P. (2025). Green nanomaterials in interior design: Advancing sustainability through nanotechnology. International Journal of Research and Innovation in Applied Science. https://www.semanticscholar.org/paper/d1cab333ac6cded204c0462fa1ff6f148d644fd8 (accessed on 25th September 2025).

52. Swapna, M., Sankararaman, S., & Korte, D. (2024). Thermal lensing and photoacoustic as potential tools for nanomaterial characterization: A review. Journal of Materials Science. https://www.semanticscholar.org/paper/00bbe1612058dd05b41010b87570f80cb840785b (accessed on 25th September 2025).

53. Trache, D., Tarchoun, A. F., Derradji, M., Hamidon, T. S., Masruchin, N., Brosse, N., & Hussin, M. H. (2020). Nanocellulose: From fundamentals to advanced applications. Frontiers in Chemistry, 8. https://doi.org/10.3389/fchem.2020.00392 (accessed on 25th September 2025).

54. Yan, D., Li, Y., Huo, J., Chen, R., Dai, L., & Wang, S. (2017). Defect chemistry of nonprecious-metal electrocatalysts for oxygen reactions. Advanced Materials, 29. https://doi.org/10.1002/adma.201606459 (accessed on 25th September 2025).

55. Yao, Q., Yu, Z., Li, L., & Huang, X. (2023). Strain and surface engineering of multicomponent metallic nanomaterials with unconventional phases. Chemical Reviews, 123, 9676–9717. https://doi.org/10.1021/acs.chemrev.3c00252 (accessed on 25th September 2025).

56. Yuan, M., Kermanian, M., Agarwal, T., Yang, Z., Yousefiasl, S., Cheng, Z., Ma, P., et al. (2023). Defect engineering in biomedical sciences. Advanced Materials, 35. https://doi.org/10.1002/adma.202304176 (accessed on 25th September 2025).

57. Zhai, W., Li, Z., Wang, Y., Zhai, L., Yao, Y., Li, S., Wang, L., et al. (2024). Phase engineering of nanomaterials: Transition metal dichalcogenides. Chemical Reviews, 124, 4479–4539. https://doi.org/10.1021/acs.chemrev.3c00931 (accessed on 25th September 2025).

CHAPTER 2

FUNDAMENTALS OF NANOSCIENCE AND NANOTECHNOLOGY

Contents

INTRODUCTION TO NANOSCIENCE AND NANOTECHNOLOGY

Nanoscience and nanotechnology are a pioneering interface in science and technology that is concerned with manipulation at the atomic and molecular levels. The discipline is comprised of the science and engineering of devices and materials on a nanoscale, where size typically ranges between 1 and 100 nanometers. Materials at this size possess unique properties that are very different from their bulk counterparts, and hence nanotechnology is an area of research that is of fundamental importance. Improvements in nanomaterials not just improve existing technology but also provide new opportunities for use. For instance, nanoscale additives significantly improve the mechanical, electrical, and thermal properties of materials, which can lead to advances in various industries, from electronics and medicine to environmental studies. The fundamental science of nanoscience is thus at the core of developing a new generation of materials, devices, and processes, underscoring the need for education and research in this dynamic field. One of the most important aspects of nanotechnology is its possible contribution to medicine, generally referred to as nanomedicine.

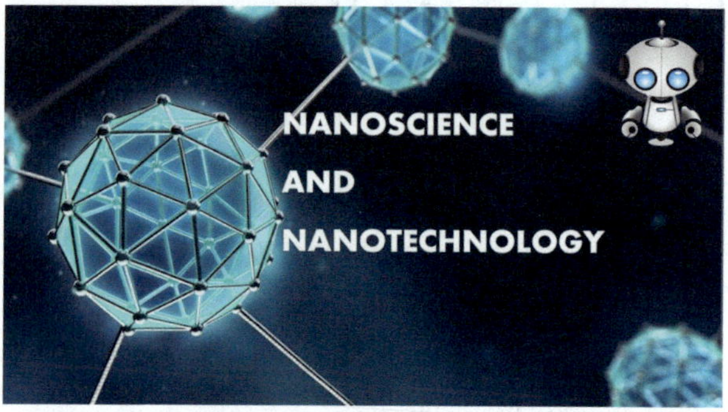

Figure 2.1. An introduction to nanoscience and nanotechnology.

Source: https://external-content.duckduckgo.com/iu/?u=https%3A%2F%2Fi.ytimg.com%2Fvi%2F2iQgTLmMNjY%2Fmaxresdefault.jpg&f=1&nofb=1&ipt=b7347dacee48e3fc7ba9a14b07e5ed6cb2a8ac2d81c05874502702b098176dfa.

This niche area applies nanoscale material to drug delivery, imaging, and diagnostics, thus revolutionizing the field of medicine. Nanoparticles can be made to be targeted at a specific cell or tissue to deliver more therapeutic effects with less harm. As highlighted in current studies, various efforts have sought to realize these possibilities by elucidating how nanoinformatics can enhance biological and clinical applications (Anguita et al., 2012). Besides, with the integration of computational methods and modeling with experimental data, researchers can be in a position to illuminate the interactions of biological systems with nanomaterials, thereby opening the way for personalized medicine approaches. As a result, the development of new nanomedical devices represents one hopeful avenue for undertaking complex health problems, which illustrates the multidisciplinary character of nanoscience. In addition, the integration of nanotechnology with traditional disciplines such as electronics illustrates its wider influence in industries. The semiconductor industry, in fact, has used nanoscale technology innovations to adhere to Moores Law, which predicts the future doubling of transistors on integrated circuits (Gupta et al., 2011). With technology miniaturizing, new methods of fabrication like bottom-up molecular self-assembly have evolved, increasing efficiency and functionality in microelectronics. This demand for smaller but more efficient devices not only enhances performance but also facilitates the dissemination of smart technologies and green electronics. With universities redesigning curriculum with courses related to nanotechnology, students find this fascinating field extremely appealing, which paves the way for the growing demand for specialized professionals in nanoelectronics and more. The integration of nanoscience with the traditional disciplines, therefore, represents a paradigm shift in the design and application of technology in the modern world (Figure 2.1).

Definition of Nanoscience

Nanoscience is literally the study of materials and phenomena at the nanoscale, typically defined as the range between 1 and 100 nanometers. This branch of science looks at how physical and chemical properties change as materials shrink to this scale. Materials of this scale will have new properties not observed in bulk, which allows for innovation in many areas such as medicine, electronics, and energy. Behaviors at this scale can be attributed to higher surface area-to-volume ratios and quantum effects, making nanoscience a rich area to study. As researchers delve further into the nature and characteristics of nanoscale materials, they uncover mechanisms that govern how such materials can be manipulated and used in a broad range of disciplines, such as developing more efficient energy solutions and novel medical therapies. The link between nanoscience and nanotechnology is pivotal in understanding how nanoscale materials can be produced, characterized, and utilized its properties. Nanotechnology is the application of the principles of nanoscience for the development of new technologies and products that take advantage of the unique characteristics of nanoscale materials. For example, the superlubricity effect where friction is minimized at the nanoscale is a highly significant field of research in this framework. In such materials, it has been shown through research that layered materials can achieve a condition of frictionlessness through specific atomic level structure compositions, which can affect their performance in materials such as lubricants and coatings (Tomlinson et al., 2012). Such interlinking of fundamental scientific research and working technological development is the strength of change brought by nanoscience and its deployment in combating modern challenges. Apart from that, nanoscience serves to connect various scientific fields to encourage interdisciplinary innovation and research. When scientists go ahead to explore phenomena at the nanoscale, they are likely to find knowledge that straddles the boundaries of traditional science fields, which sparks developments in materials science, physics, chemistry, and biology. The exploitation of plasma-specific characteristics, for example, demonstrates the manner in which the interplay between matter and energy at the nanoscale can lead to the development of new materials possessing remarkable properties, such as self-assembled nano-solids (Meyyappan et al., 2013). Research such as this, involving interdisciplinary study, not only enhances our fundamental knowledge concerning nanoscale systems but also opens up the potential for technological progress that can lead to better materials and devices. Embracing the complexity of nanoscience will be perhaps most crucial in tackling the complicated problems facing modern society.

Definition of Nanotechnology

Nanotechnology, which is a field that operates on the nanoscale, tends to involve the manipulation of material at the 1–100 nanometer scale. This unusual scale is characterized by unprecedented physical and chemical properties not exhibited by bulk material, an aspect that has spurred scientists to explore novel applications in many areas. These attributes can be enhanced strength, lower weight, or increased chemical reactivity, which are key to applications in materials science, medicine, and energy. As a sub-discipline of nanoscience, the investigation of these nanoscale phenomena allows researchers to design certain nanomaterials, leading to breakthrough technologies such as drug delivery systems and improved catalysts. The simple flexibility and increased desire for materials at the nanoscale give the push to think of nanotechnology as being a part of modern scientific exploration and innovation. In addition to traditional chemistry and biology, nanotechnology has also become a principal solution to multifaceted problems in society. Seen as a bridge between life sciences and engineering, nanotechnology has led to new uses such as nanozymes, enzyme-like nanomaterials. As outlined in recent research, the nanozymes mimic natural enzymes' catalytic function but with the evasion of limitations such as instability and high cost of their natural forms (Gao et al., 2024). The ability to harness the intrinsic properties of nanoparticles and engineer them for specific applications has made them central in a range of applications from environmental remediation to biomedical diagnostics, thereby testifying to the ubiquity of nanotechnology's impact on both scientific and practical fronts. In short, nanotechnology is a transformative science and engineering methodology with an emphasis on accuracy and functionality at unprecedented magnitudes. It not only provides ways to enhance existing technologies but also forms completely new paradigms that go beyond conventional understanding of material properties. Evolution in the theory of nanozymology mirrors the rapid progress in the discovery and development of multifunctional nanozymes such as mimicking peroxidase, catalase, and superoxide dismutase (Liang et al., 2019). These advancements demonstrate the rapidly interdisciplinary nature of nanotechnology, which takes in elements of physics, chemistry, biology, and engineering. As research continues, possible contributions of nanotechnology to society through real-world applications in health care, energy, and materials science will keep on rising, demonstrating the importance of nanotechnology in the 21st century and beyond.

Historical Context and Development

The growth of nanoscience and nanotechnology is tightly connected with past events and social transformations that have influenced scientific studies and financing. Throughout the 20th century, significant advances in physics and materials science drove the expansion of nanotechnology as a research field. Importantly, the discovery of quantum mechanics and microscopy improvements, such as the development of the scanning tunneling microscope in the 1980s, allowed for the observation and manipulation of materials at the atomic level. These technologies not only facilitated an improved understanding of the materials at the nanoscale level but also created a deluge of scientific interest. This new science would in turn affect numerous disciplines, a dynamic relationship of scientific discovery and the needs of society typical of the wider historical context that drives innovation in science and technology. Interdisciplinary collaboration cannot be overstated in the history of nanoscience. Key to making innovation possible, inter-disciplinary collaborations between engineering, chemistry, biology, and medicine have made it possible for scientists to unlock the potential of nanotechnology applications. For instance, advances in drug delivery systems with the application of nanoparticles have revolutionized the treatment of disease in medicine. As a result, disciplines such as environmental science and chemical engineering began developing huge interest, acting as 'sinks' of scientific knowledge, dynamically affecting the trajectory of research activities in the long term. This knowledge diaspora, as advanced in contemporary accounts, illustrates how shifting agendas for research create a network of innovators who stretch the boundaries of what is possible in nanotechnology. All progress in knowledge has been propelled not only by improving technology but also by urgent historical and political imperatives that necessitate such creative solutions (Arenas et al., 2016). Furthermore, the global perspective of nanoscience and nanotechnology emphasizes the economic and social impact of its developments. Countries have realized the potential competitive advantage brought about by nanotechnology development, pushing national policies towards research funding and innovation. The convergence of scientific inquiry and economic policy was at the core of the move towards applied technology from basic research. Dominant disciplines such as materials science have become an essential component of this quest for progress, changing from conventional centers of information to inherent components of contemporary technological answers. The resulting synergy of science research and policy directives illustrates how previous contexts, determined by economic imperatives and political will, inexorably set the direction for research in nanoscience and nanotechnology, affecting individual scientists and international research communities (CSIC-UPV - Centro de Física de Materiales (CFM), 2011) (Figure 2.2).

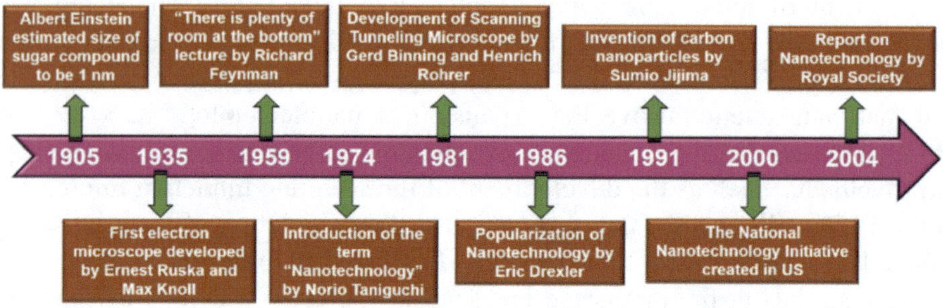

Figure 2.2. The evolution of nanoscience and nanotechnology.

Source: https://www.researchgate.net/publication/370656943/figure/fig3/AS:1143 1281157288634@1683775617013/Timeline-showing-the-milestones-in-the-history-of-nanoscience-and-nanotechnology.png.

2.1. THE SCIENCE OF NANOMATERIALS

The investigation into nanomaterials has radically redefined our thinking about material science, primarily by the examination of their unique properties at the nanoscale. Nanomaterials, as far as having structural components smaller than 100 nanometers, exhibit properties that are significantly different from the bulk. These variations are primarily brought about by the increased surface area-to-volume ratio, which generally increases reactivity and hardness, and changes optical and electrical properties. These unique features make nanomaterials the focal point of an array of uses, including catalysis, electronics, and medical diagnostics. The control and understanding of such properties allow scientists to create materials for specific purposes, thus unveiling a new paradigm of material science where imagination is fused with fundamental scientific principles (Gao et al., 2024). The invention of nanozymes is an amazing achievement in the field of nanomaterial science, as a new type of artificial enzymes with huge useful functionalities was introduced. First identified through the accidental peroxidase-like activity of ferromagnetic nanoparticles, nanozymes have proven to possess striking features such as high activity, durability, and cost-effectiveness, thus being good alternatives to traditional enzymes. From the more than 1200 various nanozymes reported in the literature, researchers have begun exploring their uses in a variety of fields ranging from biomedicine, pollution abatement, to energy conversion. The rational engineering of nanozymes, which allows for the fine-tuning of their catalytic activity, further presents their importance in accelerating research in nanotechnology. With the growth of popularity in the materials, they substantiate the inherent agility and power of transformation of nanomaterials to address complex biological and environmental challenges (Gao et al., 2024).

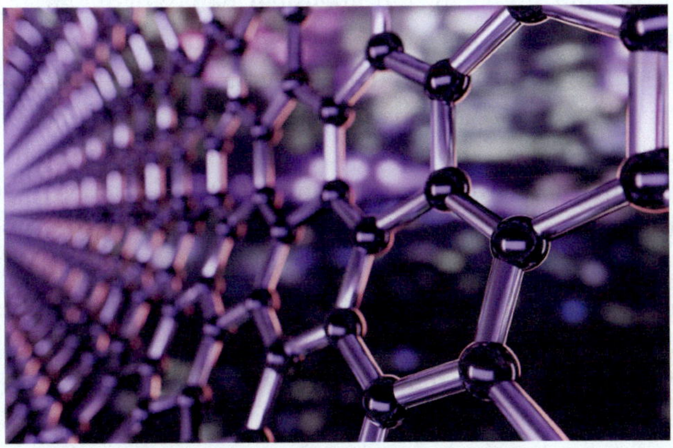

Figure 2.3. The science of nanomaterials.

Source: https://external-content.duckduckgo.com/iu/?u=https%3A%2F%2Fblog.nanochemigroup.cz%2Fcontent%2Fimages%2F2021%2F06%2FNanomaterials-in-Manufacturing-FAQs-PIC-A.jpg&f=1&nofb=1&ipt=035ca916ac3bd19504029bbc11f30a423f44f16761cf6dbe6d430d24e984b701.

The implications of nanotechnology are not only of scientific significance but also equally important in terms of ethical and societal issues that need to be carefully scrutinized. While applications for nanomaterials increase year after year in sectors such as medicine, where they are applied for the delivery of drugs as well as diagnosis, risks and benefits have come into focus as a subject of argument. Concerns regarding the environmental impact of nanomaterials, unforeseen consequences, and long-term implications on health cast a high level of urgency on legislation and public involvement. Thus, understanding the fundamentals of nanoscience and nanotechnology involves, in addition to learning about materials' properties, the creation of ethical standards and public policy frameworks that are capable of sustaining responsible innovation. In solving such intricate issues, the scientific community can see to it that innovations in nanomaterials serve society's interests (Bora et al., 2024) (Figure 2.3).

2.1.1. Types of Nanomaterials (e.g., Nanoparticles and Nanotubes)

Nanomaterials, in nanoscience and nanotechnology, have attracted lots of attention on the basis of their new properties as well as broad applications. Among various types of nanomaterials, nanoparticles (NPs) are famous for exhibiting outstanding physical and chemical behavior when scaled down to the nanometer level. These NPs can be formed from a wide variety of materials, for example, metals, ceramics, and polymers, and this leads to innovation in applications such as drug delivery, imaging, and diagnostics. Increased interaction of nanoparticles with biological systems by functionalization enables targeted therapy methods in cancer treatment. For instance, newer research concentrates on the significant roles NPs play in increasing the bioavailability of drugs and reducing cytotoxicity, thus resolving long-standing chemotherapy problems, as the literature has highlighted (Cheng et al., 2021). The second key category of nanomaterials are nanotubes, which possess extremely high mechanical, electrical, and thermal properties. Carbon nanotubes (CNTs) have, in fact, triggered a surge of interest due to their high strength-to-weight ratio and suitability for applications ranging from nanocomposites to electronics. Their structure, characterized by a cylindrical morphology, is favorably suited to conductivity and stability, and they are therefore suitable for application to nanosensors and energy storage devices. The use of CNTs in nanodevices has helped improve water purification and environmental monitoring, highlighting their potential in addressing global issues. With the rising number of studies on CNTs, continued research into the potential of CNTs to enhance sensor performance and drug delivery systems is crucial to future advancements in nanotechnology (Naresh et al., 2021). In addition to NPs and nanotubes, other nanomaterials, such as nanowires and quantum dots, are also emerging as fundamental components of most technologies. Nanowires (NWs) offer favorable features like high aspect ratios and improved conductivity and have the potential to be used in electronic devices and bio-detection sensors. On the other hand, quantum dots (QDs), or size-tunable optical properties, have been influencing imaging technologies, particularly in biomedical technology. Synthesis and characterizations of these nanomaterials are of significant research interest as they play a fundamental role in the creation of more effective diagnostic devices and biosensors. Emerging uses of these materials emphasize the importance of using nanomaterials in real-life systems practically, proving that they can enhance performance and sensitivity in every industry, and pushing the boundaries of modern nanotechnology (Naresh et al., 2021).

2.1.2. Properties of Nanomaterials (e.g., Surface Area and Quantum Effects)

Nanomaterial properties are automatically different from their bulk counterpart due primarily to their reduced size and increased surface area. When materials are reduced to the nanolevel, their volume-to-surface area ratio is significantly enhanced, leading to enhanced reactivity and interaction with the environments surrounding them. This surface area aspect is extremely critical because it controls chemical reactivity, thereby making nanomaterials valuable for numerous applications such as catalysis and drug delivery. For instance, it is possible to have more active sites in catalytic reactions with greater surface area, and this enhances reaction rates vary significantly. In addition, this aspect is critical in biomedical engineering because it makes the drug-delivery systems target effectively and reach molecular-level interactions. The transition of materials from bulk to the nanoscale not only enhances their reactivity but also enables new uses in many applications across various fields, making surface area important in nanoscience and nanotechnology (Mekuye et al., 2023). Electron confinement and resulting alteration of the electronic properties of materials enable quantum effects to be observed at the nanoscale. When materials become nanometric in size, quantum mechanics states that their electronic properties cannot be predicted by classical physics anymore. This gives rise to novel optical, electrical, and magnetic properties such as quantized energy levels and widened energy band gaps. Semiconductor nanocrystals, for instance, have size-dependent fluorescence whereby their color changes depending on their particle size, a phenomenon not found in bulk materials. These quantum phenomena have serious applications in the invention of electronic devices, optoelectronic devices, and solar cells. As a result, understanding these effects is necessary to use nanomaterials in novel technologies and applications that require atomic-scale control (Harish et al., 2022). Moreover, mechanical characteristics of nanomaterials are far superior to those of macroscopic materials. Nanomaterials possess greater strength, toughness, and durability, all of which owe a great deal to their smaller dimensions and unique structural forms. One such example is carbon nanotubes, which possess astonishing tensile strength with minimal weight, well-suited for application in material science and engineering. Mechanical properties can also be regulated by modifying the structure and organization of nanomaterials. Methods like doping or changing size and shape can yield materials with the best performance properties for certain applications, varying from aerospace engineering to biomedical implants. But this enhancement of mechanical properties also needs rigorous tests regarding their safety and environmental impact, as the toxicity profiles of nanomaterials may be very different based on the size and surface characteristics (Harish et al., 2022; Mekuye et al., 2023) (Figure 2.4).

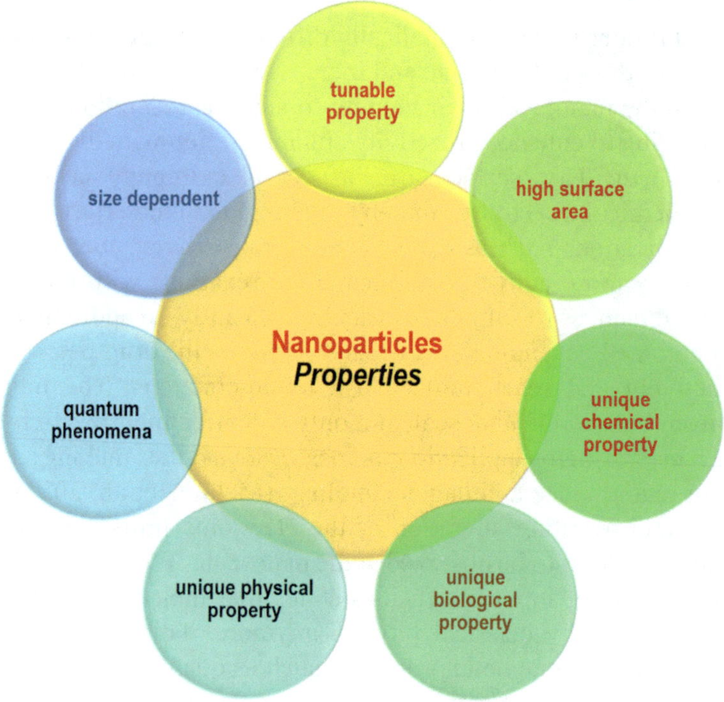

Figure 2.4. Properties of nanomaterials.

Source: https://external-content.duckduckgo.com/iu/?u=https%3A%2F%2Fwww.mdpi.com%2Fmaterials%2Fmaterials-15-05109%2Farticle_deploy%2Fhtml%2Fimages%2Fmaterials-15-05109-g001.png&f=1&nofb=1&ipt=aaf39b5492481e4faa3a87c075e0623f778167d60f8305bcac09e20090a3e7c2.

2.1.3. Synthesis Methods (e.g., Top-Down vs. Bottom-Up Approaches)

The landscape of nanoscience is basically characterized by two prominent methods of synthesis: top-down and bottom-up approaches. Top-down approach includes the process of condensing bulky large materials into nanoscale shapes using techniques such as lithography, etching, and milling. It is predominantly practiced in industries such as the semiconductor industry where precision and control are of the essence. It is also extremely energy- and resource-consuming, constraining its application to nanotechnology by sustainability. Conversely, the bottom-up strategy builds nanostructures from molecular or atomic scale, typically through chemical vapor deposition, sol-gel processes, or biomimetic synthesis. The method encompasses greater functional variability and is often associated with eco-friendly processes, which aligns with current-day sustainability goals (Perveen et al., 2025). Green synthesis, being the central part of the bottom-up process, employs indigenous materials to synthesize nanoparticles. A classical illustration is the use of plant extracts, for instance, Nigella sativa, to demonstrate the efficiency of biogenic processes towards the synthesis of silver nanoparticles (AgNPs). The advantages of utilizing this method are reduced toxicity, reduced environmental hazard, and a reduced carbon foot print when compared to conventional chemical methods. Besides, the products of the formed nanoparticles generally exhibit enhanced properties appropriate for biomedical applications and antimicrobial treatments due to their enhanced surface area to volume ratio. With the use of biological resources in the formation of nanoparticles, scientists are able to address environmental challenges pertaining to conventional production methods while simultaneously creating new materials (Chahal, 2025). Both bottom-up and top-down synthesis methods have varying advantages and disadvantages, typically depending on applications of the resultant nanomaterials. While top-down approaches provide more precision, they may lack the versatility seen in bottom-up processes, particularly those derived from biological material. The choice between synthesis methods ultimately governs the properties and functionality of resultant nanomaterials, hence their application in various fields, including electronics, medicine, and environmental science. Use of nanotechnology in these diverse areas is defined by advances in drug delivery systems and methods of environmental remediation, emphasizing the transformative position of nanoscience in modern technology (Perveen et al., 2025). With the advancement of the world further, studies to advance both methods will be essential to enhancing sustainable nanotechnology as well as addressing global challenges.

2.2. CHARACTERIZATION TECHNIQUES

Characterization techniques are of utmost importance in the area of nanoscience and nanotechnology to provide the techniques needed in determining the properties and behaviors of material at the nanoscale. Characterization techniques begin from imaging techniques, such as Atomic Force Microscopy (AFM) and Scanning Electron Microscopy (SEM), to spectroscopic techniques that analyze the structures at an atomic level. By employing these advanced techniques, researchers can retrieve such fine details of nanomaterials' surface properties, chemical composition, and crystal structure. All such fine details assist in designing materials for a specific application, thereby enhancing their functionality in various industries including medicine and electronics. In particular, nanoinformatics finds relevance in the characterization of nanomaterials, offering computer-based tools for data modeling and analysis. This is a highly dynamic field that has the ability to absorb and comprehend enormous quantities of information generated by experimental techniques. As recent studies have uncovered, the interface between nanoscience and informatics has opened up new avenues for the generation and application of nanomaterials, especially in the field of nanomedicine. The synergism between these approaches enables scientists to explore new applications, including personalized medicine, wherein information-driven wisdom guides treatment intervention based on individual patient profiling, significantly improving treatment outcomes (Anguita et al., 2012). Thin-film analysis through acoustic nanometrology is also predictive of the analytical depth achievable within this field. Surface acoustic wave techniques enable the accurate determination of elastic properties at the nanometer level, pointing to the complexity of material response to external conditions. The confinement of these waves in the region between thin films and the ambient environment gives the material a deep insight into its performance, which is of utmost importance for applications from sensors to optoelectronic devices. The innovation in the use of ultrasound techniques announces their potential for revolutionizing material characterization, thereby making room for the innovative application of nanostructured materials in new technologies (Bettucci et al., 2015).

2.2.1. Scanning Electron Microscopy (SEM)

The discovery of SEM has transformed the development of nanoscience and nanotechnology through the ability of scientists to image materials at unprecedented resolution levels. SEM operates by focusing a beam of electrons onto a specimen and the resulting interactions forming high-resolution images of the surface topography. This capability to form three-dimensional maps of nanostructures makes it possible for scientists to uncover fine details and analyze surface properties that are essential in understanding material behaviors on the nanoscale. As a result, SEM has become a critical tool in applications like materials science, biology, and semiconductor technology, where it aids nanomaterial quality assessment and characterization. High-energy electron interactions offered through SEM also facilitate many analytical methods, such as energy-dispersive X-ray spectroscopy, that can provide elemental composition data in conjunction with imaging. Besides, the use of SEM is especially significant in the study of photocatalytic nanomaterials, whose function has significant value in environmental purification.

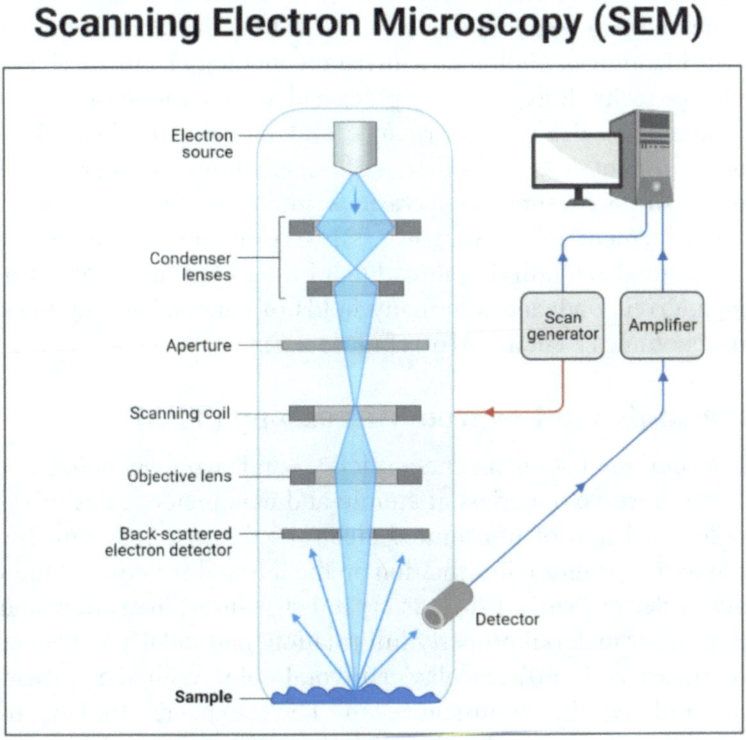

Figure 2.5. Scanning electron microscopy (SEM).

Source: https://microbenotes.com/wp-content/uploads/2021/07/Scanning-Electron-Microscope-SEM.jpeg.

According to recent literature, nanostructures such as titanium dioxide (TiO_2) are prepared and characterized via SEM in order to enhance their photocatalytic activity for use in the environment. For instance, hydrothermally prepared potassium titanate has new nanostructures whose photocatalytic adsorptivity as water treatment agents is analyzed through SEM. The minute details showcased by SEM, including surface area and porosity, are directly proportional to the materials' efficiency in degrading and adsorbing the pollutants, highlighting the applicability of the techniques to nanotechnology.

This SEM integration in synthesizing and designing nanomaterials is critical in advancing innovative water treatment and pollution management sustainable practices (Shahid et al., 2015). Besides environmental monitoring, SEM development also aligns with general education and research endeavors in the photonics field, as in organizations like the Boston University Photonics Center. Their interdisciplinarity in nanotechnology is centered on the importance of cutting-edge imaging techniques such as SEM to define and engineer photonic devices. The blending of photonics with nanotechnology facilitates innovation in medical device technology, whereby precise characterization of nanostructured surfaces maximizes device performance and biocompatibility. The ongoing studies in neurophotonics and other areas of photonics further reflect SEM's significant position in supporting various interdisciplinary projects and the broader advancement of nanoscience. SEM is therefore not only a useful analysis tool but also a unifying thread that bridges fundamental research with applications, driving advances in many fields of nanotechnology (The Boston University Photonics Center, 2016) (Figure 2.5).

2.2.2. Transmission Electron Microscopy (TEM)

TEM is a common tool in nanoscience and nanotechnology, enabling researchers to observe materials at atomic and nanometer scales of resolution. TEM employs a beam of transmitted electrons through very thin specimens, resulting in high-resolution information on the internal structure of the samples. The ability to detect features such as crystal structures, interfaces, and defects is most crucial for material property information, particularly in nanostructures where interfaces and surfaces play a pivotal role. With the growing range of nanotechnology, the significance of TEM expands further, triggering innovations in numerous applications ranging from semiconductor processing to materials science. The unique characteristics enabled by TEM are assisting in the interpretation of the fine interplay between microstructure and function in nanoscale materials, creating a solid foundation for further development. TEM's analytical capability extends beyond simple imaging, with techniques such as Electron Diffraction and Energy Dispersive X-ray Spectroscopy (EDS), offering information on crystallography and elemental composition, respectively. These complementary techniques enhance the value of TEM in nanomaterial analysis because they meet the call for extensive characterization. For example, in analyzing layer materials like doped bismuth ferrite, TEM not only assists in imaging morphology of the nanofibers but is also useful in determining their crystallinity and phase purity (Mishra et al., 2025). This information is essential for the design of material properties for technological uses in electronic devices. Integration of these sophisticated characterizations, scientists can gain a deeper insight into the behavior of the material at the nanoscale, thus facilitating the customized design of nanostructured systems aimed at particular functions. The peroxidase activity of iron oxide nanozymes has also been proven in recent research, mirroring the overall uses of nanoscale materials in biomedicine. Synthesis and characterization of iron oxide nanozymes have shown bright catalytic and targeted drug delivery functionalities, where TEM may be employed to image their structure stability and functionalization processes (Shastri et al., 2025). For example, rhamnolipid-capped iron oxide nanozymes exhibit enhanced stability and catalytic activity over conventional enzymes. While using TEM as the eyes of the scientist, researchers can see the nanoscale interactions and alterations leading to better therapeutic results. Such developments highlight the role of TEM as a vital tool in ongoing studies of nanoscale materials, promoting a greater understanding of their diverse applications across nanotechnology disciplines.

2.2.3. Atomic Force Microscopy (AFM)

AFM has emerged as a paradigm tool in nanoscience and nanotechnology, providing us with unprecedented insight into the material's structural and mechanical properties on the nanoscale. AFM is a scanning probe microscopy method that works by involving a sharp tip mounted on a cantilever that physically interacts with the sample surface. The cantilever is bent by the tip-sample forces and is quantified to create high-resolution topographical images. Importantly, AFM can be operated in various environments, i.e., air or liquid, making it an extremely useful technique for the analysis of biological specimens as well as nanostructures. Through provision of information on surface morphology, actual mechanical properties, and adhesion forces, AFM plays a crucial part in nanoscale material creation and description that is a necessity for advancement in the fields of material science and nanomedicine. Applications of AFM extend beyond mere imaging to include mechanical characterization of materials, which finds great significance in the context of nanotechnology. For instance, material mechanical properties are crucial in uses ranging from coatings to drug delivery systems. In research involving transition metal carbides and nitrides known as MXenes, AFM has been used to explore their elastic properties at the monolayer scale. Recent findings indicate that the efficient Youngs modulus of Ti_3C_2T x, a widely studied MXene, is as high as 0.33 ± 0.03 TPa, a remarkable figure for two-dimensional nanomaterials. The realization of mechanical properties by AFM not only puts in the spotlight the strength and flexibility of such nanomaterials but also paves the way to their application in composite structures for niche applications such as protective coatings and nanoresonators (Lipatov et al., 2018). In addition to mechanical properties, AFM also provides useful information on the interactions between NPs and the biological system, a fundamental aspect of nanomedicine. The mechanisms of NP uptake are significant to understand to enhance therapeutic applications, particularly for the targeting of disease states such as cancer and inflammation. AFM is a crucial tool in assessing the efficacy of various nanoparticles in cellular uptake studies, revealing how physical and chemical properties influence endocytosis. Research has indicated that controlling these properties can significantly influence NP targeting of the desirable cells while minimizing clearance by immune cells. With the advancement of nanomedicine, the combination of AFM in nanoparticle studies could potentially yield optimized drug delivery systems and therapeutic benefits (de Almeida et al., 2021). Through bridging the knowledge gap in NP interactions with cell mechanisms, AFM underscores its central role in the emerging world of nanotechnology (Figure 2.6).

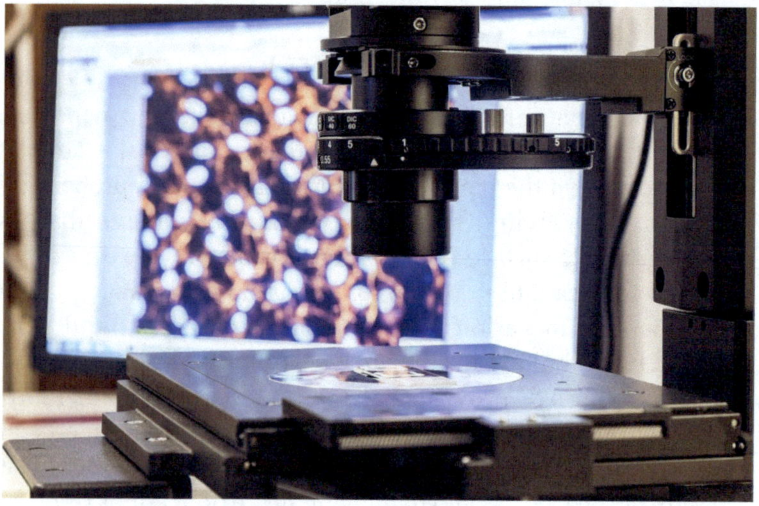

Figure 2.6. Atomic force microscopy for nanoparticles.

Source: https://www.azooptics.com/images/Article_Images/ImageForArticle_2317_16613452207212059.jpg.

2.3. APPLICATIONS IN MEDICINE

Medical applications of nanoscience and nanotechnology are among the most revolutionizing advances in contemporary biomedical research. Technology advances in this field enable the creation of new nanomaterials capable of enhancing prevention, diagnosis, and treatment of several diseases. For instance, nanoparticles have the potential to target drugs with high accuracy, significantly increasing therapy but limiting side effects. This drug targeting system is of great importance in cancer treatment, where existing therapy is prone to killing normal tissue as well as cancer tissue. Furthermore, the application of nanotechnology in imaging devices enhances the visualization of cellular and molecular biological processes, paving the way for more accurate diagnosis and personalized medicine approaches, which are imperative in the current healthcare system. Ultimately, these advancements touch on how nanotechnology can revolutionize patient treatment and therapeutic techniques in the medical field. Lastly, the newly arising science of nanoinformatics is instrumental in the formation of medical applications of nanotechnology. This emerging discipline is committed to using computational skills for simulating and modeling the interaction between nanoparticles and biological systems.

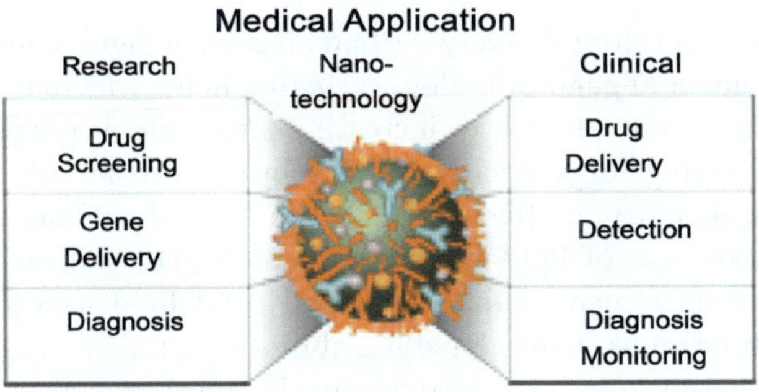

Figure 2.7. Medical applications of nanotechnology.

Source: https://www.researchgate.net/publication/228719040/figure/fig2/AS:302065235578881@1449029372744/Medical-applications-of-nanotechnology-The-size-and-tailorability-of-nanoparticles-may.png.

By merging huge quantities of data, nanoinformatics makes it possible to develop personalized drugs tailored to individual patient profiles. Interestingly, it makes clinical and experimental data integration possible, which is significant in determining toxicity and efficacy. As illustrated by recent works, the future of nanomedicine is envisioned to follow closely behind genomic medicine, particularly in terms of the integration of modeling, simulating strategies, and development of good scientific ontologies. These may enable advances in individualized health care and establish informatics' central role in making nanotechnology in medicine a reality (Anguita et al., 2012). Finally, collaboration between industry and academia is essential in converting nanoscience research into practical applications in the medical field. Centers like the Boston University

Photonics Center illustrate the interface by fostering multiparadigm research and innovation through photonics and nanotechnology. These centers are some of the pioneers in investigating the application of advanced materials in medicine, as evidenced by their work to create biophotonics for cancer treatment and customized chemotherapy regimens. Through securing key research funding and establishing educational programs, these centers encourage discovery and innovation in nanomedicine. This, in turn, the continued development and use of nanotechnology for medical practice have the potential to revolutionize the medical therapies and improve patient outcomes, reflective of the growing role that interdisciplinary collaboration is having in solving complex medical issues (The Boston University Photonics Center, 2015) (Figure 2.7).

2.3.1. Drug Delivery Systems

The advancement of drug delivery systems has been significantly influenced by the advancement of nanotechnology, ushering in the way to new therapeutic modalities. These systems seek to increase pharmacological agent safety and efficacy by delivering drugs specifically, thereby minimizing side effects as well as maximizing therapeutic effects. One of the most significant developments in this area is the use of lipid-based nanoparticles (LBNs), which contain a variety of structures such as liposomes and solid lipid nanoparticles. With an array of applications, LBNs have the ability to stabilize drugs, control the release profiles, and offer a targeted approach—most crucial in the treatment of complex diseases such as cancer. The characterization and design of these nanocarriers become central to their therapeutic efficiency, forming a critical divergence from drug delivery mechanisms (Saha et al., 2024). Furthermore, the incorporation of advanced nanostructures has idealized the role of drug delivery systems, addressing long-standing issues in medicine. Nanotechnology enables the development of drug carriers that are capable of crossing biological barriers such as cellular membranes to ensure that therapeutics are delivered to their target locations more efficiently. Dendrimers and nano-polymer nanoparticles, for instance, have been engineered to carry therapeutic agents and modify their pharmacokinetic properties. These technologies not only stabilize and enhance the solubility of drugs but also improve their absorption within cells. These advanced delivery systems have the ability to allow site-specific action, thereby achieving maximum therapeutic effect and minimizing possible side effects (Kumar et al., 2023). The use of stimuli-responsive polymers also indicates this advancement, allowing dynamic control over drug release based on environmental stimuli. Generally, the use of drug delivery systems based on nanoscience has marked a milestone shift in therapeutic approaches in medicine. The synthesis of smart polymers and nanostructures is an interface of cutting-edge technologies aimed at the delivery of drugs with extremely precise targets. It promises not just a more effective pharmacological effect but also the tailoring of treatments to individual patients' specific needs, thereby reshaping paradigms in patient care. As the technology continues to improve, further research and regulatory evaluation will be required to realize the full potential of such nanotechnological progress for clinical application. Through rigorous testing and preparation, drug delivery systems will completely redefine the landscape of therapeutic intervention and global health outcomes (Kumar et al., 2023; Saha et al., 2024).

2.3.2. Diagnostic Tools and Imaging

Advances in diagnostic tools and imaging technology have come a long way in the area of nanoscience and nanotechnology. Nanoparticles are becoming crucial components for establishing sophisticated imaging techniques with enhanced sensitivity and specificity to identify diseases. Quantum dots and gold nanoparticles are utilized, for instance, to create contrast agents much more effective than traditional imaging modalities. These nanostructured materials can be used to target specific cell types, thereby enabling precise imaging of the tumor or other pathological diseases. Utilization of such advanced nanomaterials in imaging processes not only optimizes the resolution but also paves the way towards early diagnosis, which is of paramount significance in fields like oncology and cardiology, with the ultimate outcome being better patient outcomes. Hence, the convergence of diagnostic imaging and nanotechnology brings with it a paradigm change to more effective clinical practice. In addition, the integration of nanotechnology into diagnostic devices is fundamentally rooted in the field of nanoinformatics, which maximizes the performance of such technologies through sophisticated data analysis. Through computational approaches, researchers are able to model and interpret complex interactions between biological systems and nanomaterials to best utilize them in diagnosis. Of specific interest is the informatics integration of nanoscale information into larger biological databases, where it is simpler to recognize biomarkers and improve diagnostic precision of conditions. This bidirectional stimulation of nanotechnology with data informatics drives innovation in designing personalized diagnostic platforms, as witnessed technological leaps in genomic research. As highlighted by ongoing efforts, there is immense scope for visionary research that crosses the domains of nanoinformatics and nanomedicine to enhance diagnostic performance and enhance treatment protocols (Anguita et al., 2012). Finally, the role played by academic and research institutions toward the creation of these diagnostic tools cannot be overemphasized. Facilities like Boston University Photonics Center are among the foremost institutions in developing new technologies based on light-based methods of imaging and diagnostics. Their emphasis on cross-disciplinary research provides an excellent environment for innovation, where photonics breakthroughs are being translated into health care realities. These partnerships not only lead to the creation of novel materials for the improvement of imaging methods but also address real-world medical needs, such as cancer detection and monitoring. Through the large amounts of money and research projects it provides, institutions like this guarantee the practical application of theoretical research into applied diagnostics breakthroughs, ensuring the bridging of the gap between medical use and nanoscale innovation (Figure 2.8) (The Boston University Photonics Center, 2015).

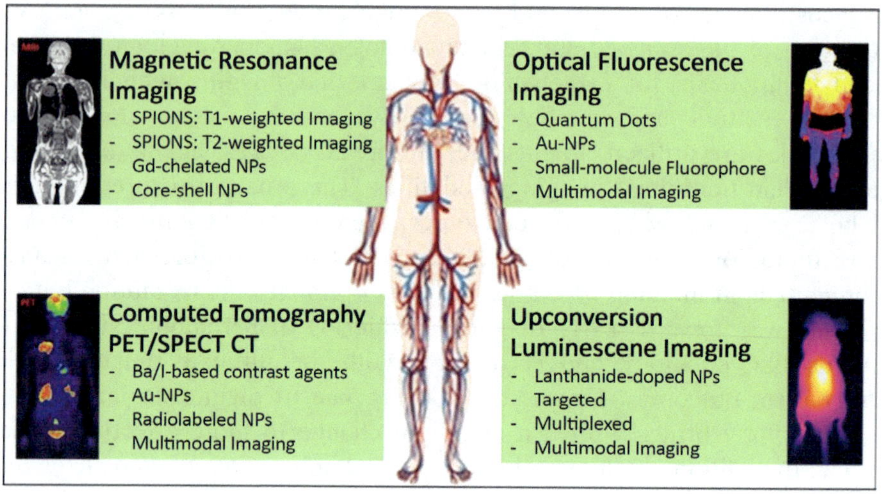

Figure 2.8. Application of nanotechnology in medical diagnosis and imaging.

Source: https://ars.els-cdn.com/content/image/1-s2.0-S0958166921002445-gr1.jpg.

2.3.3. Cancer Treatment and Therapy

The treatment of cancer has advanced significantly due to developments in nanoscience and nanotechnology. By applying the unique properties of nanomaterials, researchers are creating new therapies that are capable of making cancer treatment more effective with lesser side effects. Nanoparticles can be engineered to deliver drugs to cancer cells directly, thus enhancing localization and bioavailability, which is especially critical in the treatment of very aggressive tumors. Targeted delivery not only enhances the concentration of therapeutic agents at the site of action but also reduces the impact on nearby healthy tissues. This precision medicine demarcates the shift towards more patient-specific treatment guidelines, giving hope for improved treatment outcomes. Moreover, the application of nanotechnology in imaging techniques makes tumor visualization stronger at an earlier stage, providing timely treatment and potentially altering prognosis (Anguita et al., 2012). Nanoinformatics is also emerging as an important branch of nanotechnology, particularly in addressing cancer therapy-related problem-solving. Through focusing on the integration of computational strategies with nanomaterials, nanoinformatics aims to rationalize the design and application of nanomedicine. This inter-disciplinary field enhances our understanding of nanoparticle interactions with biologic systems and maximizes therapeutic strategies. It makes it possible to work on researching large data bases for nanoparticle toxicity and their clinical relevance. Moreover, the generation of wide-ranging databases and model approaches can facilitate finding new therapeutic drugs in addition to opening avenues towards future research. Advances in nanoinformatics, like advances in genomic science, can go a great distance in pushing cancer therapy and prevention advances further along at a faster rate, improving the pace at which new therapies are brought to the table and validated (Anguita et al., 2012). As research continues to advance, nanotechnology for the treatment of cancer is not just about drug delivery. Recent studies have demonstrated ways by which nanoparticles may be utilized in order to engage in tumor immunotherapy as delivery agents for vaccines or immune checkpoint inhibitors. Through enhancing immune responses against cancer, these methods leverage the body's own immune system to combat cancer cells more effectively. In addition, the use of nanostructured materials for photothermal and radiation therapy constitutes a new field where physical phenomena are utilized to destroy cancer cells. These new approaches not only increase treatment specificity but also provide opportunities for combination therapies, leading to synergism. Nanotechnology, thus, not only redesigns therapy paradigms but also is a force driving mixed-mode research endeavors towards cancer eradication, solidifying its status as a cornerstone for future cancer therapy (Conde et al., 2015).

2.4. APPLICATIONS IN ELECTRONICS

Nanotechnology has transformed the face of electronics to a great extent, enabling the development of devices and systems that are compact, faster, and more efficient. Materials at the nanoscale possess unique electrical properties that differ greatly from those of their bulk forms, and these have transformed various electronic components. For example, semiconductor technology is privileged with the use of nanoscale materials that support enhanced mobility of charge as well as reduced consumption of energy. By integrating these materials, manufacturers can fabricate transistors not only reduced in size but also with increased performance characteristics, paving the way for more powerful microprocessors and computing systems. Such a shift towards nanostructured electronics takes on special significance in the creation of more efficient data processing and data storage systems, highlighting the significance of nanotechnology for the future of computing and electronic devices. One of the most prominent applications of nanotechnology to electronics is in the area of single-molecule junctions, where single molecules are applied as active material within electrodes.

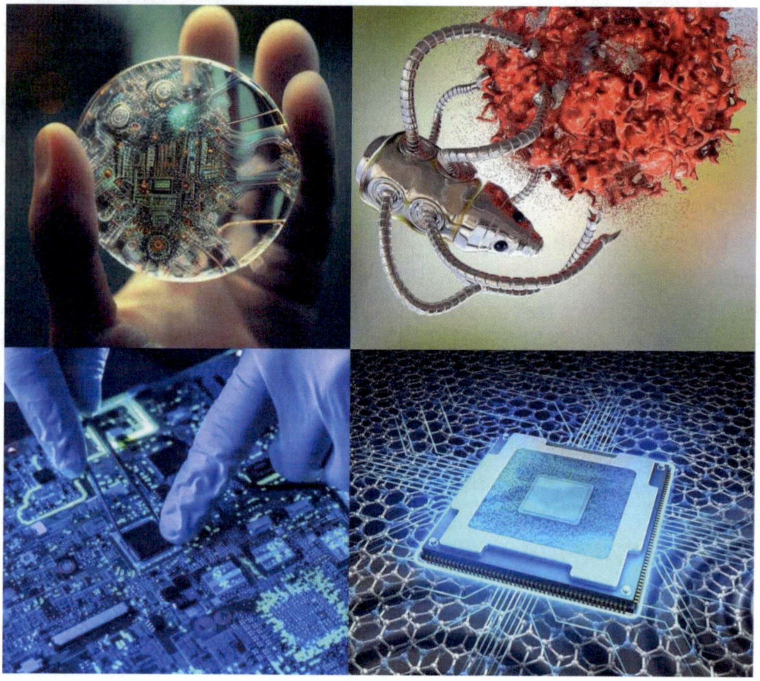

Figure 2.9. Application of nanotechnology in electronics.

Source: https://ieearc.com/wp-content/uploads/2023/10/Application-of-Nano-Technology-in-Electronics-Devices.jpg.

Recent advances have sharply leveraged the use of ionic liquids in such junctions to facilitate charge transfer and redox processes and thereby redefine our perceptions of molecular behaviors at the nanoscale. These advances are seminal pillars for advanced electronic and sensing technology, and the evaluation of ionic liquids' uses is more profound than ever before. As electrolytes and structural elements, ionic liquids can control electrochemical behavior and processes and hence maximize single-molecule device function.

These developments open a rich universe of research opportunities, including interrogating catalysis on an individual molecule's scale and the functionalization of devices for application-specific purposes, thus underscoring the significance of ongoing studies in this exciting multidisciplinary field (Zhou et al., 2024). Additionally, the overall impact of nanoscience extends far beyond mere electronic device enhancements; it indicates a broad array of applications across a myriad of fields, including energy and medicine. The distinct properties of nanoscale materials enable them to discover new applications in fields such as energy storage technologies, where nanostructured electrodes are capable of producing substantially improved capacities and efficiencies of batteries and supercapacitors. Even in the domain of nanomedicine, the technology can make drug delivery systems more efficient with targeted and controlled drug release mechanisms at the cellular level. This diversity does little more than to refer to the interdisciplinary nature of nanotechnology, and with it, accompanying social and ethical considerations regarding how it gets into consumer products and industries. As debated, the variety of applications that result from nanotechnology refers to the degree to which it can change ordinary practice and industries and thereby make significant strides in what we can do around the issues of modern technology (Bora et al., 2024) (Figure 2.9).

2.4.1. Nanoelectronics and Transistors

Development in nanoelectronics has been a benchmark achievement in the history of transistors in revolutionizing the world of contemporary electronics. With the need for more efficient and smaller-sized devices growing exponentially, the scaling down of transistors became the priority. With traditional lithography techniques hitting roadblocks, nanotechnology proves to be a significant partner by making leaps in fabrication methods. This leap not only adheres to Moores Law but also introduces a plethora of novel architectures, such as carbon-nanotube-based transistors and quantum computing. The innovation of these next-generation technologies redraws the boundaries of electronic capabilities, advancing speed and efficiency simultaneously while reducing power usage. These developments are integral to the fulfillment of continually escalating performance and sustainability requirements of the electronics industry, thereby paving the way for a high-performance and resource-aware integrated circuit future (Gupta et al., 2011). The secret to advances in nanoelectronics is at the nanoscale field-effect transistors (nanoFETs), which are significant devices in bridging the electronic-biological gap. These nanoscale devices have given an unparalleled capability for the study of cellular function at the molecular level. For instance, nanoFETs that mimic the size and operation of biological nanostructures facilitate electronic recording of cellular interaction, shedding light on cellular behavior and signal pathways. NanoFET compatibility with biological systems allows researchers to carry out multiplexed extracellular sensing, significantly enhancing the capability to acquire data in biological systems. Thus, nanoFETs are ushering in the era of revolutionizing biomedicine applications by providing the promise to revolutionize diagnostics and therapeutics through enhancing interaction with biological materials (Lieber et al., 2013). The application of nanotechnology in the development of transistors not only enhances electronic performance but also facilitates cross-disciplinary research combining biology, engineering, and materials science. With the introduction of novel transistors like spintronics and single-electron devices, the field of applications expands from quantum computing to bioelectronic circuits. These versatile developments show the importance of educational courses that equip future engineers with the capability to deal with this rapidly evolving field. Incorporating nanotechnology topics in VLSI education enables the students to gain key knowledge and skills necessary to excel in next-generation nanoelectronics. Engaging students in the research of such groundbreaking concepts allows them to contribute meaningful research to the discipline and meets industry's growing demand for competent nanoelectronics engineers, thus fortifying the all-important partnership between education and technological progress in nanoscience (Gupta et al., 2011) (Figure 2.10).

Figure 2.10. Nanoelectronics and transistors.

Source: https://news.mit.edu/sites/default/files/images/202411/MIT-Tunneling Electrons-01-press.jpg.

2.4.2. Quantum Dots (QDs) in Displays

In nanotechnology, QDs are also an innovative material used in display technology, particularly televisions and smartphone screens. Quantum dots at the nanoscale are distinctive semiconducting particles with unique optical properties, making screens extremely bright and highly colored. The size of quantum dots can be accurately defined using synthesis, and the latter affects their emission spectrum—a property known as quantum confinement. This is to the effect that through varying their size, manufacturers are able to tailor the color properties of QDs for use in a variety of applications ranging from high-definition television sets to portable electronic gadgets. The integration of quantum dots into display technology not only enhances the quality of viewing but also opens up possibilities for more efficient energy utilization, which is fast becoming an issue in a world geared towards sustainability and energy conservation. As quantum dots continue to develop, research has highlighted their future applications in display technology. One of the fields that demonstrate their utility is the use of QDs in light-emitting diodes (LEDs). Quantum dots are used as a down-conversion layer that absorbs blue light emitted from LED sources and re-emits it in other colors, essentially enhancing the color gamut of the display. This capability matters because it makes possible displays that provide better color depiction and wider color delivery, thereby improving the user experience. Furthermore, this technology conforms to the general trends in nanoscience in which it is increasingly being realized that control over nanoscale materials can become a means of providing higher-order functionalities not possible by conventional methods (Meyyappan et al., 2013). But deployment in displays has not been without its challenges. Concerns regarding the environmental impact of the materials employed in QD have arisen, especially with respect to cadmium-based quantum dots, which are poisonous. This has prompted the research towards deployment of cadmium-free quantum dots and alternative less poisonous forms. Furthermore, ongoing nanoscience research focuses on enhancing the efficiency and stability of QDs under varying light intensities, addressing functional concerns regarding lifespan in consumer devices. The drive for improvement is part of a broader movement within nanotechnology—the requirement for developing safer, more efficient materials that accommodate challenges of modern technology without creating greater ecological traces (Jensen et al., 2016).

2.4.3. Energy-Efficient Devices

The advent of energy-efficient appliances has attracted significant attention in an era of price-soaring energy demands and ecological consciousness. Underpinning this revolution is the field of nanoscience and nanotechnology, which has opened up new materials and techniques that maximize energy efficiency in numerous fields. Nanomaterials, particularly nanoparticles, have unique physical and chemical properties which differ substantially from materials in bulk and therefore enable improved energy transfer and storage options. For example, application of nanostructured materials in solar cells enhances their efficiency through enhanced light absorption and electron transfer. In extension, these technologies not only promise to reduce energy consumption but also pave the way for sustainable technologies that symbolize environmental responsibility and economic viability, encapsulating the call for modern engineering practice to support the challenges mounted to energy consumption at the global scale (Joudeh et al., 2022). In consumer electronics, the application of nanotechnology to energy-saving devices has brought spectacular levels of performance and sustainability.

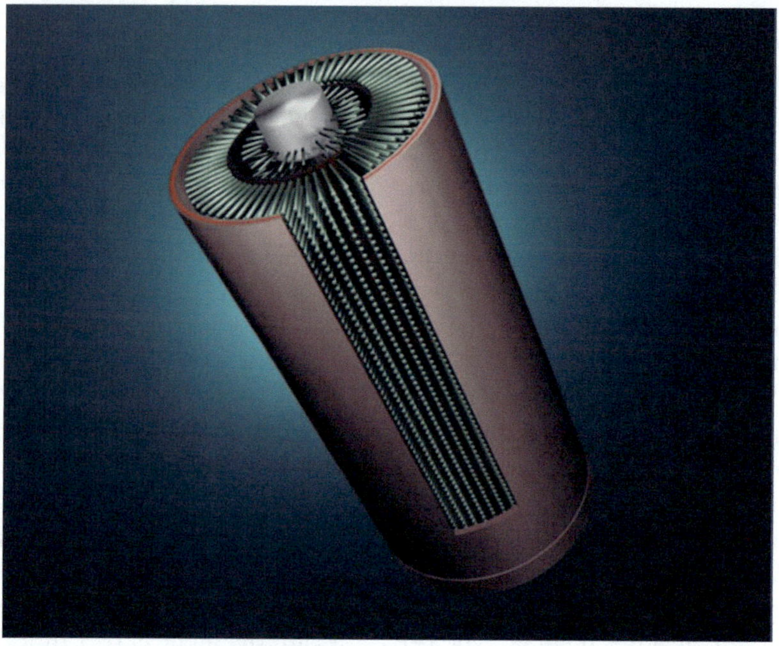

Figure 2.11. Energy-efficient devices.

Source: https://www.mouser.com/blog/Portals/11/Nano_EnergyStorage.jpeg.

Liquid crystals in various display technologies demonstrate the benefit of nanostructured materials. These materials can be transformed into mesophases that enhance the responsiveness and efficiency of devices such as televisions, computers, and smart phones. Secondly, the incorporation of metallomesogens in display technology shows a promising avenue for the creation of multifunctional materials with higher quality properties, including photo and electroluminescence. This advancement not only leads to lower energy consumption but also widens the ranges of use in energy-saving devices, particularly in optoelectronics where energy efficiency and performance are critical (Cristián Cuerva et al., 2021). As such technologies evolve, they open the way for a new line of energy-saving products that achieve balance between performance and least damage to the environment. A move towards the use of biogenic nanoparticles in energy-saving products is a revolutionary measure towards sustainable nanotechnology. By building nanoparticles by plant and microorganism-mediated methods, researchers have developed a cleaner alternative to current technologies, which are energy-wasting and ecologically damaging. These nanoparticles possess unique properties that can be exploited for various applications, including catalysis in chemical reactions to improve the efficiency of energy conversion. Advances in the characterization of such materials make it increasingly feasible to introduce such materials into commercial products applied for energy wastage prevention. The thorough understanding of the physicochemical properties of nanoparticles, as reported in recent reviews, also highlights the importance of this cross-disciplinary approach in catalyzing the development of energy-efficient technologies that perform efficiently and are environmentally friendly (Joudeh et al., 2022) (Figure 2.11).

2.5. ENVIRONMENTAL IMPLICATIONS

The assessment of environmental impacts related to nanoscience and nanotechnology carries with it opportunities and threats that need to be taken into account in a thoughtful manner. Nanomaterials, due to their unique properties at the nanolevel, are useful for enhancing environmental protection measures, including the creation of more efficient catalysts for the degradation of pollutants and very efficient filters. The same features that make nanomaterials attractive also carry huge risks. Their presence in the environment necessitates that their toxicological consequences on ecosystems and human health become a concern. As scientists and policymakers tread this new frontier, they will need to take heed of this balancing act between ecological stewardship and technological advancement, more or less weighing the benefit of nanotechnology against potential environmental risk and devastation (Lakhtakia et al., 2005). When examining public perception of nanotechnology's ecological footprint, the stakeholder communication tactics must be taken into consideration. The variety of sites for discourse—scientific communities through to environmental interest groups—has an effect on the diversified understanding of how nanotechnology interacts with nature. Engaging the public in debate regarding nanotechnology's benefits and risks is essential for guaranteeing decision-making based on informed choice. These interactions create room for deliberation by which the trajectory of nanoscience research and development may be instructed by the issues at stake for citizens. Responding to citizens' opinions and practicing tact to the ethicalities at play, the creation of trust in nanotechnology is a possibility and necessity as it ultimately decides policy regulation and investment in sustainable processes (Murphy et al., 2009). Ultimately, the environmental effect of nanotechnology and nanoscience will be significantly dependent on future regulation and research anticipating future issues. As the technology evolves, it is essential that well-rounded lifecycle analyses of the nanomaterials are conducted. This includes a review of the manufacturing process, their application, and their eventual recycling or disposal. Such combined analyses can inform best practices for reducing adverse environmental impacts while securing social gains. Furthermore, coordination among scientists, entrepreneurs, government agencies, and non-state actors is essential in establishing robust regulatory systems. By adopting a comprehensive perspective on caring for the environment in nanotechnology, stakeholders can navigate the complexities of innovation without compromising ecological integrity and public health (Lakhtakia et al., 2005).

2.5.1. Nanotechnology in Pollution Control

Nanotechnology has emerged as a pioneering technology in the fight against environmental pollution, leveraging the unique properties of nanostructured materials to develop innovative solutions. Perhaps the most significant area of application is the cleaning of water and land from heavy metals and organic pollutants. Nanoparticles, particularly those made of metals such as titanium and iron, possess a high surface area-to-volume ratio, which enhances their reactivity and adsorptivity. These features enable them to precipitate and bind toxic substances effectively, hence enabling their removal from polluted environments. For instance, it has been proven that superparamagnetic iron oxide nanoparticles can remove heavy metals from aqueous solutions, thereby being effective agents for environmental cleanup. These breakthroughs not only highlight nanotechnology's prospect but also the necessity for integrating these methods into integrated pollution control programs. In addition to water treatment, nanotechnology is also revolutionizing air quality management through the synthesis of nanoparticles with the ability to adsorb airborne contaminants. Titanium dioxide (TiO_2) and zinc oxide (ZnO) nanoparticles are nanomaterials with embedded photocatalytic activity that degrades volatile organic compounds (VOCs) by the action of ultraviolet light. This renders it possible to convert harmful contaminants into less toxic substances, hence reducing air pollution in urban environments. Besides, the nanomaterials can be added to building materials, and this leads to the creation of self-cleaning surfaces as well as air cleaning structures. Utilization of such materials in everyday surroundings promotes a sustainable way of controlling pollution that is not just concerned with the efficiency of nanotechnology but also with its capacity to enhance public health and protection of the environment. Finally, nanotechnology is also involved in controlling pollution in the area of monitoring and assessment of environmental contaminants. New nanosensors have been developed to detect a range of pollutants with high sensitivity and specificity. Nanoparticles made of metals or carbon are typically used as sensors with special optical or electronic properties when they come in contact with specific chemicals. For example, certain nanosensors can detect the presence of heavy metals or organic pollutants at concentrations much lower than present protocols, while offering real-time data that can inform policy-making in regulation and remediation efforts. Moreover, as recognized in emerging studies, the incorporation of nanotechnology can aid in the understanding of toxicological effects related to exposure, in turn facilitating a more precise risk assessment and management strategies (Baatrup et al., 2012; Hanna et al., 2017). Collectively, all these advancements demonstrate the multifaceted capabilities of nanotechnology in mitigating pollution problems, hence solidifying its incomparable role in environmental science (Figure 2.12).

Figure 2.12. Nanotechnology in pollution control.

Source: https://www.echemcom.com/data/ecc/coversheet/991642841926.png.

2.5.2. Nanomaterials in Water Treatment

Applications of nanomaterials in water treatment have emerged as a revolutionary advance in preventing water quality problems globally. Nanotechnology enhances traditional purification processes by a variety of new methods such as adsorption, photocatalysis, and membrane filtration. In particular, nanomaterials like titanium dioxide (TiO_2) nanoparticles have generated much interest for their photocatalytic use, which can enable the decomposition of organic pollutants by UV light. By making use of the unique features at the nanoscale, scientists can be able to make pollutant removal more effective and hence supply cleaner and safer water. This innovation addresses not only the immediate demand for effective wastewater treatment methods but also minimizes environmental pressures typically associated with conventional methods, highlighting the central role of nanotechnology in water management technologies (Meyyappan et al., 2013). Another innovation in nanotechnology application for water treatment is the development of nanostructured adsorbents that have increased surface area and enhanced adsorption capacity. Such adsorbents can be engineered to trap certain pollutants, such as heavy metals and organic contaminants. For instance, potassium titanate nanofibers synthesized using hydrothermal processes have shown outstanding ability to both adsorb and photocatalytically degrade toxic compounds in water media.

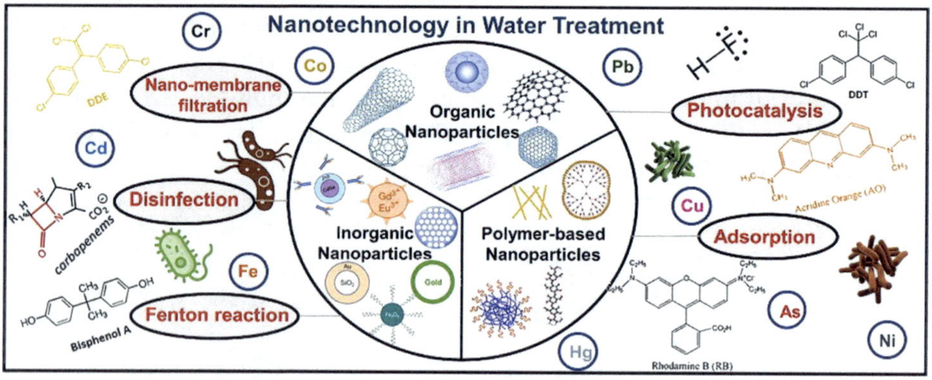

Figure 2.13. Nanomaterials in water treatment.

Source: https://ars.els-cdn.com/content/image/1-s2.0-S0960852421013420-ga1.jpg.

The unique morphological and compositional features of such nanometer-scaled materials allow the engineering of specific interactions with pollutants, resulting in better removal efficiency in treatment operations. This efficacy and abstractness are important in obtaining cost-effective solutions to meet rigorous water quality requirements, which is a reflection of the significance of integrating nanomaterials into water treatment technologies (Meyyappan et al., 2013). Besides, the addition of nanomaterials not only increases treatment efficiency but also provides new mechanisms such as solar photocatalysis. Solar radiation as an environmentally friendly energy source for driving chemical reactions represents a clean solution compared to conventional energy-scratching mechanisms. Recent studies have indicated that titanium dioxide-based photocatalysts have been effective in degrading organic pollutants like methylene blue under solar irradiation. This not only decreases the dependence on chemical treatments but also improves photocatalyst regeneration, ultimately making the water treatment process more sustainable. With ongoing advances in the field, interdisciplinary collaborations between nanoscience and environmental engineering will become important to design robust, efficient, and sustainable solutions to water treatment problems (Shahid et al., 2015) (Figure 2.13).

2.5.3. Potential Risks and Safety Concerns

As the number of advances in nanotechnology dispersed across various industries grows, there is an acute need to address the risks and safety concerns of its applications. The unique properties of nanomaterials (NMs), such as their large surface area-to-volume ratio and novel physicochemical characteristics, can lead to unanticipated interactions with biological systems and the environment. Although these properties can improve product performance in areas such as medicine, food safety, and environmental management, they can also pose toxicological threats whose implications remain to be fully understood. This complexity in the behavior of nanoscale materials demands rigorous assessment practices for establishing potential risks of their use throughout production and consumption processes, pointing toward the necessity of transitioning to systemic safety control in the nanotechnology sector (Malik et al., 2023). Application of nanotechnology in food and agribusiness industries, being promising but potentially having severe safety concerns as well, similarly necessitates investigation. Engineered NMs are increasingly being applied in food packaging, preservation, and quality enhancement, but there is concern regarding their compatibility with biological systems in terms of toxicity and bioaccumulation. For instance, certain nanomaterials have the potential to inadvertently enter the food chain and cause some unknown effect on human health and the ecosystem. Although some regulations exist within some developed countries to assess these risks, the majority of developing countries have no regulations at all, which creates gaps in the safety standards. The lack of adequate understanding of the toxicodynamic and toxicokinetic properties of NMs also highlights the necessity to create powerful research and regulation systems to ensure consumer and environmental safety (Hossain et al., 2021). The environmental impact of nanotechnology also cannot be ruled out. While businesses apply nanomaterials at a growing level, issues about their potential to cause environmental pollution and impact sustainability hang in the air. The fate of such materials in natural ecosystems remains controversial, and they may pose potential adverse effects on water and soil quality and biodiversity. The unique nature of NMs may lead to new pathways of persistence and toxicity that earlier materials have never encountered, rendering remediation difficult. Therefore, it requires a concerted approach through collaboration among researchers, policymakers, and business executives for designing safer nanomaterials and minimizing environmental risks. Guidelines and safety standards will be essential in ensuring accountable nanotechnology development and application along with encouragement of sustainable practices (Hossain et al., 2021).

2.6. ETHICAL AND SOCIETAL CONSIDERATIONS

In spite of the continuous growth of nanotechnology, ethical issues regarding its applications must be examined carefully. Innovations in nanoscience have the potential to revolutionize industries such as medicine, energy, and materials science but with dangers of safety issues, environmental impact, and unequal distribution. Ethical concerns must consider not only the potential for good from nanotechnology but also the potential for harm through misusing or misinterpreting it. For example, nanomaterials can enhance drug delivery systems significantly, but the risk of nanotoxicity cannot be discounted. Stakeholders like policymakers, scientists, and the public must balance the conversation about these double edges of innovation. Effectiveness of interaction relies on accountability and transparency because public trust is based on perceived advantages and hazards of scientific advancements (Anderson et al., 2012). Thus, promoting an ethical system is a priority to guarantee effective decision-making in nanotechnology's development. Social dimensions of nanotechnology extend beyond individual risk to encompass broader socio-economic issues. Equity issues are raised especially, as profits from nanotechnology may not be shared fairly. Poor countries may be denied access, widening the disparity if it remains within wealthier countries, where they have more resources. Thus, ethical reflection must include thinking about how nanotechnology may be induced and utilized in a way that serves social justice. Engagement in decision-making within nanotechnological advancements is crucial. In Europe, for instance, the emphasis on government responsiveness to citizens as part of participatory governance models (Anderson et al., 2012) becomes more critical. Such an approach not only makes scientific discourse more democratic but also aligns technological progress with the interest and values of society in contrast to the notion that technological progress is a linear path to progression. Furthermore, the ethical framework guiding the development of nanotechnology needs to be built on a reflection of the values forming the basis of scientific research. Scientists are increasingly being understood as citizens who owe not only to the science they do but also to the societies in which they live. Recognizing the values built into their work can be used to empower scientists to transcend technical concerns and speak about broader ethical concerns, examining how their research aligns with the common good. This dimension challenges the aim of scientific activities on behalf of advancing social welfare. The idea of a common good must be made integral to the nanotechnology discourse so that development would be made with a vision that emphasizes ethical responsibility and accountability. Through this, stakeholders are made to deal with the complexities of technology developments while encouraging an open dialogue that integrates ethical stewardship.

2.6.1. Public Perception of Nanotechnology

The public's perception of nanotechnology plays a crucial role in shaping its development and use in medicine, environmental studies, and consumer goods. With further progress in this science, individuals are likely to be confused with the complexity and repercussions of nanotechnology. Most of what they know is usually obtained from media coverage and public discourse, which may be either positive or negative. Misconceptions about the safety and moral issues of nanoscale material mainly outweigh public opinion, forming a base for skepticism and resistance against proposed innovation. A critical examination of such beliefs explores immense differences in cognition, which requires evaluating public communication methods to bridge the gap between scientific facts and public information. Public outreach to nanotechnology involves not only discussing down risk and uncertainty but also concern for potential benefit to society. Effective communication strategies must emphasize the beneficial effect of nanotechnology, such as advancement in medical treatment and green energy technology. Balancing scientific investigation with the public interest will guarantee the development of a more positive image, thereby making acceptance on a broader scale easier. This approach is consistent with philosophy's questions concerning technoscience, particularly as scientists push the boundaries of ethical and social dimensions of their activities. As suggested in the literature, the critical concerns of nanoscientific research are in alignment with values in society, which lead scientists to consider what reception their findings will have and how this impacts public confidence in technology (Nordmann et al., 2008). Besides, the effect of education initiatives and outreach activities cannot be overemphasized in shifting public attitudes towards nanotechnology. Initiatives that provide freely available information and interactive experience can de-mystify technological terminologies pertinent to nanoscience while emphasizing the relevance of its application in everyday life. Public workshops, seminars, and internet forums can serve as important avenues in carrying out exchanges among scientists, policymakers, and the public. These activities should be based on freely debating risks and ethical concerns, thereby strengthening public confidence in scientific advancement. Ultimately, crafting an informed public is important for the advancement of nanotechnology since public support is dependent on reciprocal understanding of both its promise and limitations (University of New Mexico, 2012) (Figure 2.14).

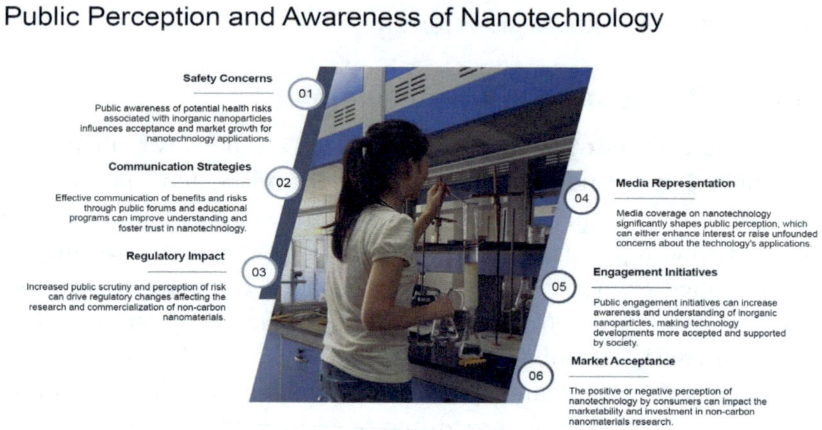

Figure 2.14. Public perception of nanotechnology.

Source: https://www.slideteam.net/media/catalog/product/cache/1280x720/p/u/public_perception_and_awareness_of_nanotechnology_inorganic_nanoparticles_ppt_slides_st_ai_ss_slide01.jpg.

2.6.2. Regulatory Frameworks and Policies

The interplay of nanoscience and nanotechnology with regulatory regimes and policies becomes increasingly significant as the technology advances. Because of the unique characteristics of nanomaterials, traditional mechanisms of regulation generally cannot handle the specific risks and benefits of these technologies. Effective regulation should take into account the advanced interactions of nanoparticles with the environment and biological systems. Without such targeted guidelines, the risk exists of inhibiting innovation or, worse still, the uncontrolled spread of risky practices. Furthermore, as technology becomes more developed, it makes sense to establish dynamic policies that can adapt themselves, keeping pace with scientific advances and public health issues. In agricultural applications, regulatory systems are essential to facilitating the safe use of nanotechnology. Agriculture faces immediate challenges, like food security and pest control, which call for creative responses. Nanotechnology has the ability to transform the methods of farming today by employing nanoparticles for increased effectiveness of pesticides and fertilizers. The regulatory channels, however, need to guarantee that these technologies do not sacrifice safety or ecological integrity. Policymakers need to thoroughly analyze the consequences of nanoparticle-based products and their long-term implications on soil fertility and ecosystems. By giving priority to scientific research and the involvement of stakeholders, regulation can ensure that nanotechnology can remain within agricultural systems safely as highlighted through a critical examination of current practices and their impacts on productivity and food safety (Mittal et al., 2020). International cooperation will come in handy to provide harmonized regulation that comes into terms with the global implications of nanotechnology. Since most countries are funding nanoscience research, there is the imperative of having coordinated policies that can cross-cut the regulatory landscape across borders. Coordination efforts of nanotechnology projects with the United Nations Sustainable Development Goals are particularly important since they underscore the role of innovation in responding to global challenges. Mutual partnerships among high-profile nanocenters can help in terms of knowledge and best-practice exchange in regulation to enhance the overall nanotechnology framework (Pokrajac et al., 2021). With the promotion of an entrepreneurial climate coupled with stringent safety measures, policymakers can unleash the potential of nanotechnology to enhance sustainable development initiatives across sectors, thereby making progress trickle down to society in general.

2.6.3. Ethical Implications of Nanotechnology Research

The expansion of nanotechnology research is filled with profound ethical concerns which have to be weighed with caution. While scientists advance the boundaries of controlling materials at the nanoscale, there emerges a chance for beneficial applications and collateral damage. Possibly the most severe concern of government and society is that of nanotoxicity. Nanomaterials have been engineered with properties that are distinct from their bulk materials counterparts, and those differences may eventually find their way into detrimental effects on human health and the environment. Without careful assessments of such risks, the use of nanomaterials in consumer products would subject populations to harmful substances unknowingly. This requires a robust regulatory framework that features systematic test and evaluation procedures to determine safety and efficacy, thus preventing the ethical issue of prioritizing innovation over public well-being. In addition, the social effects of nanotechnology raise significant ethical issues of equity and access. Nanotechnology holds the ability to transform numerous fields, including medicine and environmental remediation, such that these technologies become central to the solution of global challenges. But disparities in exposure to such technology can exacerbate current inequalities. For instance, rich countries and citizens are able to benefit disproportionately from cutting-edge nanomedicines and nanotechnology-enabled agricultural breakthroughs at the cost of excluded groups. Such inequality underscores the ethical imperatives for equal access to nanotechnology benefits, to the point of not allowing nano-assisted advancements to widen socio-economic differences.

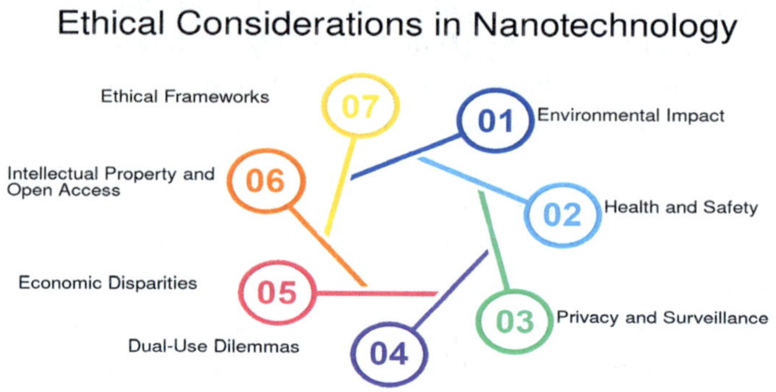

Figure 2.15. Ethical considerations in nanotechnology.

Source: https://fastercapital.co/i/Daio-sNanotechnology--Redefining-Possibilities-at-the-Atomic-Level--Ethical-Considerations-in-Nanotechnology.webp.

To correct such issues, inclusive policies from policymakers are required to prioritize inclusive approaches that broaden the advantages of nanotechnology to poor communities, promoting a more balanced approach to innovation. Lastly, the ethical dimensions of nanotechnology research also include its impact on education and professional qualifications. The integration of nanoscience into curricula necessitates teachers' relevant digital competencies to appropriately educate on such cutting-edge technologies. As the study attesting to outstanding improvements in teachers' digital skills upon training in applications of nanotechnology suggests, an effort should be made to bridge the gap between education customs and the evolving scientific world. Incorporating ethical considerations in the education of future scientists and teachers ensures they are not only skillful in employing nanotechnology but also knowledgeable regarding its potential implications. Focusing on ethics in addition to technical education shapes a generation of professionals who are able to handle the intricate ethics embedded in the practice of nanotechnology, leading to responsible innovation that supports societal well-being and environmental sustainability (Al-Muqbil, 2024; Kumar et al., 2025) (Figure 2.15).

SUMMARY

In drawing conclusions on the Fundamentals of Nanoscience and Nanotechnology, it is understood that it is not just an expansion of existing scientific paradigms but a paradigm change in how things are known at the nanolevel and how materials interact with one another. The judicious synthesis and characterization of nanostructures bring unprecedented possibilities for innovation across all fields of application, from medicine to electronics. By combining outcomes from theoretical and experimental studies, scientists are beginning to discover the intricate interactions that define nanoscale effects. They don't only contribute scientific advancement but also set the stage for applications which can transform industries. The finding of new materials and technologies creates fertile ground for solving contemporary challenges, enhancing the role of nanotechnology as a basis for the development of technology in the future. The methods employed in nanoscience continuously evolve, indicating the cross-disciplinary interaction between such sciences as chemistry, biology, and physics. One of the most crucial elements driving this process is the ability to characterize materials with unprecedented accuracy. Nanometrology applications of surface acoustic waves, for instance, emphasize the necessity of advanced characterization techniques. The elastic characteristics measured by such high-frequency waves provide valuable information for evaluating thin films that can potentially enhance the operation of a wide variety of devices and systems where nanostructured materials have a decisive influence. Alongside this demand, the newly developed field of acoustic nanometrology facilitates higher quality control and design optimization, hence advancing the utility of nanotechnology in practical use as exemplified by BETTUCCI et al. (2015). Moreover, the study of biological systems such as diatom cultures also reflects the intersection of ecological study and technological advancements with a new perspective in nanotechnology. The use of diatoms in synthesizing intricate nanostructures through biological processes is a definitive sign of what technological innovation nature-based systems can translate. The method, besides offering a method of synthesis of complex structures, also promises to lead the way towards nanotechnological sustainability. With experiments being conducted in emerging new environments like the International Space Station revealing new discoveries, one realizes the possibility of wider directions of research that could pave the way to improved understanding of biological nanostructures. These advancements can ultimately lead to the development of a nanotechnology toolkit that harnesses biodiversity for utilitarian ends, based on the expansive applications of this field as discussed in Camp et al. (2007).

In nanoscience and nanotechnology, it is important to know the distinctive nature of nanoscale materials. Such materials have certain properties that are made possible by their smaller size, which alters their behavior significantly compared to bulk materials. For instance, nanoscale materials such as quantum dots and nanowires make possible novel electronic and optical effects that open up breakthroughs in fields ranging from materials science to electronics. Perhaps most notably, the influence of shape and size on properties such as conductivity and reactivity allows for specially designed materials to be produced. With these unique characteristics, scientists can create new devices and systems that benefit from the improved performance feature of nanomaterials, and which reflect a substantial shift in the technological capability level in areas like energy storage and conversion, and drug delivery systems (Meyyappan et al., 2013). Another essential component of nanoscience is the intersection of nanotechnology and biological systems, which has the potential to transform medical and biological sciences. The advent of nanoscale field-effect transistors (nanoFETs) is a paradigmatic technology that bridges electronic engineering and biology. NanoFETs are devices that can be employed as probes to directly couple with cellular systems and hence learn about biological processes at an unprecedented spatial scale. The advances of nanoFETs in the same size range as biological nanostructures enhance our abilities to investigate cellular communication and signal transduction events. In addition, multiplexed extracellular sensing by planar nanoFET arrays enables the simultaneous observation of several biological signals, and hence, our insights into complex biological systems and new diagnostic tools capable of enhancing patient outcomes (Lieber et al., 2013). Moreover, the integration of the electronic and biological systems at different levels offers significant opportunities for new advances in tissue engineering and regenerative medicine. The use of macro-porous nanoelectronic scaffolds is an exemplary example of the merger of nanotechnology and three-dimensional cellular structure, which allows for a more conducive environment for growth and function of tissue. With advancements in this direction, there is an urgent need to address the new challenges that come with merging these seemingly unrelated technologies, most importantly maintaining biocompatibility and functional integration. The continued advancement and development of nanoFET-based technologies will truly develop paradigms of breakthroughs that not only enhance our appreciation of fundamental biological processes but also get implemented in clinical environments, thereby demonstrating the revolutionary potential of nanotechnology across the science and medicine of today (Lieber et al., 2013).

REFERENCES

1. Al-Muqbil, N. S. M. (2024). The impact of a nanotechnology-based training program on the development of digital competencies among high school biology teachers. Journal of Curriculum and Teaching. https://www.semanticscholar.org/paper/77d3fecbe764166142c8cb881ea40e679876c800 (accessed on 25th September 2025).
2. Anderson, A., Mnyusiwalla, A., Shvedova, A. A., Spagnolo, A. G., McKibben, B., Trouiller, B., Buzea, C., et al. (2012). Global ethics and nanotechnology: A comparison of the nano ethics environments of the EU and China. https://core.ac.uk/download/16414418.pdf (accessed on 25th September 2025).
3. Anguita, A., Pazos, A., Geissbuhler, A., Smith, B., Kim, B. Y., Kulikowski, C., Rosse, C., et al. (2012). Nano informatics: Developing new computing applications for nanomedicine. https://core.ac.uk/download/28969094.pdf (accessed on 25th September 2025).
4. Arenas, A., De Domenico, M., & Omodei, E. (2016). Quantifying the diaspora of knowledge in the last century. https://core.ac.uk/download/pdf/81184937.pdf (accessed on 25th September 2025).
5. Baatrup, E., Berret, J. F., Bie, P., Fresnais, J., et al. (2012). Superparamagnetic iron oxide polyacrylic acid coated γ-Fe_2O_3 nanoparticles does not affect kidney function but causes acute effect on the cardiovascular function in healthy mice. http://arxiv.org/abs/1401.1847 (accessed on 25th September 2025).
6. Bettucci, A. (2015). Surface acoustic waves in thin films nanometrology. https://core.ac.uk/download/54531668.pdf (accessed on 25th September 2025).
7. Bora, S., Pooja, D., & Kulhari, H. (2024). Introduction of nanoscience and nanotechnology. https://doi.org/10.1007/978-981-99-5314-1_1 (accessed on 25th September 2025).
8. Camp, P. J., deLuis, J., Gordon, R., Hoover, D., et al. (2007). Testing prospects for reliable diatom nanotechnology in microgravity. https://core.ac.uk/download/pdf/10539495.pdf (accessed on 25th September 2025).
9. Chahal, P. (2025). Precision at the nanoscale: Exploring the synthesis, applications, and challenges of nanotechnology in modern chemistry. Resmilitaris. https://www.semanticscholar.org/paper/834ebee97c7148b826cd873bd42157f13868e7e6 (accessed on 25th September 2025).
10. Cheng, Z., Li, M., Dey, R., & Chen, Y. (2021). Nanomaterials for cancer therapy: Current progress and perspectives. Journal of Hematology &

Oncology, 14. https://doi.org/10.1186/s13045-021-01096-0 (accessed on 25th September 2025).

11. Cuerva, C., Cano, M., & Lodeiro, C. (2021). Advanced functional luminescent metallomesogens: The key role of the metal center. Chemical Reviews, 121, 12966–13010. https://doi.org/10.1021/acs.chemrev.1c00011 (accessed on 25th September 2025).

12. de Almeida, M. S., Sušnik, E., Drašler, B., Taladriz-Blanco, P., Petri-Fink, A., & Rothen-Rutishauser, B. (2021). Understanding nanoparticle endocytosis to improve targeting strategies in nanomedicine. Chemical Society Reviews, 50, 5397–5434. https://doi.org/10.1039/d0cs01127d (accessed on 25th September 2025).

13. Gao, L., Wei, H., Dong, S., & Yan, X. (2024). Nanozymes. Advanced Materials, 36. https://doi.org/10.1002/adma.202305249 (accessed on 25th September 2025).

14. Gupta, N., Hmurcik, L. V., Xiong, X., Zhang, Y., et al. (2011). From microelectronics to nanoelectronics: Introducing nanotechnology to VLSI curricula. https://core.ac.uk/download/52955977.pdf (accessed on 25th September 2025).

15. Hanna, S. K., Heller, D. A., Horoszko, C. P., Johnson, D. R., et al. (2017). Redox-active nanomaterials for nanomedicine applications. https://core.ac.uk/download/215479341.pdf (accessed on 25th September 2025).

16. Harish, V., Tewari, D., Gaur, M., Yadav, A. B., Swaroop, S., Bechelany, M., & Barhoum, A. (2022). Review on nanoparticles and nanostructured materials: Bioimaging, biosensing, drug delivery, tissue engineering, antimicrobial, and agro-food applications. Nanomaterials, 12, 457–457. https://doi.org/10.3390/nano12030457 (accessed on 25th September 2025).

17. Hossain, A., Skalický, M., Brestič, M., Mahari, S., Kerry, R. G., Maitra, S., Sarkar, S., et al. (2021). Application of nanomaterials to ensure quality and nutritional safety of food. Journal of Nanomaterials, 2021, 1–19. https://doi.org/10.1155/2021/9336082 (accessed on 25th September 2025).

18. Jensen, P., Kreimer, P. R., & Levin, L. G. (2016). Does size matter? The multipolar international landscape of nanoscience. https://core.ac.uk/download/159288861.pdf (accessed on 25th September 2025).

19. Joudeh, N., & Linke, D. (2022). Nanoparticle classification, physicochemical properties, characterization, and applications: A comprehensive review for biologists. Journal of Nanobiotechnology, 20. https://doi.org/10.1186/s12951-022-01477-8 (accessed on 25th September 2025).

20. Kumar, D., & Tanwar, R. (2025). Nanoscience and the future: Transforming technologies for sustainable development. Current Engineering Letters and Reviews. https://www.semanticscholar.org/paper/a18ba8b9a333ad66bb2580cb455d21899a47a6d7 (accessed on 25th September 2025).
21. Kumar, V. N. S., Sunkar, S., K. S., & Nellore, J. (2023). Nano drug delivery systems: A mini-review. Nanoscience & Nanotechnology–Asia. https://www.semanticscholar.org/paper/b8fbe10ec4c39507e2d84a6ece501e3cc34fe536 (accessed on 25th September 2025).
22. Lakhtakia, A., et al. (2005). A map of the nanoworld: Sizing up the science, politics, and business of the infinitesimal. http://arxiv.org/abs/physics/0510209 (accessed on 25th September 2025).
23. Liang, M., & Yan, X. (2019). Nanozymes: From new concepts, mechanisms, and standards to applications. Accounts of Chemical Research, 52, 2190–2200. https://doi.org/10.1021/acs.accounts.9b00140 (accessed on 25th September 2025).
24. Lieber, C. M., & Tian, B. (2013). Synthetic nanoelectronics probes for biological cells and tissues. https://core.ac.uk/download/28944253.pdf (accessed on 25th September 2025).
25. Lipatov, A., Lu, H., Alhabeb, M., Anasori, B., Gruverman, A., Gogotsi, Y., & Sinitskii, A. (2018). Elastic properties of 2D $Ti_3C_2T_x$ MXene monolayers and bilayers. Science Advances, 4. https://doi.org/10.1126/sciadv.aat0491 (accessed on 25th September 2025).
26. Lorusso, L. (2013). Nanotechnology, aims, and values. https://core.ac.uk/download/576784170.pdf (accessed on 25th September 2025).
27. Malik, S., Muhammad, K., & Waheed, Y. (2023). Nanotechnology: A revolution in modern industry. Molecules, 28, 661–661. https://doi.org/10.3390/molecules28020661 (accessed on 25th September 2025).
28. Mekuye, B., & Abera, B. (2023). Nanomaterials: An overview of synthesis, classification, characterization, and applications. Nano Select, 4, 486–501. https://doi.org/10.1002/nano.202300038 (accessed on 25th September 2025).
29. Meyyappan, M., Neyts, E. C., & Ostrikov, K. (2013). Plasma nanoscience: From nano-solids in plasmas to nano-plasmas in solids. http://arxiv.org/abs/1306.6711 (accessed on 25th September 2025).
30. Mishra, J., Swain, S., & Sundaray, B. (2025). Study of structural and magnetic properties of nickel doped bismuth ferrite electrospun nanofibers. Advances in Natural Sciences: Nanoscience and Nanotechnology. https://

www.semanticscholar.org/paper/84f7e528180bc4a3b839a521c3cc9be3cf969e4f (accessed on 25th September 2025).

31. Mittal, D., Kaur, G., Singh, P., Yadav, K., & Ali, S. A. (2020). Nanoparticle-based sustainable agriculture and food science: Recent advances and future outlook. Frontiers in Nanotechnology, 2. https://doi.org/10.3389/fnano.2020.579954 (accessed on 25th September 2025).

32. Murphy, P. (2009). The challenges of "upstream" communication and public engagement for Irish nanotechnology. https://core.ac.uk/download/11309134.pdf (accessed on 25th September 2025).

33. N/A. (2011). CSIC-UPV–Centro de Física de Materiales (CFM). Memoria CFM 2011. https://core.ac.uk/download/45449033.pdf (accessed on 25th September 2025).

34. N/A. (2012). University of New Mexico. Nanoscience & Microsystems 2012 APR self-study & documents. https://core.ac.uk/download/151599127.pdf (accessed on 25th September 2025).

35. N/A. (2015). The Boston University Photonics Center Annual Report 2014–2015. https://open.bu.edu/bitstream/2144/29753/1/Photonics-Center-Annual-Report-2015-final.pdf (accessed on 25th September 2025).

36. Naresh, V., & Lee, N. (2021). A review on biosensors and recent development of nanostructured materials-enabled biosensors. Sensors, 21, 1109–1109. https://doi.org/10.3390/s21041109 (accessed on 25th September 2025).

37. Nordmann, A. (2008). Philosophy of nano technoscience. https://core.ac.uk/download/151257283.pdf (accessed on 25th September 2025).

38. Osorio De Castro Conde, J. (2015). The golden age in cancer nanobiotechnology: Quo Vadis? https://core.ac.uk/download/78064917.pdf (accessed on 25th September 2025).

39. Perveen, F., Ullah, F., Rehman, A., Zaman, H., Abbas, W., Khan, M., Bilal, M., et al. (2025). A review on nanotechnology-driven green synthesis of silver nanoparticles using Nigella sativa. Insights–Journal of Health and Rehabilitation. https://www.semanticscholar.org/paper/efc1c2ec04f3ea1d9c428a95d3bb1a981cbf2022 (accessed on 25th September 2025).

40. Pokrajac, L., Abbas, A., Chrzanowski, W., Dias, G., Eggleton, B. J., Maguire, S., Maine, E., et al. (2021). Nanotechnology for a sustainable future: Addressing global challenges with the International Network4Sustainable Nanotechnology. ACS Nano, 15, 18608–18623. https://doi.org/10.1021/acsnano.1c10919 (accessed on 25th September 2025).

41. Saha, A., Chauhan, S. B., & Singh, I. (2024). Targeted drug delivery with lipid-based nanoparticles: Innovations for improved cancer therapy. Nanoscience & Nanotechnology–Asia. https://www.semanticscholar.org/paper/51dd15306ef2385f24fab9f59c9570f78f61c232 (accessed on 25th September 2025).
42. Shahid, M. (2015). Photo desorption and higher-order assemblies of potassium polytitanate for water treatment. https://opus.lib.uts.edu.au/bitstream/10453/36988/1/01front.pdf (accessed on 25th September 2025).
43. Shastri, J., & Kelkar-Mane, V. (2025). Rhamnolipid-capped iron oxide nanozyme-substrate combination as promising antibacterials of the new generation. Advances in Natural Sciences: Nanoscience and Nanotechnology, 16. https://www.semanticscholar.org/paper/84bd4e7aedb4b5c596ec807bcc2a79c3da502486 (accessed on 25th September 2025).
44. Tomlinson, G. A., Singer, I. L., Müser, M. H., Robbins, M. O., Hod, O., Bragg, W. H., & Frenkel, Y. I. (2012). Superlubricity–A new perspective on an established paradigm. http://arxiv.org/abs/1204.3749 (accessed on 25th September 2025).
45. Zhou, L., Zhang, M., Huo, Y., Bai, L. P., He, S., Wang, J., Jia, C., et al. (2024). Application of ionic liquids in single-molecule junctions: Recent advances and prospects. Green Energy & Environment, 9, 1784–1801. https://doi.org/10.1016/j.gee.2024.01.003 (accessed on 25th September 2025).

CHAPTER 3

TYPES OF NANOMATERIALS

Contents

INTRODUCTION

Nanomaterials, i.e., materials whose structure constituents are at the nanoscale (typically 1–100 nanometers), have been the subject of extensive scientific study and industrial use due to their unique properties and phenomena. They have higher strength, improved electrical conductivity, and higher chemical reactivity compared to their bulk counterparts. Nanomaterials exhibit novel and fascinating applications in a vast array of fields including medicine, electronics, environmental science, and energy generation. Consequently, the study of the types of nanomaterials is paramount in unlocking their promise for technology innovation and addressing the issues of today. The discovery of the materials not only determines their potential to revolutionize but also to propel the justification for continuing research on their properties and application. Nanomaterials can be categorized based on their origin, composition, and functionality. In general, they fall into two general categories: organic and inorganic nanomaterials. Organic nanomaterials primarily involve carbon-based systems like nanocarbon structures, such as graphene and carbon nanotubes, whereas inorganic nanomaterials involve a wide range of materials such as metal oxides, noble metals, and semiconductor nanoparticles. Both have distinctive properties that render them suitable for specific applications. Silver nanoparticles, for instance, are prominently known for their antimicrobial properties and therefore are extremely useful in healthcare as well as environmental technology. Meanwhile, quantum dots, as unique semiconductor nanomaterials, have transformed optoelectronic devices due to their tunable optical properties. Thus, understanding these groups is necessary for complete optimization of nanomaterial functionalities in various industries. The development of nanomaterials has large-scale implications for the regulation patterns and governance, particularly with their possible harm and ethical aspects becoming the center of debate. A mixed mode of governance is where Europe has advanced, blending adaptive techniques with traditional regulatory practices, based on the ongoing discussions regarding safety and effectiveness (Ruggiu et al., 2015). This regulatory context establishes the requirement for a systematic means of monitoring the production and application of nanomaterials, ensuring safety without discouraging innovation. Issues in the industry demand critical discourse relating to responsible research and official guidelines, as regulation must expand with technology development. As mentioned in relevant literature, the integration of human rights and ethical principles into governance systems is essential to ensure that responsible practices of nanotechnology are encouraged (N/A, 2017). With proper governance and informed policy-making, the potential of nanomaterials can be utilized while addressing issues of society with respect to health concern and environmental impacts.

Definition of Nanomaterials

The etymology of nanomaterials' definition is rooted in their peculiar structural properties, normally defined as materials of a size within the nanoscale, i.e., between 1 and 100 nanometers. At this dimension, materials possess discernible physical and chemical properties, which are appreciably different from the bulk phase. The explanation for this size-dependent effect is the increased surface area-to-volume ratio, which leads to increased reactivity and interaction with living systems. As a result, nanomaterials have found applications in diverse fields, including electronics, medicine, and environmental rehabilitation. More than 1600 consumer products based on nanotechnology that are available today are evidence of the growing significance of the materials in modern application (Abhyankar et al., 2015). The definition and characteristics of nanomaterials are thus critical to understand in preparation for studying their diverse types and applications. Moreover, the determination of nanomaterials involves consideration of their engineered nature and integration of nanoscale building blocks into matrices.

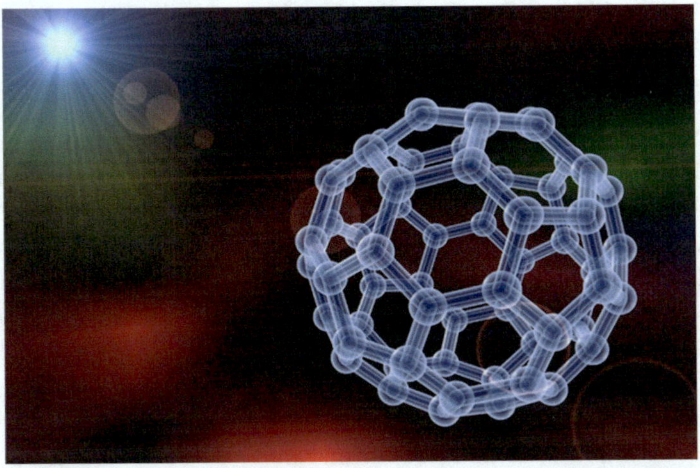

Figure 3.1. Nanomaterials.

Source: https://www.azonano.com/images/Article_Images/ImageForArticle_ 4147(1).jpg.

The materials themselves can be broadly classified into three types: nanoparticles, nanocomposites, and nanostructured materials. Nanoparticles are individual particles having dimensions on a nanoscale with specific optical, electrical, and magnetic properties. Nanocomposites, however, are materials that integrate nanosized particles into a matrix in order to enhance mechanical and thermal properties. Nanostructured materials, engineered at the atomic or molecular level, have modified mechanical and electrical properties and thus find uses in emerging technological devices. Despite the favorable benefits of these materials, the interaction between their engineered properties and potential health hazards necessitates rigorous test practices (Adamec et al., 2019). At the heart of the science on nanomaterials is the life-cycle assessment of their safety and environmental risk at each step—production, through disposal. The risks associated with nanoparticle release, particularly in occupational contexts, attest to the need for strict regulation. Current chemicals and environment legislation has a tendency to exclude specific reference to nanomaterial issues. In addition, the possible physio-chemical transformations that are likely to occur in nanoparticles through their life cycles are not well understood, and that fuels concerns regarding their safety and effectiveness. Development of standardized risk evaluation procedures is essential for providing information to consumers and industries regarding implications of nanotechnology-based product usage. The importance of addressing these issues is underscored by the need for a robust methodology to evaluate release and impact of nanoparticles in the workplace (Abhyankar et al., 2015; Adamec et al., 2019) (Figure 3.1).

Importance of Nanomaterials in Modern Science

No overestimation of the contributions of nanomaterials in modern science can be made, as they represent a forefront of innovation beyond traditional material science. Their unique properties, owing to their sizes and their structure at the nanoscale, make them interesting for advances in a range of different areas, from electronics to biotechnology. Nanomaterials, such as carbon nanotubes and metal nanoparticles, possess better electric, thermal, and optical properties than the bulk material. This allows us to imagine a new era of uses, such as more efficient solar cells and delivery systems for targeted drugs. These discoveries, besides highlighting the revolutionizing potential of nanomaterials, also underscore the importance of in-depth research and development to optimally harvest their potential and integrate them into practical applications (Grimpe et al., 2025). Regional innovation systems play a pivotal role in the production and use of nanomaterials and underline the importance of collaboration between public research institutions and private companies. The development of R&D-friendly spaces, such as science parks and nanoclusters, forms the cornerstone of encouraging innovation in nanotechnology. Data from the European Patent Office reveal that the resultant synergistic R&D activity from clustered regions significantly contributes to the creation of patents on nanomaterials. By gaining insight into and optimizing the dynamics of collaborative systems, actors can enhance economic growth and technological advancement. These efforts contribute to the regional economy, as well as generate a steady flow of technological innovations that can transform industry standards and societal applications (Grimpe et al., 2025). The interdisciplinary scope of nanomaterials offers advantages beyond the conventional limits, since it impacts various scientific and engineering fields. For instance, in medicine, nanoparticles are revolutionizing medicines with improved bioavailability and targeted delivery systems to eliminate side effects of conventional methods of treatment. In environmental science, nanomaterials in the treatment of water and wastes are providing new solutions to acute environmental issues. These applications demonstrate the relevance of nanomaterials to sustainable development and addressing global issues. Continued investment and research into different types of nanomaterials are required, as their development and use in novel technologies over the future have the potential to redefine future scientific frontiers (Figure 3.2).

Figure 3.2. Nanomaterials in modern science.

Source: https://lniconference.com.au/wp-content/uploads/2021/05/march-how-nanomaterials.jpg.

3.1. CLASSIFICATION OF NANOMATERIALS

Classification of nanomaterials is required in terms of grasping their characteristic properties and applications, primarily in relation to their size-dependent properties. Nanomaterials are typically categorized as zero-dimensional (0D), one-dimensional (1D), two-dimensional (2D), and three-dimensional (3D) materials. Zero-dimensional nanomaterials, such as nanoparticles, possess particle diameter less than 100 nanometers in all directions. Nanomaterials tend to exhibit enhanced properties, leading to a variety of uses in the area of medicine, electronics, and catalysis. Alternatively, one-dimensional nanomaterials such as nanowires or nanotubes have one dimension several orders of magnitude larger than the other two, allowing them to be more readily used in electronic devices on a nanoscale due to their unique electrical and thermal properties. This hierarchical system of classification simplifies it for scientists to design materials specifically well-adapted for particular functionalities. Further specialization within these categories offers detailed exploration into specific nanomaterial types, such as carbon-based, metal-based, and polymer-based nanomaterials. Carbon nanomaterials, graphene, and carbon nanotubes are extremely mechanically robust and electrically conductive and are exceedingly valuable for numerous industrial applications. Metal nanoparticles, due to their unique optical properties, are being heralded as useful entities in areas like biomedical imaging and photothermal cancer treatment. In addition, nanomaterials from polymers, including nanocellulose, are of interest as a result of their biocompatibility and low toxicity, especially in biomedical applications. The multi-functionality of such materials emphasizes the necessity of their accurate classification since it creates opportunities for innovations from technology fields ranging from electronics to environmental remediation, thus extending the frontier of nanotechnology. Among the most promising nanomaterial classes is the manufacturing of biosensors that subsequently enhance medical diagnostic capabilities. As described in a series of recent reviews, nanocellulose is one of the best potential candidates for the application of biosensing technologies due to its extremely desirable characteristics like biocompatibility and low toxicity and remarkable mechanical properties (Langari et al., 2022). Such classification promotes a systematic understanding of how different types of nanocellulose can be integrated into diverse sensor platforms, enhancing their functionality. Besides, the individual properties of nanoparticles, by virtue of their size and surface area, allow for new possibilities in medicine and electronics, becoming fundamental components in new technological developments (Joudeh et al., 2022). Classification of nanomaterials therefore not only improves theoretical data but also gives inputs towards practical use in various scientific disciplines.

3.1.1. Based on Dimensionality: Zero, One, Two, and Three-Dimensional

Dimension-based classification of nanomaterials—zero, one, two, and three-dimensional—gives a systematic outlook on their unique character and anticipated applications. Zero-dimensional nanomaterials, such as nanoparticles, possess no length, width, or height but do possess a significant surface area relative to volume. This characteristic makes them more reactive and responsive to the environment around them and renders them particularly useful in drug delivery and catalysis. Functionalization of such nanoparticles can condition their behavior to provide improved specificity and efficiency in biomedicine and in environmental cleaning. Zero-dimensional nanomaterials are also used as building blocks for the construction of higher-order structures

(Sahu et al., 2023). One-dimensional nanomaterials, including nanowires and nanotubes, have a greater length with a minimum cross-sectional diameter, thus obtaining unique electrical and mechanical characteristics. Carbon nanotubes, for example, have excellent electrical conductivity and tensile strength, making them highly valuable across various technologies, from sensors to flexible electronics. The possibility of manipulating and conducting electricity along such nanostructures has drawn interest in integrating them into nanoelectronics and energy storage devices.

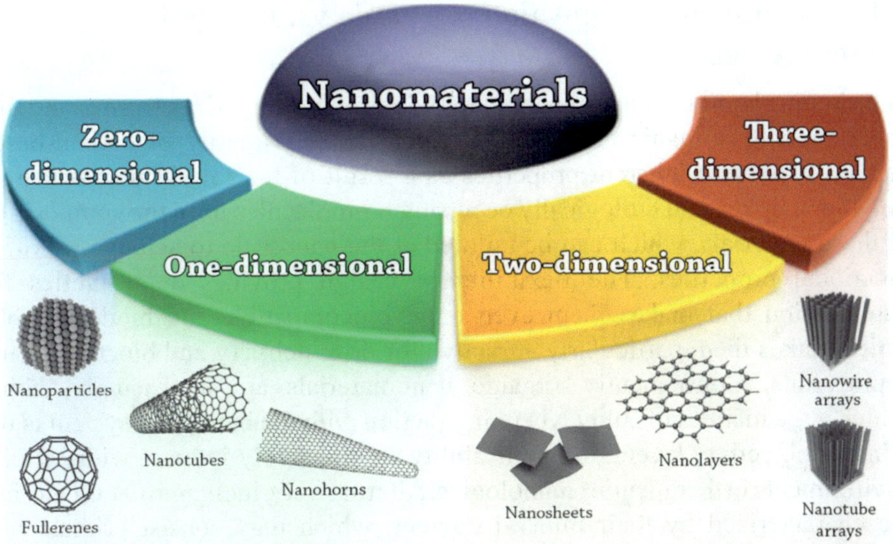

Figure 3.3. Nanoparticle classification based on dimensionality.

Source: https://media.springernature.com/lw1200/springer-static/image/art%3A10.1186%2Fs12951-022-01477-8/MediaObjects/12951_2022_1477_Fig1_HTML.png.

The one-dimensional nanomaterials' extended structure enhances their surface area to volume ratio and, at the same time, promotes electrons' mobility, which makes it crucial for photovoltaic as well as electrochemical cell applications. These attributes put one-dimensional nanomaterials in keen focus within the comprehensive field of nanotechnology (Naresh et al., 2021). Two-dimensional nanomaterials, particularly graphene and graphene derivatives, are of great interest due to their exceptional properties, arising from the unique structure of the atoms in one layer. Graphene, with its highest electrical and thermal conductivity on record, offers unmatched potential in electronic devices, transistors, and biosensors. Also, the ability to scale up these materials into large sheets without losing their useful properties has opened up possibilities for composite materials and flexible electronics. The inherent two-dimensionality offers high surface area with low weight, making the materials suitable to use in such applications as energy storage devices and filtration systems. The present study emphasizes the addition of two-dimensional nanomaterials to various technologies, considering their potential to revolutionize various domains (Naresh et al., 2021; Sahu et al., 2023) (Figure 3.3).

3.1.2. Chemical Composition: Organic vs. Inorganic Nanomaterials

The chemical composition of nanomaterials simply categorizes them as organic and inorganic. Organic nanomaterials, consisting primarily of carbon-based materials, possess inherent properties as a result of their molecular structure. Polymers, lipids, and biologically occurring biomolecules form the composition of these materials, which can be tailored at the nanoscale to achieve maximal functional properties. The flexibility of design provides opportunities for optimization that makes them even more biocompatible and biodegradable, which makes them particularly attractive for drug-delivery and bioengineering applications. Consequently, organic nanomaterials are increasingly taking center stage in areas of study where interaction with the biological system is the primary objective, leveraging their ability to effectively interact with cellular environments in therapeutic technologies. Alternatively, inorganic nanomaterials are characterized by their mineral content, which may consist primarily of metals, metal oxides, and other non-carbon substances. Examples include gold and silver nanoparticles, which are favored for their stability and their optical properties. Synthesis of these nanoparticles has become a critical field of study, with the purpose of tailoring size, morphology, and stability, which has a direct impact on their performance in a range of applications, such as sensors and medical imaging (Fratoddi et al., 2017). Inherent physical and chemical properties of inorganic nanomaterials, including conductivity and magnetic properties, present opportunities for their applications in future electronics and catalysis. This chemical difference thus emphasizes the multivariant functional uses of nanomaterials and their roles in various technological uses. Additionally, the functional performance of organic and inorganic nanomaterials introduces significant implications with regard to their chemical composition. The reaction of such materials to their environment generally leads to size-dependent effects that are remarkably dominating in nanostructured structures. For instance, acoustic cavitation-based sol-gel techniques highlight the way inorganic nanomaterials can benefit from high-pressure environments to become porous and homogeneous during gelation in order to enhance their characteristics including strength and surface area (Fedriani et al., 2009). This innovation leads us to believe that it is possible to optimize the chemical composition of nanomaterials to create innovations for applications in biomaterials for tissue engineering and environmental remediation. Understanding the chemical distinctions between organic and inorganic nanomaterials is therefore essential to adapt their properties to meet particular working requirements in a broad range of applications (Figure 3.4).

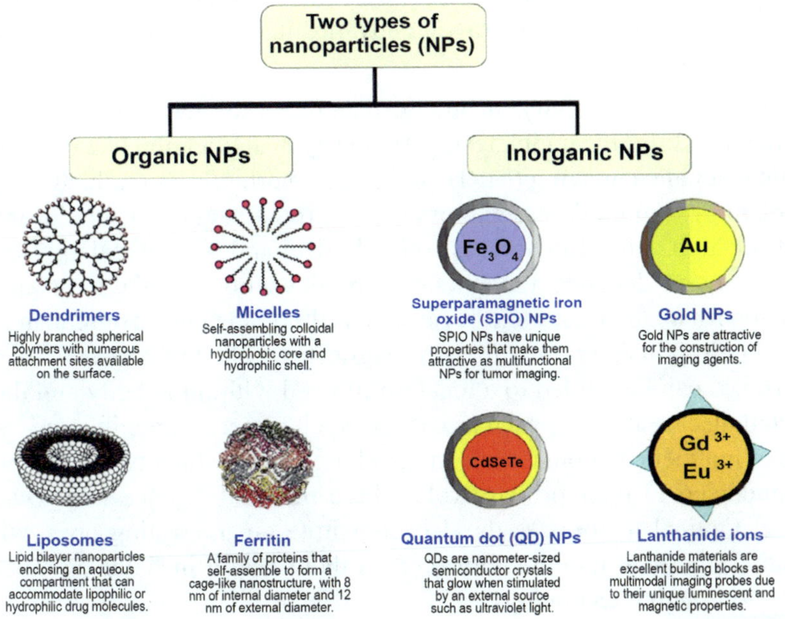

Figure 3.4. Nanomaterials chemical composition.

Source: https://www.researchgate.net/publication/350775063/figure/fig1/AS:1021040251633664@1620446374259/Two-fundamental-types-of-nanomaterial-including-organic-and-inorganic-nanoparticles.png.

3.1.3. Functional Properties: Conductive, Magnetic, and Optical Nanomaterials

The investigations on functional properties of nanomaterials have unveiled tremendous advances in conductive materials, particularly those employed in the field of electronics and energy. Conductive nanomaterials such as graphene and carbon nanotubes exhibit high electrical conductivity and excellent mechanical strength, and they are well-positioned for use in flexible electronic devices and energy storage devices. Such nanomaterials are capable of producing efficient and light batteries and supercapacitors better than the traditional energy storage devices. Moreover, the conductance of these nanomaterials is capable of being tuned, and hence they can be introduced into sensors and wearable devices, extending the boundaries of current technology. Such a capability makes conductive nanomaterials central to meeting the growing demand for advanced electronic components that are functional and efficient, rendering them fundamental in the ever-growing advancement of electronic material and device development.

Apart from conductivity, magnetic nanomaterials have special functions that have gained significant interest, particularly in application like data storage, magnetic resonance imaging (MRI), and targeted drug delivery. Such materials, often cobalt or iron oxide based, possess superparamagnetic behavior that can be used for high control in various technological and biomedical applications. They exhibit size-dependent magnetic properties with potential to enhance the performance of magnetic storage devices as well as contribute to the enhancement of biotechnology using more efficient diagnostic tools. For instance, magnetic nanoparticles can be guided to a location of need within the body and they aid in targeted disease therapy, such as cancer. Such targeted targeting underscores the role of magnetic nanomaterials in novel medical treatments and technology, and it underscores their primary role where material science intersects with medicine. Optical nanomaterials also constitute an interesting area, wherein their native photonic functionality renders them useful in fields as diverse as sensors and imaging technology.

The nanomaterials, generally composed of semiconductor nanocrystals, are capable of manipulating light at the nanoscale to enable more efficient performance in devices like solar cells and photodetectors. Their ability to exhibit phenomena such as surface plasmon resonance allows them to make very sensitive biosensors with an ability to detect biomarkers at very small concentrations, which is vital in the detection of diseases at early stages. Moreover, progress in optical nanomaterials enhances telecommunication by making the light-based communication system more efficient. In general, the study of conductive, magnetic, and optical nanomaterials is paving the way for paradigmatic applications across various fields, underscoring their role in scientific research as well as practical innovation, evident in the comprehensive overviews of their activities in future technology and medicine treatments (Altintas et al., 2019; Anderson de la LLana et al., 2014).

3.2. CARBON-BASED NANOMATERIALS

Carbon-based nanomaterials constitute a special subgroup within the large clan of nanomaterials, primarily due to their remarkable physical and chemical properties. Carbon nanotubes, graphene, and fullerenes, for example, exhibit remarkable strength, electrical conductivity, and heat stability. Their nanometer scales not only facilitate these superior properties but also provide multiple applications across several disciplines such as electronics, materials science, and biomedical engineering. For instance, carbon nanotubes are coveted for their promise to stiffen composite materials, while graphene—a two-dimensional lattice of single-layer carbon atoms—holds revolutionary promise in electronics due to the high charge carrier mobility. This variety of applications indicates the importance of carbon-based nanomaterials in the composite ecosystem of nanotechnology. The applications of carbon-based nanomaterials are also followed by growing concern regarding their environmental and health consequences. With these nanomaterials being integrated into market commodities and industrial production, greater concern about their possible risks requires an effective regulatory framework to safeguard public health. Innovations are numerous, but the requirement to acknowledge uncertainties regarding nanomaterials life cycles cannot be overstated. Current regulation measures, as cited in recent criticisms, demand greater control of nanotechnology applications to avert risks to public health and environmental conservation. Specific recommendations include increasing risk research funding and inviting greater public participation in the regulatory process. These measures will balance technological innovation with safety to enable the safe development of carbon-based nanomaterials. Regulation of carbon-based nanomaterials is all about managing complex regulatory regimes that are driven by evolving scientific knowledge and societal demands. The European Commission, for instance, has adopted adaptive and flexible regulatory approaches to address such challenges, illustrated through initiatives across sectors like cosmetics and biocides. The EU's embraced hybrid model of regulation comprising consultations, self-regulation, and traditional regulatory measures reflects a method to keep the dynamic aspects of nanotechnology in check. However, this models reliance on an ad-hoc model causes coherence and anticipation problems for emerging risks. Critical analysis posits that the current framework may lack proactive measures, highlighting the importance of governance that considers empirical evidence as well as proactive ethical issues such as human rights. Carbon-based nanomaterial regulation thus relies on innovation versus steps to establish public trust and safety, ensuring a responsible nanotechnology future (Figure 3.5).

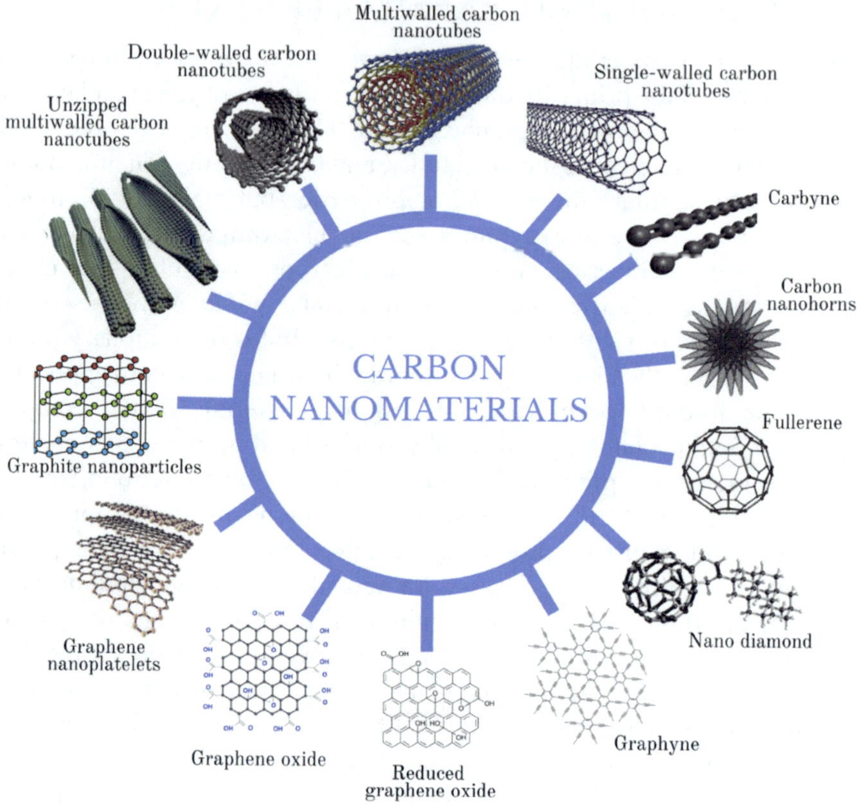

Figure 3.5. Carbon-based nanomaterials.

Source: https://media.springernature.com/lw1200/springer-static/image/art%3A10.1007%2Fs11051-024-06006-2/MediaObjects/11051_2024_6006_Figa_HTML.png.

3.2.1. Graphene: Properties and Applications

Graphene, a single atomic layer of carbon atoms forming a two-dimensional lattice, has witnessed tremendous attention among the nanomaterials fraternity due to its exceptional properties. With its high electrical conductivity, mechanical toughness, and thermal characteristics, graphene comes as a contender for use in numerous diverse industries. The combination of these properties makes graphene overcome traditional materials, particularly in electronics, whose outstanding carrier mobility has created more efficient and faster transistors. Its high surface area also enhances its application in energy storage devices like batteries and supercapacitors. Thus, the unique characteristics of graphene not only position it as a prime player in advancing nanotechnology but also identify its potential utility in various industries, providing new material science and engineering solutions. In biomedical applications, integration of graphene holds huge promise, especially in healthcare biosensing technology. Due to its excellent biocompatibility and ability to facilitate detection of a wide range of biologically significant molecules, graphene can be tailored for specific diagnostic applications. Application of graphene-based sensors facilitates sensitive detection of nucleic acids, proteins, and pathogens, hence significantly enhancing early detection and monitoring of diseases (Altintas et al., 2019). The incorporation of graphene into biosensing platforms has revolutionized the functionality of point-of-care testing, thereby improving patient outcomes. As graphene can be blended with other nanomaterials, the resulting synergy effects arising from such blending form the basis of new opportunities for further enhancing highly advanced sensing technologies. This seamless blending of characteristics from diverse nanomaterials reflects the very promising field of application of nanotechnology in medicine. Researches on the biocompatibility of graphene materials further solidifies its useful role in a wide range of applications. For instance, nanosheets of Graphene Oxide (GO) have shown improved biocompatibility compared to pristine Single Wall Carbon Nanotubes (p-SWCNTs) in *in vitro* models in various human cell lines. GO was found to have minimal cytotoxicity, which was established by observation of the cultures as well as by quantitative assays, and hence demonstrated its capability as a drug delivery and molecular imaging nanocarrier (Calcaterra et al., 2018). The favorable uptake profile of GO into cells indicates that it can be employed in targeted treatments and hence enhance the efficacy of treatment modalities. Generally, the biocompatibility and functional versatility of graphene materials assure their status as key players in the evolving landscape of nanomaterials, with functionalities extending from electronics to medicine.

3.2.2. Carbon Nanotubes (CNTs): Structure and Uses

The typical structure of CNTs is a structural feature with strong influence on the vast array of applications of the material across various domains. Composed of rolled-up graphene sheets, CNTs possess superior properties attributed to their unique geometry. These tube architectures have different diameters and lengths, facilitating the changes in mechanical, thermal, and electrical behavior. The high aspect ratio of the CNTs gives them superior tensile strength, greater than that of steel, and excellent electrical conductivity, rendering them highly efficient in numerous applications, ranging from composite material to nanoelectronics. The ability to engineer their structures at the nanolevel enhances their integration into next-generation technologies with enhanced performance for sensors and transducers, which are increasingly relevant in modern electronic devices, robots, and biotechnology, as it points to the promise of CNTs for future technological breakthroughs. Adding carbon nanotubes to materials enhances their physical and functional properties, particularly in sensing devices. CNTs have been seen to possess the ability to serve as conductive fillers for polymer nanocomposites, which enhances the electrical performance and mechanical strength of the materials. By such inclusion, there is a possibility to develop sensitive and flexible sensor devices that can detect parameters such as pressure, concentration of gases, and humidity with high sensitivity and rapid response times.

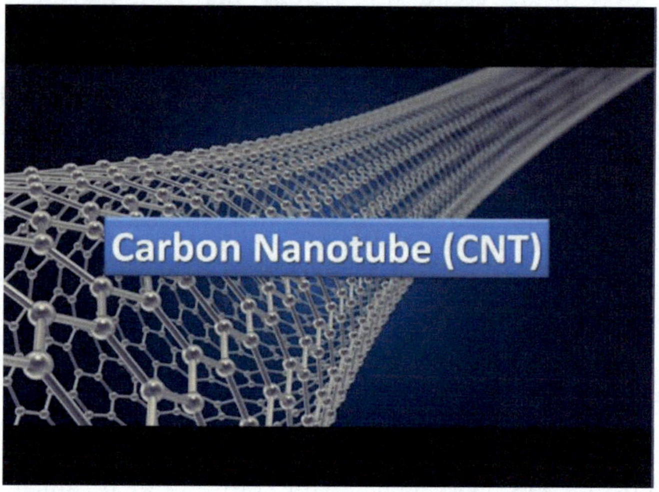

Figure 3.6. Carbon nanotube (CNTs).

Source: https://external-content.duckduckgo.com/iu/?u=https%3A%2F%2Fi.ytimg.com%2Fvi%2Fmf5wPBpnRnQ%2Fhqdefault.jpg&f=1&nofb=1&ipt=24502e30dc02139ae52f3d1e72729f02334edb01cb36d9b36af9c6af202bfccc.

The combination of CNTs with conducting polymers has been the basis for the emergence of conducting polymer nanocomposites (CPNCs) with enhanced performance features in sensor devices. This combination can also enable one to create lightweight and flexible sensors that are so essential for wearable devices and smart technology, further establishing the position of CNTs as a cornerstone for driving sensing technologies for every-day use (Sharma et al., 2023). Their ontogenetic nature also extends to their uses in biomedicine, particularly in biosensor design. Using their high surface area and conductivity, CNTs can be functionalized with biomolecules to make highly sensitive diagnostic probes. Functionalization through this means makes it possible to achieve fast and accurate identification of numerous biomarkers in medical diagnosis ranging from glucose detection to cancer markers. Besides, the use of CNTs in biosensing devices assists in overcoming harsh obstacles in the transduction process, hence enhancing sensitivity and reducing response time. As scientists continue to elaborate on the potential of CNTs in biosensors, the health diagnosis implications grow more vivid, making CNTs an integral factor in the emerging epoch of nanomaterials and its implications in modern science (Naresh et al., 2021) (Figure 3.6).

3.2.3. Fullerenes: Unique Characteristics and Potential Applications

Fullerenes are a particular group of carbon allotropes that are characterized by their unusual molecular shapes, which are spherical, ellipsoidal, or tubular. Such spherical structures consist completely of carbon atoms in the form of polycyclic structure and possess characteristics that differ significantly from their bulk carbon counterparts. Apparently, fullerenes possess an outstanding stability, chemical reactivity, and electron-accepting ability, which make them suitable for a broad spectrum of applications, particularly in nanomedicine. Their structural variety makes them suitable for use as drug delivery systems, where their drug-entrapment capability facilitates bioavailability and targeted delivery, offering examples of promise in the remodeling of treatment modalities for diseases like cancer and neurodegenerative diseases. The fullerenes have applications beyond drug delivery, with their unusual electronic properties making them significant in nanotechnology and high-tech materials. Because they conduct electricity and have a high surface area, fullerenes are employed as key elements in the fabrication of organic photovoltaic devices and sensors. Specifically, they are utilized to enhance the efficiency of solar panels by facilitating charge transport. Moreover, fullerenes' incorporation into high-electron-mobility environments with transistors' applications presents their notable properties that fulfill the demands of modern electronic and optoelectronic devices. Studies of fullerenes in such areas symbolize a critical milestone towards green energy solutions and novel sensory functions capable of enhancing real-time monitoring in diverse environments. Moreover, fullerenes' integration into biosensing application further underlies the wide utility of fullerenes in diagnosing health conditions. Because fullerenes are able to identify a multitude of biologically important molecules, fullerenes can potentially facilitate the sensitivity and specificity of diagnostics tests. By virtue of their unique redox properties, as characterized in recent work, fullerenes can facilitate nucleic acid, cancer biomarker, and pathogen sensing, and thus find themselves at the forefront of nanomaterials for medicine. Such multipurpose application has enormous promise for early disease detection and point-of-care diagnostics, leading to enhanced patient outcomes through enhanced speed and accuracy of diagnosis. The current research in fullerenes is a classic illustration of the dynamic potential of nanomaterials to transform healthcare diagnostics and therapies, culminating in pathfinding innovations in the medical technology discipline (Altintas et al., 2019; Hanna et al., 2017).

3.3. METAL-BASED NANOMATERIALS

Metal-based nanomaterials are at the forefront of the use of nanotechnology due to the unique physical and chemical properties that diverge significantly from their bulk material counterpart. These materials, like nanoparticles, nanowires, and metal oxides, possess greater electrical conductivity, catalytic activity, and optical properties, making them facile to many different applications, from electronics to medicine. The larger surface area-to-volume ratio in metals of nanosize, for instance, facilitates more interaction with the surrounding environment and thus greater reactivity. Although these advantages have promising prospects, however, they also present potential risks. The life cycle of metal nanoparticles, from release into the environment to interaction with biological systems, remains poorly understood, requiring comprehensive research to ensure safety and compliance with regulation in their application (Abhyankar et al., 2015). Nanomaterials based on metals have revolutionized the manufacturing and functioning of electronic devices. Silver and gold nanoparticles, for example, are widely utilized in applications that vary from sensors to transparent conductive films to photovoltaic cells because of their superior electrical properties and plasmonic characteristics. Recent developments have illustrated that incorporation of metal-based nanomaterials into polymers can also enhance the performance characteristics of composite materials further. Researchers have investigated the synthesis of piezoelectric nanocomposites comprising polymers with incorporated metal salts to achieve drastic improvements in electromechanical properties. This breakthrough not only paves the way for new wearable technology but also points to the importance of selecting appropriate metal compounds in order to optimize the functional properties of resulting nanocomposites (Ballirano et al., 2018). Although very promising, extensive use of metal-based nanomaterials is limited by fears regarding environmental and health hazards. Concerns exist regarding the potential toxicity of nanoparticles released at the manufacturing, use, and waste treatment processes, which could pose a risk to human health and the environmental system. Current regulation lacks specific guidelines for assessing the safety of nanomaterials, hindering industries from being compliant. To address such challenges, one needs to devise standardized methods for assessing the risk and knowledge transfer to consumers on safe nanoproduct handling.

Through openness and regulation, the nanotechnology sector can foster public trust, leading to broader public acceptance of metal-based nanomaterials in the market (Abhyankar et al., 2015) (Figure 3.7).

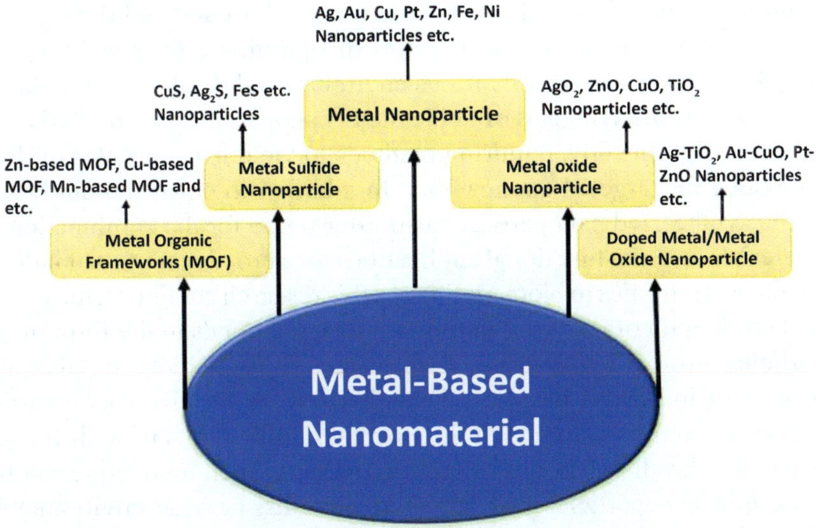

Figure 3.7. Metal-based nanomaterials.

Source: https://external-content.duckduckgo.com/iu/?u=https%3A%2F%2Fwww.frontiersin.org%2Ffiles%2FArticles%2F528583%2Ffchem-08-00341-HTML%2Fimage_m%2Ffchem-08-00341-g001.jpg&f=1&nofb=1&ipt=d55c9f77baceaea19c792492bcfeaef754be00083432ed190e01f03fc33d41b1.

3.3.1. Silver Nanoparticles (AgNPs): Antimicrobial Properties

The application of AgNPs as antimicrobial therapeutics is now a leading area in nanotechnology based, to a great degree, on their physicochemical properties. AgNPs are highly potent towards a wide range of microbes including bacteria, viruses, and fungi. This antimicrobial action is primarily due to the silver ions released from the nanoparticles which interact with the microbial cell membrane, leading to cell lysis and resultant death. The variability of AgNPs renders them compatible for incorporation into various matrices, which guarantees enhanced stability and extended antimicrobial activity in applications as varied as healthcare devices and fabrics. From the infection control perspective, application of the nanomaterials is a cutting-edge approach that may be superior to conventional antibiotics in a world with escalating antimicrobial resistance.

However, thorough examination of their effect and environmental persistence continues to be imperative in development and application strategies. Recent innovations in the synthesis of AgNPs have disclosed a multitude of ways for the customization of their size, shape, and surface properties, which directly affect their antimicrobial efficacy. Studies have revolved around stabilizing AgNPs by a wide range of ligands with the aim of optimizing their stability while ensuring their bioactivity under various environmental conditions. On the basis of recent surveys, the synthesis of noble metal nanoparticles, particularly silver, has increased significantly, requiring methods to yield nanoparticles of desired characteristics for targeted applications. In addition to enhanced antibacterial action, the synthesized nanoparticles also offer scope for the combination with other materials in multi-functional applications for a broad spectrum of industries ranging from electronics to biomedicine. As this research continues, the potential for mass production of powerful antimicrobial compounds in the form of silver nanoparticles grows, highlighting their potential for addressing the global health issue of infectious diseases. Though evidence for silver nanoparticles' antimicrobial activity is strong, environmental ramifications of their longevity necessitate the development of greener alternatives. Such innovative solutions, such as lignin nanoparticles filled with silver ions, provide environmentally friendly options that are antimicrobial and have a low ecological impact. Such formulations prove that using biodegradable materials can reduce the harmful effects commonly brought about by metallic silver nanoparticles. Particularly, the prolific release of silver ions that become attached to bacterial membranes is made possible for cell killing without having much added toxicity compared to classical silver-containing drugs. This strategy is in line with the guidelines of green chemistry by promoting the use of renewable materials in the synthesis of nanomaterials. Further research into environmentally friendly alternatives is a promising future direction to take advantage of the antimicrobial properties of silver without succumbing to sustainability hurdles in material science and public health applications, ensuring the best of nanomedicine and nanotechnology in the years to come (Duval et al., 2015; Fratoddi et al., 2017).

3.3.2. Gold Nanoparticles (AuNPs): Applications in Medicine and Electronics

AuNPs have also gained significant attention because of their extensive medical applications, for instance, biosensing and drug delivery. The improved optical properties are due to surface plasmon resonance, making them useful for enhanced imaging and detection capabilities in the biological system. In clinical usage, AuNPs serve as good biosensors for various biomarkers in the diagnosis of disease such as cardiovascular disease (CVD) for which early treatment is required. The ability of biosensors to detect various biomarkers is important because CVD poses a huge health hazard globally. As stated in reports, the study of cheap and effective biosensing devices is indispensable in a bid to assist in reducing reliance on expensive laboratory tests in clinics, illustrating the groundbreaking potential of gold nanoparticles in medicine (Gurbuz et al., 2012).

Apart from application in medicine, gold nanoparticles also have a profound role in electronics, i.e., the development of sophisticated microchips and sensors. Their high electrical conductivity and stability make AuNPs appropriate for the creation of ultra-sensitive electronic devices. Recent advances have made it possible to introduce gold nanoparticles in biochips, enabling the continuous real-time monitoring of human disease biomarkers, a possibility of supreme significance for personal medicine.

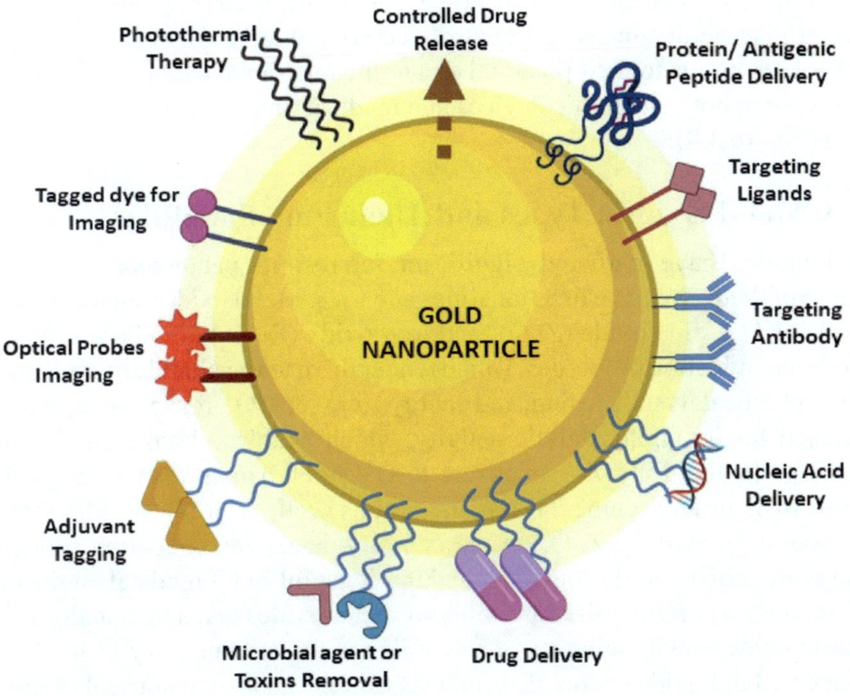

Figure 3.8. Gold nanoparticles (AuNPs).

Source: https://external-content.duckduckgo.com/iu/?u=https%3A%2F%2Fwww.mdpi.com%2Fvaccines%2Fvaccines-10-00505%2Farticle_deploy%2Fhtml%2Fimages%2Fvaccines-10-00505-g001.png&f=1&nofb=1&ipt=bbb250c763bc1ee7ce8cd09af2f0afcacf567028a45c8f5a6083fdf8df482b54.

The biochips not only enable real-time monitoring of health but also minimize the extent to which invasive procedures must be used, representing an unprecedented technological development in the healthcare industry. The inherent properties of AuNPs enhance the effectiveness and reliability of such devices as they can be fabricated to impart high specificity and sensitivity, thereby overcoming the current limitations in clinical diagnostic methods (Baj-Rossi et al., 2012). The convergence of electronics and nanomedicine through the integration of gold nanoparticles holds great promise for effective and biocompatible medical solutions. Designing fully implantable, minimally invasive devices is underway with AuNPs being a prime to overcome existing technological hurdles. These advances are predicted to ensure long-term biocompatibility, data transmission security, and high reliability in the monitoring of human metabolites. Gold nanoparticles paving the way for new health care applications, further studies will target enhancing their functionalization for efficient disease diagnosis and control. The ongoing study of the dual use of AuNPs later points to their potential as a multi-functional nanomaterial that can revolutionize both electronic devices and medical diagnostics (Baj-Rossi et al., 2012) (Figure 3.8).

3.3.3. Metal Oxides: Types and Their Functionalities

Metal oxides have garnered significant interest in nanomaterials research because of their extensive functionalities and uses. Metal oxides such as titanium dioxide (TiO_2), zinc oxide (ZnO), and iron oxide (Fe_2O_3) are characterized by their semiconducting properties, which render them very valuable in applications such as photocatalysis, sensing, and energy storage. TiO_2, for instance, is highly renowned for its photocatalytic activity, which enables photo-degradation of organic pollutants following exposure to UV light. This property has enabled the use of it in environmental remediation as well as in the development of self-cleaning materials. ZnO, on the contrary, has semiconducting properties along with antibacterial properties, making it useful in biomedical engineering and coatings. Therefore, the variability of metal oxides takes on a major role in accessing new functionalities that are in line with contemporary technological demands. Metal oxide nanoscale synthesis, which can lead to unusual electronic and optical properties, is a feasible route for tuning their functional properties. Different methods of fabrication such as sol-gel techniques, hydrothermal synthesis, and chemical vapor deposition have the ability to control their morphology and surface properties and thus their reactivity and performance. For example, the controlled synthesis of iron oxide nanoparticles can produce a wide range of magnetic properties, which are extremely critical for applications ranging from targeted drug delivery to magnetic resonance imaging. Apart from that, recent reports indicate that metal oxides can be incorporated into composite scaffolds with carbon-based materials and enhance the performance as a whole. By combining the strengths of different materials, researchers can develop nanocomposites with synergistic functionality and hence tailor functionalities for sensor, catalyst, and energy device applications. An extension of the study of metal oxides is their interaction with biological systems, where metal oxides are emerging as strong contenders for biocompatible applications. Recent evaluations of graphene oxide (GO) depict promising biocompatibility features, with low morphological alterations of treated cell lines in biocompatibility tests. In contrast, single-wall carbon nanotubes (SWCNTs) were cytotoxic in varying concentrations, which indicates a critical distinction if utilized in therapy. The findings point out that GO nanosheets have the potential to be excellent nanocarriers in drug delivery and molecular imaging with favorable biointeraction profiles. Also, functionalizability of metal oxides can make it possible to design advanced drug delivery systems. Such systems can tailor drug release profiles and target individual cells, which is critical in enhancing therapeutic responses while avoiding possible side effects, demonstrating the advanced applications of metal oxides for contemporary medicine and biotechnology.

3.4. POLYMER-BASED NANOMATERIALS

The application of polymer-based nanomaterials takes center stage in the realm of nanotechnology, where their diverse uses range across medicine, environmental science, and materials engineering. The small size and unique properties of these nanomaterials facilitate enhancement in drug delivery systems, diagnostics, and therapeutics. Their low cytotoxicity, combined with high functionalization capacity, qualifies them as excellent candidates for biomedical applications. Polymer nanomaterials, particularly nanohydrogels, have been in the limelight of recent research as they can stabilize and deliver bioactive agents efficiently. With advancements in polymer chemistry, it becomes possible for such materials to address complex biological challenges as the efficiency of the treatments enhances and side effects are minimized (Quazi et al., 2022). This cross-talk between polymer science and nanotechnology points to the revolutionary promise of polymer-based nanomaterials in contemporary therapeutics. From their biomedical applications, polymer-based nanomaterials also present opportunities of larger scope in food packaging and sustainability. Nanotechnology-derive advanced packaging strategies have revolutionized the food industry by increasing food product safety, shelf life, and quality. By incorporating polymer-based nanomaterials into packaging, manufacturers are able to create intelligent systems that monitor levels of contamination and freshness, offering confidence in food integrity throughout the supply chain. Bio-based nanomaterials are also gaining traction as a green alternative for traditional plastics. They have the promise of biodegradable and eco-friendly packaging alternatives that minimize the environmental problems caused by plastic litter (Chausali et al., 2021).

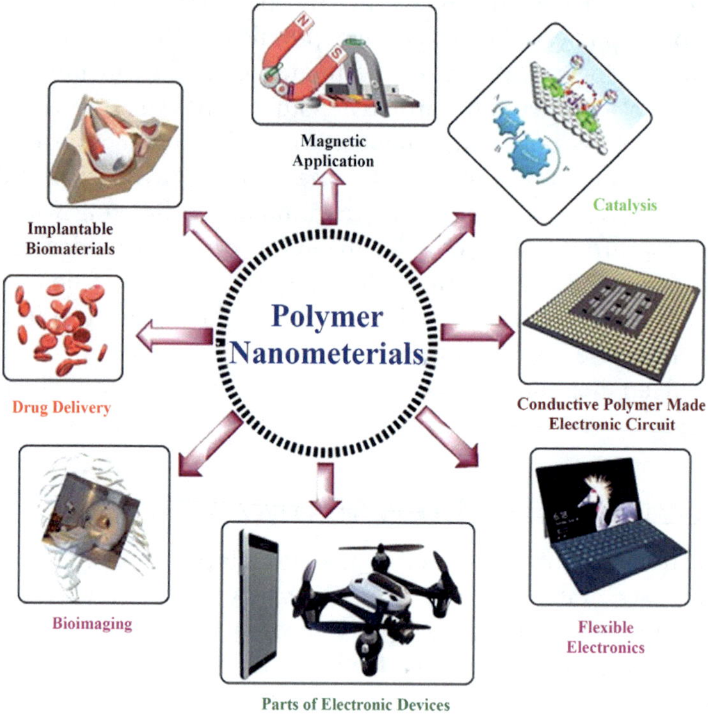

Figure 3.9. Polymer-based nanomaterials.

Source: https://external-content.duckduckgo.com/iu/?u=https%3A%2F%2Fwww.pharmaexcipients.com%2Fwp-content%2Fuploads%2F2021%2F07%2FPolymer-Nanomaterials.jpg&f=1&nofb=1&ipt=a722c4aa518cd72c95e578ba94d3c91502c352a2a073ed7579cff9a2892b682d.

Polymer nanomaterial addition to food packaging not only enhances function but also plays a role in increasingly consumer-directed demand for more sustainable methods within food production. Polymer-based nanomaterial design and synthesis present particular challenges and opportunities in the materials science field. Formulating targeted properties typically requires a delicate interplay among polymer composition, molecular structure, and processing conditions. Advancements in the techniques such as 3D printing and electrospinning have enabled high-precision control over the structure and morphology of these nanomaterials. Furthermore, studies are also being carried out to enhance biodegradability and biocompatibility of polymer-based systems to enable their safe use in applications. With further studies on these materials, expectations are generated about overcoming existing limitations in terms of cost, scalability, and performance. The prospects for future development of polymer-based nanomaterials are bright, as they offer keys to innovative solutions for problems of long standing in healthcare, conservation of the environment, and sustainable development (Chausali et al., 2021; Quazi et al., 2022) (Figure 3.9).

3.4.1. Nanocomposites: Enhancing Material Properties

Nanocomposites are leading the way in materials science because they have the ability to significantly enhance mechanical, thermal, and electrical properties. Through the incorporation of nanoscale fillers, such as carbon nanotubes, metal nanoparticles, or clay minerals into a matrix material, researchers are able to create composites with exceptionally improved performance compared to traditional products. This enhancement comes from the unique properties of nanoscale materials, which possess the ability to stabilize the matrix, strengthen interfacial adhesion, and lend reinforcing functionality at a molecular level. For instance, the incorporation of carbon nanomaterials has been shown to increase tensile strength and elasticity as well as confer conductivity. These benefits extend beyond industry, however, to consumer products since they can lead to lighter, stronger, and more resilient materials that can meet the higher demands of modern technology and sustainability initiatives. In ecological uses, nanocomposites have drawn interest because of their potential for creating high-tech material with the ability to achieve significant pollutant removal. The integration of nanomaterials in composite architectures allows it to become possible to improve reactivity and surface area, which are important to the operations like adsorption and catalytic elimination of pollutants. Nanocomposites, especially, are capable of targeting and eliminating a variety of pollutants ranging from heavy metals to organic chemicals in water and air purification systems. Utilization of nanoscale material and polymers in composite membrane development further enhances filtering efficiency, which poses an intriguing possibility for water treatment technologies. Such new applications illustrate the multifaceted utility of nanocomposites, revealing their key roles in addressing environmental problems of pressing concern as well as achieving improved material performance. Besides, the study of nanocomposites has extended to energy conversion technologies where their integration into thermoelectric materials has also been very promising. For instance, in the case where Half-Heusler (HH) alloys, when nanostructured, exhibit highly enhanced thermoelectric performance, particularly in high-temperature operating regimes of 700 K. Such nanocomposites harness the unique properties of structures at the nanoscale to reduce thermal conductivity while increasing electrical conductivity, hence improving overall energy efficiency. The flexibility of such materials lays open the door to new approaches in harvesting waste heat from industrial processes, emphasizing the future of nanomaterials in sustainable energy prospects. With future developments in nanocomposite technology, the potential for developing next-generation materials is a key point of intersection between materials science, energy efficiency, and environmental stewardship (Altintas et al., 2019; Poon et al., 2012).

3.4.2. Dendrimers: Structure and Applications

Dendrimers are branched tree-shaped macromolecules with a definite structure, made of a central core, branching units in the middle, and terminal groups of functions. Their unique structure is the reason behind their vast diversity of properties such as large surface-to-volume ratios and precise molecular size. Dendrimer synthesis occurs by a process of repetitive branching, which enables one to control size and functionality precisely. These properties make dendrimers extremely versatile to be applied to a wide range of applications in numerous fields, testifying to their importance in nanomaterial research and the way they can impact technology and medicine. Further, their ability to encapsulate medications and provide targeted delivery systems puts dendrimers center stage in nanomedicine, where they can enhance the effectiveness of medication delivery and eliminate undesirable side effects that are inherent in conventional therapies. The application of dendrimers is not only in medicine; they also form a subject of research in biosensing technology. Dendrimers may improve the efficiency of biosensors by acting as transduction components due to their ability to interact with biological molecules at the molecular level. Owing to their tunable functionality and high surface area, dendrimers can potentially be used to target selective biomolecules in order to make biosensors in diagnostics as well as environmental monitoring more sensitive and specific. For instance, the use of dendrimers within biosensor platforms allows for effective transduction of biological recognition events into quantifiable signals for the detection of low-abundance biomarkers. As highlighted in current studies, the use of dendrimers in biosensors not only improves signal transduction but also facilitates miniaturization and the overall performance of these vital devices (Naresh et al., 2021). Although encouraging outcomes are associated with dendrimers, challenges continue to surround their widespread application within the clinic and commercial viability. Biocompatibility, toxicity, and scaling concerns of production must be resolved so that they can be shifted from research in the laboratory to practical application. Current work is focused on optimization of dendrimer formulations to further enhance their safety profile without sacrificing their potential function. Along with this, research is being carried out to explore new synthetic methods that could lead to the synthesis of dendrimers with specific functionalities aimed at addressing issues of targeted drug delivery and biosensing. As per recent articles, the development of deeper insights into dendrimer interactions at the molecular level will aid in the overcoming of current bottlenecks and the realization of their full potential for therapeutic and diagnostic applications (Cheng et al., 2021) (Figure 3.10).

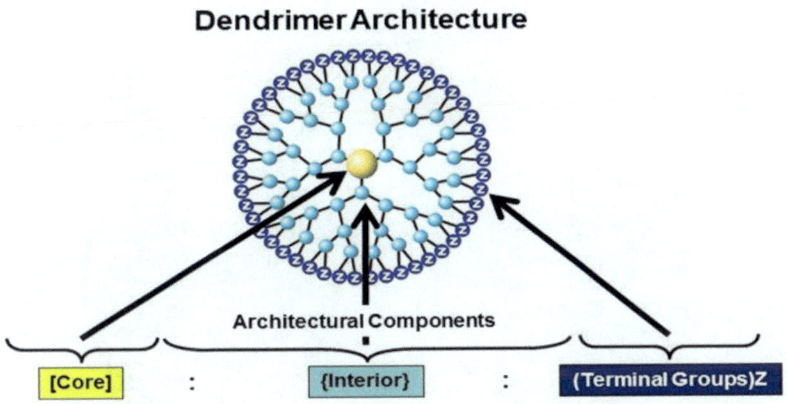

Figure 3.10. Dendrimers.

Source: https://external-content.duckduckgo.com/iu/?u=http%3A%2F%2Fwww.orientjchem.org%2Fwp-content%2Fuploads%2F2014%2F09%2FVol30_No3_Dend_Singh_fig6.jpg&f=1&nofb=1&ipt=276fc4112a038d92993e880c14df4a766ef3c883016697ee7db11b2483bba606.

3.4.3. Biodegradable Nanopolymers: Environmental Impact and Uses

Biodegradable nanopolymers are a new area of innovation in the world of nanomaterials that have proven highly promising to address key environmental challenges compared to traditional plastic waste. These nanopolymers, which degrade naturally over time, possess the potential to mitigate the impact of long-lasting pollutants on the ecosystem. Their molecular structure, normally derived from biodegradable renewable natural polymers, is biodegradable through microbial activity and therefore environmentally friendly as an alternative to conventional plastics. The integration of nanotechnology into biodegradable polymers enhances their properties, such as mechanical resistance and heat resistance, necessary for packaging and agricultural films applications. By minimizing the environmental footprint of such materials, biodegradable nanopolymers show the way that highly advanced materials can be in line with green goals without sacrificing functional advantages. Applications for biodegradable nanopolymers are diverse and transcend industries, aiming at consumer goods as well as medical devices.

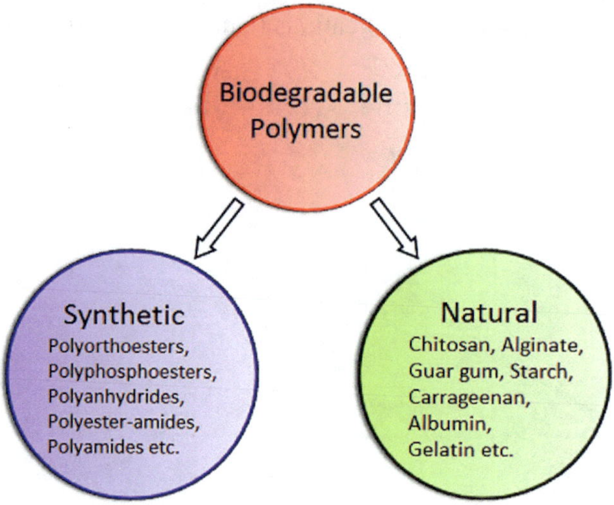

Figure 3.11. Biodegradable nanopolymers.

Source: https://external-content.duckduckgo.com/iu/?u=https%3A%2F%2Fd3i71xaburhd42.cloudfront.net%2Fd53c1fb42ce2784f601a32cc43597b118b3d9769%2F2-Figure1-1.png&f=1&nofb=1&ipt=04ae39cc856a3a548969b6c736c16f3b5e062d19bce9c4d5a6a52fb5d79d5aea.

In agriculture, they are used to produce controlled-release fertilizers and pesticides that promote sustainable agriculture with reduced chemical runoff and healthier soils. In medicine, biodegradable nanopolymers are applied as drug delivery systems, delivering targeted therapies with reduced side effect compared to traditional methods. These developments are particularly noteworthy in the context of nanotechnology's role in improving the bioavailability and efficacy of drugs, as addressed in the thorough research on nanoscale drug delivery systems. These applications do not merely illustrate the divergent versatility of biodegradable nanopolymers but also propose a paradigm shift towards more sustainable processes in product development and manufacturing across disciplines. Though their potential properties are high, the environmental impact of biodegradable nanopolymers is complex and demands stringent monitoring of their lifecycle and degradation. Mechanisms like the rate of biodegradation, release of by-products, and reaction with the environment must be strictly analyzed to prevent generating unforeseen environmental problems through the application of these materials. In addition, the knowledge of mechanisms of toxicity of various types of nanoparticles is very significant since their increased chemical reactivity relative to bulk counterparts may prove to be harmful to human health and to the ecosystem. Holding these challenges in mind will be important as scientists continue to work toward creating safer, more effective biodegradable nanopolymers aligned with environmental sustainability goals, notes recent research highlighting their implications for health as well as the environment (Abbasi et al., 2023; Liu et al., 2024) (Figure 3.11).

3.5. CERAMIC NANOMATERIALS

Among many types of nanomaterials, ceramic nanomaterials are most notable for their unique properties and vast applications. These inorganic, non-metallic nanomaterials possess exceptional mechanical strength, thermal stability, and chemical resistance. The nanoscopic nature of these ceramics, ranging from 1–100 nanometers, enhances their reactivity and applicability to high-end products such as biomedical devices, catalysts, and environmental remediation. The increasing need for ceramic nanomaterials is primarily driven by their capacity to improve product performance and durability, and consequently, they are suitable for emerging technologies applied in sectors as disparate as electronics and construction. Production of ceramic nanomaterials is carried out with a variety of methods, including sol-gel process, hydrothermal synthesis, and the microwave route. All of these possess varying advantages, allowing the engineering of targeted properties such as porosity, density, and crystalline structure. For instance, sol-gel processing is well suited for regulation of the chemical composition and inclusions of several elements, which makes it easy to create multifunctional nanoceramics. The recent advances also include the application of ceramic nanomaterials within composite materials with the material's significant enhancement of its mechanical and thermal properties. These advances demonstrate the working versatility of ceramic nanomaterials in solving the technological deficiencies in most fields. In addition, collaborative projects such as the 2BFUNTEX project demonstrate the diversity of these materials used in functional textiles and amplify their significance in various industries. The project brings together research institutes, universities, and small-and-medium-sized enterprises (SMEs) to develop the potential of functional textile structures using innovative ceramic materials that are previously unexplored. This interdisciplinary approach not only provides the door to rapid industrial uptake but also focuses on training and cooperation in setting up new functions in ceramic nanomaterials. By engaging industry stakeholders at every stage of the process, the project aims to promote innovation and delineate technological gaps, emphasizing the greater use of ceramic nanomaterials in revolutionary applications, as discussed in analogous technology sectors highlighted in current endeavors outlined in analogous technological fronts (Kiekens et al., 2012a, 2012b) (Figure 3.12).

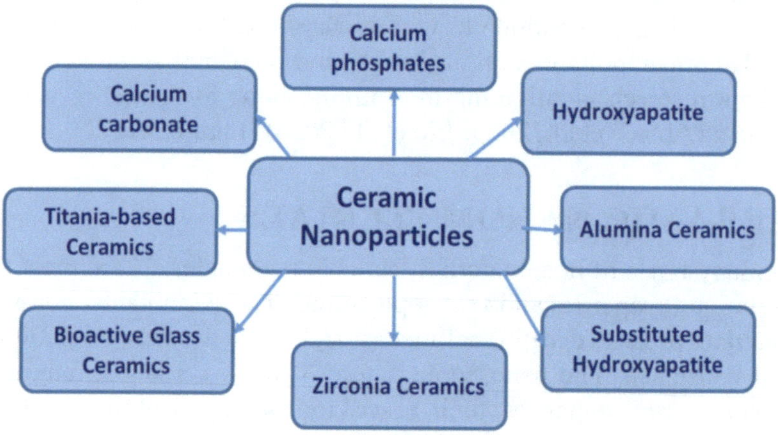

Figure 3.12. Ceramic nanomaterials.

Source: https://media.springernature.com/lw685/springer-static/image/chp%3A10.1007%2F978-3-030-40513-7_39/MediaObjects/472188_1_En_39_Fig6_HTML.png.

3.5.1. Types of Ceramic Nanomaterials: Oxides, Nitrides, and Carbides

Ceramic nanomaterials are a prominent group of nanomaterials these days, comprising oxides, nitrides, and carbides. They are characterized by their enhanced mechanical, thermal, and electrical properties and thus come to be used across various industries. Oxides, for example, zirconia and alumina, possess excellent chemical stability and wear resistance and hence come out as ideal choices as biomedical implants and cutting tools. Their thermal resistance and chemical conditions resistance are essential for industries ranging from aerospace to automobile. Ceramic oxides' strength and flexibility also find application in catalysis and sensing, where their high activity and large surface area can be employed for enhancing performance. Nitrides, a second type of ceramic nanomaterial, possess special characteristics that are different from oxides. Silicon nitride and boron nitride, for example, are highly hard and thermally stable and find their applications in machining tools and components of high-performance electronic devices. The structural integrity of nitride ceramics under adverse conditions gives them usefulness in difficult applications like nuclear reactors and aerospace. Moreover, advancements in methods of synthesis, such as the sol-gel process, have made it easier to synthesize homogeneous mixed nitrides that have improved physical properties and broader opportunities for application, hence widening their scope of use across different industries (Mahata et al., 2023). Another significant subclass of ceramic nanomaterials are the carbides, primarily due to the fact that they are highly hard and conduct heat very well. Materials like silicon carbide and titanium carbide are also extensively employed in cutting tools and abrasives, further enhancing the manufacturing process productivity. The hard wear-resistant nature of carbides leads to their extended life under harsh working conditions, reducing replacements. In addition, ongoing research into carbide nanocomposites is highly promising for their future, particularly in their advancement through designed microstructures and combinations with other nanomaterials (Levashov et al., 2017). As carbide ceramics are further researched, their applications in various technologies will expand, proving their relevance in the field of nanomaterials.

3.5.2. Applications in Electronics and Catalysis

Nanomaterials have revolutionized the electronics world by drastically increasing the performance and efficiency of devices. Owing to their quantum effects and high surface area to volume ratio, the materials offer room for enhancement in device components such as transistors, sensors, and batteries. For instance, the application of carbon-based nanomaterials like carbon nanotubes and graphene has enhanced charge mobility and signal sensitivity by much greater extent compared to traditional materials. In addition, the development of two-dimensional materials like boron and carbon nanosheets has excellent electronic and optical properties that can be incorporated into next-generation electronics. This is particularly evident with how they can be incorporated into flexible electronics, which not only have increased efficiency but also the ability to create light and compact devices. The multidisciplinary approach to nanomaterials promotes innovation in electronics, leading to revolutionary technologies and innovative consumer products (Moreno et al., 2017). In catalysis, nanomaterials demonstrate impressive properties that broaden reaction rates and selectivity as a result of their high surface area and active sites. Nanocatalysts typically deliver superior catalytic features compared to their bulk nature, rendering them vital in chemical processes. For example, platinum nanoparticles alter reaction pathways, which can reduce activation barriers and cause reactions to occur under less severe conditions. Carbon nanomaterials, particularly graphene derivatives and metal-organic frameworks, have also been noted for their ability to support catalytic processes and induce charge transfer, resulting in significantly improved catalytic efficiency. The versatility of the nanomaterials allows site-specific utilization, from the management of pollution through photocatalytic degradation to energy conversion in fuel cells, making them important in environmental as well as energy applications (Makaremi et al., 2018). The contribution of the nanomaterials to energy storage devices, particularly batteries and supercapacitors, cannot be overemphasized. The enhancement of rechargeable batteries to enhance storage capacity and lifetimes increasingly depended on integration with nanostructured materials. Recent studies demonstrate that two-dimensional nanosheets, e.g., hydrogen boride (HB), have remarkable performance features for application as anode materials with much larger capacities than their traditional counterparts like graphite. These improvements are attributed to the nanosheets unique structural features and electrical properties that enable efficient ion intercalation. Therefore, the confluence of energy storage with nanotechnology has the potential not only to surpass present limitation but also to bring about opportunities for sustainable energy uses. Such uses reflect the disruptive role that nanomaterials have in spearheading electronic advances as well as catalytic performance, heralding a new era of material science and engineering (Makaremi et al., 2018).

3.5.3. Advantages of Ceramic Nanomaterials Over Traditional Materials

Ceramic nanomaterials are a key advancement that surpasses conventional materials in most applications due to advantages they present. Reducing the size of material to the nanometer scale significantly enhances the physical and chemical properties of ceramics, such as strength, toughness, and thermal resistance. Unlike brittle bulk ceramics, nanoscale ceramics possess improved mechanical properties, resulting in improved durability and resistance. Besides, these materials have decreased porosity and increased surface area, leading to enhanced performance in uses like catalysts and structural elements. This shift in properties allows ceramic nanomaterials to keep pace with modern engineering and technology's high demands, making them a preferred choice in industries like aerospace, biomedical devices, and automotive industries, thereby sustaining their use in new applications compared to traditional alternatives (Joudeh et al., 2022). Besides their improved mechanical properties, ceramic nanomaterials have extraordinary thermal stability and thermal resistance, characteristics not commonly found in traditional materials. Their thermal stability is high, making them particularly advantageous for uses requiring toughness against extreme conditions such as thermal barrier coatings and aerospace. Notable is that their inherent resistance to heat decreases thermal degradation, and the life of critical equipment and tools is extended. Supporting this thermal advantage is that they can be corrosion- and chemically resistant, making them suitable for use in challenging environments.

Figure 3.13. Advantages of ceramic nanomaterials.

Source: https://ars.els-cdn.com/content/image/3-s2.0-B9780323996204000075-f07-01-9780323996204.jpg.

That they can function effectively where common materials fail positions ceramic nanomaterials at the forefront of application fronts from industrial manufacturing to energy solutions, thus bridging gaps opened by conventional materials and maximizing operational efficiency (Ateş et al., 2020). In consideration of the environmental impact of material production, ceramic nanomaterials also contribute towards sustainability. Being capable of being synthesized using biorenewable resources, they offer a green alternative to conventional materials to satisfy current social demand for less polluting production techniques. Utilizing sustainable sources and advanced nanotechnology techniques, ceramic nanomaterials not only reduce fossil fuel reliance but also restrict wastes. The use of nanomaterials in composite matrices is now improving their application across other industries while providing massive improvements in mechanical and thermal properties. With scientists continuing to learn and innovate in the field, the expanding roles of ceramic nanomaterials could lead to advancements in sustainability, addressing environmental challenges and material performance simultaneously (Ateş et al., 2020) (Figure 3.13).

3.6. APPLICATIONS OF NANOMATERIALS

The applications of nanomaterials cross multiple industries, reflecting their versatility and potential to revolutionize. In medicine, for instance, nanomaterials have revolutionized drug delivery, allowing targeted therapies to reduce side effects while increasing the rate of successful treatments. The targeting ensures active pharmaceutical ingredients are delivered to sites of action, enhancing patient outcomes. Secondly, nanomaterials find applications in diagnostics as well to facilitate extremely sensitive detection methods that can identify disease at an earlier stage. Their potential to upgrade the medical devices' function is impressive in that innovation like imaging agents and nanoscale sensors has been responsible for advancements in personalized medicine. This intricate combining of nanomaterials with medicine reflects their core position in current medical technologies and speaks to the urgent need for comprehensive safety assessments on such materials during their life cycle, as viewed by increasing problems in the field (Abhyankar et al., 2015). Nanomaterials are also having significant implications in the science of materials, in addition to medicine. They are used to improve the properties of conventional materials, leading to the development of nanocomposites with increased strength, lightness, and durability. For instance, the incorporation of nanofillers, i.e., nano-silica, into polymers can greatly improve the mechanical strength and heat stability and make these materials suitable for demanding applications within the automotive and aviation industries. Apart from that, nanomaterials help in energy-efficient technologies, such as the advancement of solar cells and batteries. Breakthroughs in energy storage, driven by nanotechnology, promise enhanced efficiency and longevity, which is crucial given the increasing worldwide demand for renewable energy. The complementarity of enhanced performance and energy efficiency evokes the revolutionary capability that nanomaterials have in not only enhancing existing technologies but also addressing existing environmental issues (Grimpe et al., 2025). Additionally, the use of nanomaterials in the environment underscores an intriguing pollution combat and resource management frontier. Nanomaterials have also found successful use in the field of water treatment, where they can be employed to clean up contaminants at a record level. Nanoparticles, for example, are used effectively in filtration devices to extract heavy metals and pathogens from water sources. Nanomaterials aid environmental remediation activities as well, assisting in the breakdown of harmful substances and the improvement of soil quality. However, their widespread application is weighed against safety concerns regarding processing and disposal of nanomaterials, and hence the need for developing standardized risk assessment tools. Of note will be ensuring to have appropriate regulatory mechanisms in place as society becomes increasingly reliant on nanotechnology as a way of pursuing sustainable endeavors, compelling the debate on the interface of innovation and safety (Abhyankar et al., 2015) (Figure 3.14).

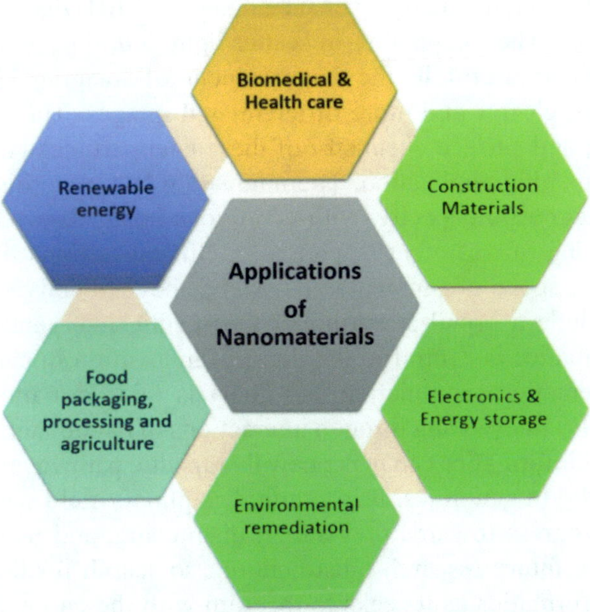

Figure 3.14. Applications of nanomaterials.

Source: https://media.springernature.com/lw685/springer-static/image/chp%3A10.1007%2F978-3-030-62761-4_1/MediaObjects/480036_1_En_1_Fig4_HTML.png.

3.6.1. Medicine: Drug Delivery and Imaging

Medical application of nanomaterials, in particular drug delivery and imaging, is a significant milestone in medicine. Targeted therapy is facilitated by nanocarriers such as liposomes, dendrimers, and polymeric nanoparticles in terms of the solubility, stability, and bioavailability of therapeutic agents. These nanoscale systems can be engineered to react to specific biological signals such that site-specific release of drugs minimizes side effects and maximizes therapeutic action. Current advances signal that nanomaterials are capable of traversing biological barriers such as cellular membranes to enable more effective delivery of conventional drugs and biologics. This capability is especially beneficial in the therapy of challenging illnesses such as cancer, where targeted delivery to tumors will drastically enhance therapy while reducing damage to normal tissue. Medical imaging techniques have also been significantly advanced by the application of nanomaterials, particularly in the development of new contrast agents. Gold and silver nanoparticles, for instance, have also shown tremendous promise in enhancing imaging modalities such as computed tomography (CT) and magnetic resonance imaging (MRI) due to their superior optical properties. Their capability to scatter light and impact electromagnetic fields makes them useful in the improvement of imaging signal contrast, hence producing clearer and more differentiated images. The ability to tailor the size, shape, and surface chemistry of these nanoparticles enables tailoring for a specific imaging application. The innovation has, according to research, led to better diagnostic capacity, with conditions being detected at an earlier stage and tracking of treatment responses in clinical settings better (Fratoddi et al., 2017). The scope of nanomaterials extends beyond delivery of drugs and imaging to include a broad spectrum of therapeutic interventions. Coupling with nanoinformatics is central to the drive of innovation in nanomedicine as it enables systematic accumulation and analysis of information pertinent to nanomaterials. By intervening through interactions of nanomaterials with living organisms, nanoinformatics can make new therapeutic pathways more efficient, just as -omics did in genomics. It is a multidisciplinary field with the capacity to accelerate progress towards preventing, diagnosing, and treating all forms of diseases. The future research must continue to search for the promise and limit of nanoinformatics as it seeks to meet up with the gap existing between experimental discovery and clinical application, hence transforming medical therapeutics and diagnostics into a new horizon (Anguita et al., 2012).

3.6.2. Energy: Solar Cells and Batteries

Application of nanomaterials in solar cells is a fundamental step towards maximizing energy absorption and efficiency of conversion. Traditional solar cells, which are composed of silicon, are constrained in their energy conversion efficiency largely by the material's intrinsic properties. That is something that can be changed with the addition of nanostructured materials such as quantum dots or nanoparticles of titanium dioxide, with which light-harvesting functionality can be significantly increased. By using control over nanoscale composition, researchers can make the surface more responsive to sunlight and, consequently, more energy can be harvested. Furthermore, technological development in this area provides a possibility of fabricating solar panels more compact, flexible, and deployable in an array of contexts, thus broadening the application of solar technology in renewable energy schemes (McManus et al., 2010). Batteries similarly demonstrate revolutionary properties with the discovery of nanomaterials, leading to enhanced cycling stability and energy density. Traditional battery technologies are handicapped when addressing charge capacity and life span. Nanomaterials such as silicon nanowires and conductive polymers improve the electrochemical conductivity and surface area of electrodes. This promotes higher electron and ion mobility with charge and discharge cycling, consequently converting to accelerating charging speeds and longer life operation. The integration of these new materials not only improves performance but also plays a role in the development of lighter and more efficient batteries, which are pivotal in new applications such as electric vehicles and solar energy storage systems (MORENO et al., 2017). Also, the integration of nanotechnology is not employed solely for the improvement of performance; it is pivotal in the sustainability of solar cells and batteries. For solar technology, enhanced nanomaterial-strengthened panels can be more resilient and have a longer lifespan, thus minimizing waste and the need for replacements. This longer lifespan helps to reduce the environmental footprint entailed in solar panel production and disposal. For batteries, nanomaterial improvements can further minimize the use of rare and toxic components through the exploitation of more readily available materials that also provide high performance. Therefore, the use of nanomaterials throughout the energy spectrum not only improves efficiency and versatility but also contributes to sustainability objectives critical for emerging energy alternatives (McManus et al., 2010).

3.6.3. Environmental Remediation: Water Purification and Pollution Control

Environmental remediation using nanomaterials has been recognized as a key focus area of research to help counteract the negative impacts of pollution, especially in water cleaning processes. Nanotechnology presents a succession of new options, with increased efficiency in the decontamination of water bodies. For instance, engineered nanomaterials such as carbon nanotubes and metal oxides are being researched for their ability to adsorb heavy metals and organic pollutants. These materials possess unique properties, including elevated surface area and reactivity, which enable them to bind and extract noxious compounds effectively. As highlighted in research, application of nanomaterials in environmental processes may revolutionize water quality to a greater extent by enabling the degradation and removal of harmful agents with fewer secondary effects of pollution, thereby addressing the imperative challenge posed by polluted sources of water (Bacchini et al., 2013). Additionally, the effect of nanomaterials on existing ecosystems is also of paramount research focus, particularly the possibility of inducing toxicological effects. It has been indicated through research that certain nanoparticles, despite being useful in remediation processes, can be dangerous to biological communities within aquatic ecosystems. For example, nano-titanium dioxide (nano-TiO_2), which is common in most uses, has been found to alter microbial community compositions when it is introduced into systems. Results of research demonstrate that exposure to nano-TiO_2 decreases bacteria populations in aquatic bodies temporarily, showing higher sensitivity of certain organisms to nanomaterial exposure (Ozaki et al., 2013). This indicates necessity for proper assessment of nanomaterial application in environmental purification so that pollutant removal benefits are achieved at the expense neither of ecological well-being. Lastly, strategic integration of nanotechnology in water purification systems is a green alternative to pollution control. By taking advantage of the ability of nanomaterials such as photocatalysis and specific remediation, researchers are discovering novel methods to control current and future pollutants. For example, utilization of composite materials composed of various nanomaterials can enhance pollutant degradation rates through synergistic effects. Application of these technologies not only makes cleaning contaminated water more effective but also offers a template for designing more sustainable environments. Increasing environmental issues confronting the world, further research in applying nanomaterials in environmental remediation will play a critical role in supplying efficient and sustainable technologies for safeguarding public health and ecological integrity (Bacchini et al., 2013).

SUMMARY

Our research into various types of nanomaterials, it is evident that their extensive properties and utilizations carry extensive consequences in a broad range of scientific fields. Nanomaterials' interaction with biological systems indicates their prospect in nanomedicine. Magnetic nanowires, for instance, have been discovered to have the capability of being taken up by cells with no acute toxicity. Such a property holds promising potential for drug delivery and cell manipulation, the way to new treatments. The advancement of nanotechnology relies not only on a comprehensive understanding of individual nanomaterials but also on the science that elucidates their mechanisms of action, toxicology, and degradation routes, e.g., in the study described on magnetic nanowires (Baeza-Squiban et al., 2011). Furthermore, the one-dimensional nanomaterial behavior, such as that of silicon nanowires, offers better understanding of the cellular interaction kinetics that are vital to their application in nanomedicine. The membrane wrapping by physical contact phenomenon exhibited by amine-functionalized silicon nanowires indicates their potential for application in controlling the mechanics and behavior of cells. So does it make necessary that nanomaterials be well designed and functionalized so as to tailor them to intended applications in biomedical applications. As the SiNW-NH_2 study has pointed out, visualization of these interactions can become vital in informing and directing the further development of nanomaterials that are efficient and biocompatible (Huang et al., 2017). It is important to realize the interaction between biological entities and nanomaterials in creating nanotechnology. These results not only add to the rational construction of novel nanomaterials but position them as breakthrough drugs in the clinic. The degradation behavior of nanowires, as outlined in their interaction with mouse fibroblasts, shows that their toxicity risk over the long term could be mitigated. This is very important in maintaining long-term safety in nanomedicine, assuaging regulators and the general public. Thus, as we keep unraveling the complications of nanomaterials, their extensive applications and inherent risks must be balanced against one another in order to unlock their full potential towards developing technology and enhancing health results (Baeza-Squiban et al., 2011).

The topic of types of nanomaterials encompasses their extensive applications and inherent nature. Central to this argument is the fact that nanomaterials exist in a variety of shapes and forms, either as zero-dimensional (0D), one-dimensional (1D), two-dimensional (2D), or three-dimensional (3D) structures. Each of these exhibits unique properties that drive their application and functionality for specific uses. For instance, 0D nanoparticles such as gold or silver nanoparticles are widely renowned for their optical properties, making them valuable in drug delivery and imaging. One-dimensional nanomaterials like nanowires, however, exhibit high aspect ratios to increase their use in electronics and sensors. The increasing multidimensionality of such materials prompts researchers to explore the inherent mechanisms behind their different functionalities. An astonishing question in nanomaterials research is the interaction of these nanomaterials with biological systems and, in particular, toxicity. Research has emphasized the issue of understanding how nanoparticles might affect human health alongside the environment. For example, mass toxicology studies have provided evidence of the harmful effects of air pollution, notably particulate matter (PM), which is usually made up of nanoparticles from industrial emission and automotive exhaust. The morphology of such particles through advanced techniques such as scanning electron microscopy (SEM) and energy dispersive spectroscopy (EDS) has proved to reveal their biochemical interactions in human cells. Studies have indicated that some nanoparticles are recognized to exert toxicity on pulmonary epithelial and immunological cells, and there exists a necessity for thorough risk assessments of their application in medicine and industry (Wilkinson et al., 2013). Moreover, mechanisms of cellular interactions with various nanomaterials have garnered much attention, mainly in the field of nanomedicine. For instance, physical wrapping of amine-functionalized silicon nanowires (SiNW-NH_2) during internalization in cells has a primary pathway for targeted drug delivery. Using optical microscopy and confocal fluorescence microscopy, researchers have unveiled the steps of cellular uptake and highlighted the involvement of physical stress imposed on the cell membrane (Huang et al., 2017). This information is a fundamental one in the intelligent design of nanomaterials for medical use, as it presents the bridge between nanotechnology and biological systems. Continuous identification of such interactions provides the basis for the safe and efficient use of nanomaterials in environmental and medical applications.

REFERENCES

1. Abbasi, R., Shineh, G., Mobaraki, M., Doughty, S., & Tayebi, L. (2023). Structural parameters of nanoparticles affecting their toxicity for biomedical applications: A review. Journal of Nanoparticle Research, 25. https://doi.org/10.1007/s11051-023-05690-w (accessed on 25th September 2025).
2. Abhyankar, H., Blackburn, D., Brighton, J., Gendre, S., et al. (2015). Measurement of nanoparticles release during drilling of polymer nanocomposites. https://core.ac.uk/download/42144135.pdf (accessed on 25th September 2025).
3. Adamec, V., Köbölová, K., Pavlíková, S., Procházková, M., et al. (2019). Current approaches of occupational and safety health management in work environments containing nanoparticles. https://core.ac.uk/download/287648860.pdf (accessed on 25th September 2025).
4. Ago, H., Bachtold, A., Bahr, J., Baughman, R., Bockrath, M., Bonnell, D., Brown, L., et al. (2011). Site-specific forest-assembly of single-wall carbon nanotubes on electron-beam patterned SiOx/Si substrates. http://arxiv.org/abs/1105.1598 (accessed on 25th September 2025).
5. Altintas, Z., & Pirzada, M. (2019). Nanomaterials for healthcare biosensing applications. https://core.ac.uk/download/287885382.pdf (accessed on 25th September 2025).
6. Anguita, A., Pazos, A., Geissbuhler, A., Smith, B., Kim, B. Y., Kulikowski, C., Rosse, C., et al. (2012). Nano informatics: Developing new computing applications for nanomedicine. https://core.ac.uk/download/28969094.pdf (accessed on 25th September 2025).
7. Ateş, B., Köytepe, S., Ulu, A., Gürses, C., & Thakur, V. K. (2020). Chemistry, structures, and advanced applications of nanocomposites from bio renewable resources. Chemical Reviews, 120, 9304–9362. https://doi.org/10.1021/acs.chemrev.9b00553 (accessed on 25th September 2025).
8. Bacchini, F., Greppi, G., Mura, S., Roggero, A., et al. (2013). Advances of nanotechnology in agro-environmental studies. https://core.ac.uk/download/16749559.pdf (accessed on 25th September 2025).
9. Baeza-Squiban, A., Auffan, M., Berret, J. F., Brunner, T. J., Cappallo, N., Champion, J. A., Chanteau, B., et al. (2011). Interactions between magnetic nanowires and living cells: Uptake, toxicity and degradation. http://arxiv.org/abs/1108.2081 (accessed on 25th September 2025).
10. Baj-Rossi, C., Burleson, W., Carrara, S., Cavallini, A., et al. (2012). Fully integrated biochip platforms for advanced healthcare. https://core.ac.uk/download/286729684.pdf (accessed on 25th September 2025).

11. Ballirano, P., Chandraiahgari, C. R., De Bellis, G., Fortunato, M., et al. (2018). Piezoelectric effect and electroactive phase nucleation in self-standing films of unpoled PVDF nanocomposite films. https://core.ac.uk/download/188825344.pdf (accessed on 25th September 2025).
12. Calcaterra, A., Mardente, S., Mari, E., Orlandi, A., et al. (2018). Metal-free graphene oxide (GO) nanosheets and pristine-single wall carbon nanotubes (p-SWCNTs) biocompatibility investigation: A comparative study in different human cell lines. https://core.ac.uk/download/158327534.pdf (accessed on 25th September 2025).
13. Chausali, N., Saxena, J., & Prasad, R. (2021). Recent trends in nanotechnology applications of bio-based packaging. Journal of Agriculture and Food Research, 7, 100257–100257. https://doi.org/10.1016/j.jafr.2021.100257 (accessed on 25th September 2025).
14. Cheng, Z., Li, M., Dey, R., & Chen, Y. (2021). Nanomaterials for cancer therapy: Current progress and perspectives. Journal of Hematology & Oncology, 14. https://doi.org/10.1186/s13045-021-01096-0 (accessed on 25th September 2025).
15. Davies, J. C. (2008). Nanotechnology Oversight: An Agenda for the New Administration. https://core.ac.uk/download/71353341.pdf (accessed on 25th September 2025).
16. de la LLana, S. A., Anderson, I. S., Capitán Vallvey, L. F., Cecchini, M., et al. (2014). A general perspective of the characterization and quantification of nanoparticles: Imaging, spectroscopic, and separation techniques. https://core.ac.uk/download/186617942.pdf (accessed on 25th September 2025).
17. Duval, A., Kumar, A., Panáček, A., Panáček, A., Verma, A., Nel, A. E., Richter, A. P., et al. (2015). An environmentally benign antimicrobial nanoparticle based on a silver-infused lignin core. https://core.ac.uk/download/151156916.pdf (accessed on 25th September 2025).
18. Elliott, K. (2008). Varieties of exploratory experimentation in nanotoxicology. https://core.ac.uk/download/11921904.pdf (accessed on 25th September 2025).
19. Esquivias Fedriani, L. M., Morales Flórez, V., Piñero, M., Rosa Fox, A., et al. (2009). Nanostructured sonogels. https://core.ac.uk/download/322844972.pdf (accessed on 25th September 2025).
20. Fratoddi, I. (2017). Hydrophobic and hydrophilic Au and Ag nanoparticles: Breakthroughs and perspectives. https://core.ac.uk/download/153366182.pdf (accessed on 25th September 2025).

21. Grimpe, C., & Patuelli, R. (2025). Knowledge production in nanomaterials: An application of spatial filtering to regional systems of innovation. https://core.ac.uk/download/pdf/6548125.pdf (accessed on 25th September 2025).
22. Gurbuz, Y., Gürbüz, Y., Kolkar, M., Niazi, J. H., Qureshi, A., et al. (2012). Biosensors for cardiac biomarkers detection: A review. https://core.ac.uk/download/pdf/11743545.pdf (accessed on 25th September 2025).
23. Hanna, S. K., Heller, D. A., Horoszko, C. P., Johnson, D. R., et al. (2017). Redox-active nanomaterials for nanomedicine applications. https://core.ac.uk/download/215479341.pdf (accessed on 25th September 2025).
24. Hernandez Moreno, S., Hernandez Moreno, S., Solache De La Torre, S. C., Solache De La Torre, S. C., et al. (2017). Nano-technological products in architecture and construction. https://core.ac.uk/download/154796653.pdf (accessed on 25th September 2025).
25. Huang, Y., McNear, K., & Yang, C. (2017). Understanding cellular internalization pathways of silicon nanowires. https://core.ac.uk/download/215928782.pdf (accessed on 25th September 2025).
26. Joudeh, N., & Linke, D. (2022). Nanoparticle classification, physicochemical properties, characterization, and applications: A comprehensive review for biologists. Journal of Nanobiotechnology, 20. https://doi.org/10.1186/s12951-022-01477-8 (accessed on 25th September 2025).
27. Kiekens, P., & Van der Burght, E. (2012). ENMat international projects: FP7 NMP coordination action: 2BFUNTEX. https://core.ac.uk/download/55859985.pdf (accessed on 25th September 2025).
28. Kiekens, P., & Van der Burght, E. (2012). ENMat international projects: FP7 NMP large collaborative project: 3D-LightTrans. https://core.ac.uk/download/55811497.pdf (accessed on 25th September 2025).
29. Levashov, E., Loginov, P., Mishnaevsky, L., Sidorenko, A., et al. (2017). Nanocomposites for machining tools. https://core.ac.uk/download/112373012.pdf (accessed on 25th September 2025).
30. Liu, Y., Liang, Y., Yuhong, J., Peng, X., Han, J., Du, Y., Yu, X., et al. (2024). Advances in nanotechnology for enhancing the solubility and bioavailability of poorly soluble drugs. Drug Design Development and Therapy, 18, 1469–1495. https://doi.org/10.2147/dddt.s447496 (accessed on 25th September 2025).
31. Mahata, D., Mahata, S., Mahato, S. S., Panda, S., et al. (2023). Perspective chapter: Sol-gel science and technology in context of nanomaterials–Recent advances. https://core.ac.uk/download/588401014.pdf (accessed on 25th September 2025).

32. Makaremi, M., Mortazavi, B., & Singh, C. V. (2018). 2D hydrogenated graphene-like borophene as a high-capacity anode material for improved Li/Na ion batteries: A first principles study. http://arxiv.org/abs/1803.07137 (accessed on 25th September 2025).

33. McManus, S., Przybyt, K., Ribela Rodríguez, R., Riera Bauzá, A., et al. (2010). Nanotechnology for energy saving: Nano-product (system)-concept design. https://core.ac.uk/download/41803544.pdf (accessed on 25th September 2025).

34. Mousavi Langari, M., Nikzad, M., & Labidi, J. (2022). Nanocellulose-based sensors in medical/clinical applications: The state-of-the-art review. Carbohydrate Polymers, 304, 120509–120509. https://doi.org/10.1016/j.carbpol.2022.120509 (accessed on 25th September 2025).

35. Naresh, V., & Lee, N. (2021). A review on biosensors and recent development of nanostructured materials-enabled biosensors. Sensors, 21, 1109–1109. https://doi.org/10.3390/s21041109 (accessed on 25th September 2025).

36. Ozaki, A. (2013). Assessing the effects of titanium dioxide nanoparticles on microbial communities in stream sediment using artificial streams and high throughput screening. https://core.ac.uk/download/48602050.pdf (accessed on 25th September 2025).

37. Poon, J., Tang, X., Tritt, T. M., Weidenkaff, A., et al. (2012). Recent advances in nanostructured thermoelectric half-Heusler compounds. http://arxiv.org/abs/1307.2156 (accessed on 25th September 2025).

38. Quazi, M. Z., & Park, N. (2022). Nanohydrogels: Advanced polymeric nanomaterials in the era of nanotechnology for robust functionalization and cumulative applications. International Journal of Molecular Sciences, 23, 1943–1943. https://doi.org/10.3390/ijms23041943 (accessed on 25th September 2025).

39. Ruggiu, D. (2015). The consolidation process of the EU regulatory framework on nanotechnologies: Within and beyond the EU case-by-case approach. https://core.ac.uk/download/53525097.pdf (accessed on 25th September 2025).

40. Sahu, A., Dosi, R., Kwiatkowski, C. O., Schmal, S., & Poler, J. C. (2023). Advanced polymeric nanocomposite membranes for water and wastewater treatment: A comprehensive review. Polymers, 15, 540–540. https://doi.org/10.3390/polym15030540 (accessed on 25th September 2025).

41. Sharma, S., Verma, A., Mavinkere Rangappa, S., Siengchin, S., & Ogata, S. (2023). Recent progressive developments in conductive-fillers based polymer nanocomposites (CFPNC's) and conducting polymeric

nanocomposites (CPNC's) for multifaceted sensing applications. Journal of Materials Research and Technology, 26, 5921–5974. https://doi.org/10.1016/j.jmrt.2023.08.300 (accessed on 25th September 2025).

42. Umbrello, S. (2019). Explorative nanophilosophy as tecnoscienza: An Italian perspective on the role of speculation in nanoindustry. https://core.ac.uk/download/158370840.pdf (accessed on 25th September 2025).
43. Wilkinson, K. (2013). Particulate airborne impurities. https://core.ac.uk/download/11698736.pdf (accessed on 25th September 2025).

CHAPTER 4

SYNTHESIS AND FABRICATION TECHNIQUES OF NANOMATERIALS

Contents

INTRODUCTION

Synthesis and fabrication of nanomaterials are now crucial fields in the field of materials science, driven by the unique characteristics nanomaterials exhibit at the nanoscale domain. Being objects with dimensions in the range of 1–100 nanometers, nanomaterials possess distinct physical, chemical, and biological properties that have led to their applications in numerous industries including electronics, catalysis, and energy storage. Their enhanced surface area-to-volume ratio makes them more reactive, and their quantum mechanical character can give rise to very distinct behaviors compared to their bulk counterparts. As such, understanding diverse synthesis methods is significant because it has direct bearing on the material characteristics and potential applications of the resulting nanostructures. Effective synthesis methods can tune the performance parameters and make nanomaterials workable for industrial purposes, making them adapt to growing demands of different technological advancements. Among numerous methods adopted in nanomaterial synthesis, the approach of flame spray pyrolysis (FSP) is a most potential method due to its scalability and efficacy.

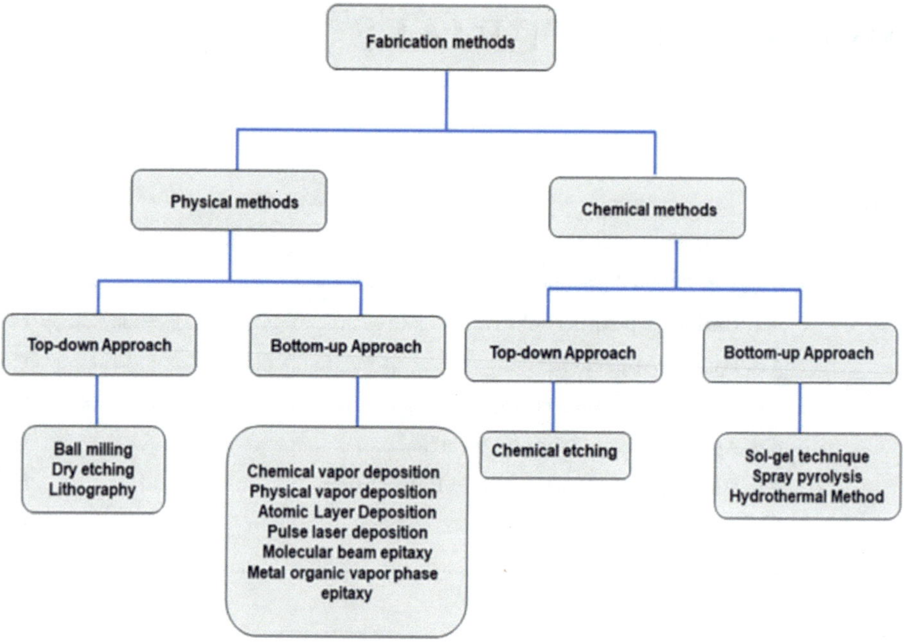

Figure 4.1. Fabrication of nanomaterials.

Source: https://ars.els-cdn.com/content/image/3-s2.0-B9780128228586000017-f01-02-9780128228586.jpg.

By facilitating fast nanoparticle formation with controlled morphologies and compositions, FSP provides a path to achieve high catalytic activity in such materials as TiO_2-based compounds. Conventional synthesis paths are often reproducibility-acerbated and scalability-bounded, which restricts large-scale applications. FSP circumvent these constraints and allow continuous manufacture while maintaining fine control over the process parameters. Recent work indicates fruitful use of FSP in the synthesis of catalytically active nanomaterials with diverse applications, for example, photocatalysis and thermocatalysis. The ongoing studies on FSP highlight the significance of detailed consideration of its benefits and limitations to obtain a comprehensive idea regarding its implications in nanomaterial synthesis. Aside from traditional catalytic materials, there has been increased focus on sustainable processes, which have resulted in a new generation of nanomaterials in the form of nanocellulose, which has enormous potential as an energy storage material. With the present world demand for green technology, nanocellulose emerges as a viable choice through its unique structural features, being cost-effective, and biodegradability. The creation of energy storage devices in the form of supercapacitors with nanocellulose or its composites is an example of an increasing focus on integrating environmentally friendly materials into technology. Energy storage is essential as society has significant hurdles to overcome with the conversion and storage of renewable energy, and the use of nanocellulose can help deliver significant greener, more sustainable solutions. Future study in this area will be essential in addressing many issues associated with renewable energy systems, thereby propelling the global transition to sustainable energy consumption (Figure 4.1).

Definition of Nanomaterials

Nanomaterials definition is a wide range of materials that are defined by their unique physical and chemical characteristics at the nanoscale, that is, dimensions ranging from 1–100 nanometers. This is the range of size since materials exhibit behaviors that vary from their bulk owing primarily to the increased surface area to volume ratio. Such properties developed at this scale include such as enhanced strength, lighter weight, higher chemical reactivity, and unique optical properties. The definition of nanomaterials thus relies both on size as well as thc ensuing properties, which make possible innovations in several fields such as electronics, medicine, and biotechnology. These attributes not only facilitate novel applications but also require innovations in synthesis and fabrication techniques tailored to control matter at the nanoscale, thus remorselessly pushing the frontiers of technological potential. Nanomaterials are generally classified under two categories: top-down and bottom-up strategies, and these indicate respective synthesis methods. The top-down approach involves the breakdown of large materials to nanoscale structures physically or chemically, while the bottom-up approach involves the construction of nanoscale structures atomically or molecule-by-molecule. Each approach has the potential for inherent strengths and weaknesses in terms of efficiency, scalability, and uniformity of the final product material. For instance, the top-down process can create defects and non-uniform particle sizes, whereas the bottom-up process can be employed with greater control over composition and morphology of nanomaterials. Understanding these variations in nanomaterial preparation is extremely crucial for scientists and engineers who seek to tailor materials for specific purposes, thereby illustrating the importance of opting for appropriate processes to achieve desired characteristics and functionalities of materials. The potential applications of nanomaterials, as exemplified by nanocellulose, provide an indicator of their success in the advancement of various industries. A plant cellulose derivative, nanocellulose, has very desirable characteristics, including high mechanical strength, biocompatibility, and renewability, and is therefore at the pinnacle of green materials. In the past few years, there has been an explosion of enthusiasm for functionalizing nanocellulose to create a broad spectrum of products, from nanocomposites to biomedical applications, that meet essential requirements for the uses of renewable energy and environmental remediation. As mentioned, further studies in nanocellulose properties and application further facilitate learning about nanomaterials and opens opportunities for innovation in the various fields. However, to maximize their potential also means addressing the issue of production scalability and uniformity of materials, thereby establishing the future scope of nanomaterial research and development (Trache et al., 2020; Zhi-Wu et al., 2020).

Importance of Nanomaterials in Modern Technology

The importance of nanomaterials in modern technology cannot be overstated with their transformative effect cutting across electronics, energy, and biotechnology sectors. Nanoscale materials exhibit new properties, such as enhanced electrical, thermal, and mechanical properties relative to bulk material. Such unique behavior comes about as a result of increased surface area and quantum effects that enable innovations such as improved speed and efficiency in electronic devices, ultra-lightweight but strong materials, and new energy storage technologies. Nanomaterials are the essential materials of these advances, which step up the size, velocity, and efficacy of the device, and all these are the essential components of today's technology-driven world. Continued advancements in the synthesis and fabrication processes further increase the versatility and breadth of use of nanomaterials so that certain engineered suit approach can solve specific technological problems. Particularly, the use of fluidic methods in nanomaterial synthesis offers unprecedented flexibility for precise control and modification. The combination of fluid mechanics with engineered systems facilitates the development of complex nanoplatforms that have tunable material properties suitable for various applications, especially in energy research and storage media. As is being witnessed with MXenes, a next-generation class of two-dimensional materials, fluidic techniques offer rapid sample processing and precise fluid management, essential in discovering their full potential in the area of energy (Richard et al., 2023). These advances allow scientists to properly design nanomaterials that are specifically designed to address the specific demands of current technologies, leading to more efficient and sustainable processes across industries. The field of soft materials, especially the photoresponsive ones, once more demonstrates the paradigm-shifting impact of nanomaterials, highlighting their multitasking capability for technology. Photoresponsive functional materials, derived from nature, have emerged as a leading class for next-generation optical technologies and soft robots. Techniques for their synthesis, allowing for dynamic regulation and control of their properties, open new avenues in the design of smart nanostructured systems (Li et al., 2019). From the advancement of bio-interface applications to the capability of shape-morphing, these advances point towards the possibility of building sophisticated structures with precise control over their functionality. As synthesis and fabrication methods for nanomaterials continue to advance, so does their significance for the development of many areas of technology, which serves to underscore the necessity for research and development in this vital field.

Overview of Synthesis and Fabrication Techniques

Synthesis and fabrication of nanomaterials comprise a vast array of methods that form the central point of designing their characteristics and maximizing their applications. One of the most advanced methods is high-temperature shock (HTS) synthesis, which is of interest due to its efficacy and environmental benefits. The process allows for quick creation of metastable nanomaterials with abundant defects, which are crucial in energy-related application. The ability to control the kinetics of nanostructure creation offers tremendous advantages, such as the capacity to create complex nanostructured materials, including bimetallic nanostructures and high entropy alloys. Additionally, application of HTS in the synthesis of support films for transmission electron microscopy grids demonstrates its importance in direct imaging and characterization of such materials, with the aim of revealing its unique structural characteristics and furthering understanding in the field of nanotechnology (Cui et al., 2024). Besides HTS, synthesis of nanofibrous materials has moved into the limelight, owing primarily to its simplicity and versatility. These smart nanofibers, with their high specific surface area and multi-dimensionality, have wide-ranging applications across industries. They may be synthesized using techniques such as solution spinning and electrospinning, which not only yield fibers of excellent mechanical properties but also allow for the incorporation of functional materials during synthesis. Thus, such nanofibrous scaffolds are able to react favorably to environmental stimuli, and thus are apt for applications in healthcare, energy, and environmental systems. The rising relevance of such smart materials emphasizes the need for novel fabrication methods, which can meet the growing demands for smart solutions in daily applications (Chamanehpour et al., 2023). The future of nanomaterial synthesis and manufacturing techniques is ready for more advancement as researchers move toward novel approaches and technologies for existing processes. Breakthroughs in high-throughput and liquid-phase synthesis technologies should enhance scalability and efficiency to meet numerous kinds of applications aside from the energy-related applications that currently prevail. Further, the marriage of characterization techniques with synthesis techniques will unlock more about the performance of nanomaterials at the atomic level, paving the way towards better control of material properties. The interdisciplinary nature of the field fosters cross-disciplinary collaboration among chemists, physicists, and engineers, driving innovations in response to global needs. As demand is emerging for efficient and sustainable production processes, studying these dimensions of nanomaterial synthesis will be crucial in driving innovation and enhancing the performance of next-generation materials (Chamanehpour et al., 2023; Cui et al., 2024) (Figure 4.2).

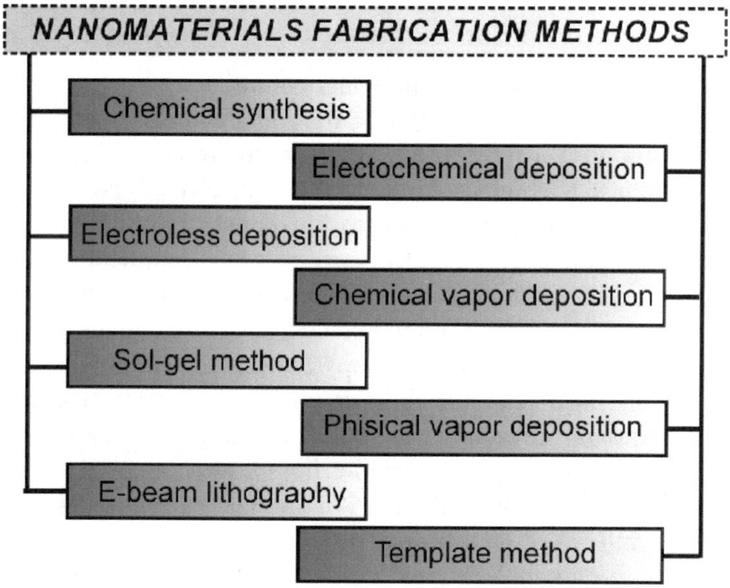

Figure 4.2. Synthesis and fabrication techniques of nanomaterials.

Source: https://www.researchgate.net/publication/286875083/figure/fig1/AS:614318471139351@1523476345953/Exemplary-methods-for-fabrication-of-zero-and-or-the-one-dimensional-nanomaterials-The.png.

4.1. TOP-DOWN APPROACHES

Top-down approaches have emerged as key methodologies for synthesizing structures at nanometric scales in nanomaterials. These processes involve the physical reduction of bulk to nanoscale particles through lithography, milling, and etching. Among the advantages of top-down approaches is that they are able to generate material of desired size and uniformity, which is critical in electronics and biotechnology applications. For instance, the controlled fabrication of nanostructures permits devices to be built that can recognize a range of biochemicals, opening doors to innovative diagnostics in medicine (Altintas et al., 2019). This level of precision can enable optimization of the performance of devices through the ability of materials to exhibit reproducible behaviors, a vital necessity for safe and trusted applications in integrated nanosystems. Thus, the top-down methods are highly promising for researchers who prefer to use the unique aspects of nanomaterials to their full extent. Even though top-down techniques share numerous benefits, they are not entirely free of difficulties. The high-energy processes involved with such methods have the potential to create defects and structural imperfections, which can adversely affect the target properties of nanomaterials. Moreover, scalability is also an issue since some of the top-down methods are not well-suited for being efficiently transferred from lab contexts to large-scale manufacturing. Nevertheless, improvements in processing technologies, including three-dimensional printing and nanolithography, are increasingly finding ways to overcome such limitations. Some recent developments have thus proved promising in generating ordered nanostructures with superior functions than their bulk counterparts. For example, three-dimensional graphene networks illustrate how the development of new fabrication techniques can yield multifunctional structures, well beyond energy storage to other fields such as bioelectronics (Mattevi et al., 2016). Finally, continuous refinement of top-down methods must occur to achieve the full potential of nanomaterials across most areas. Top-down and bottom-up synthesis techniques in nanomaterials often culminate in hybrid strategies, which take the best from both techniques. By integrating the precision offered by top-down strategies with the self-assembly characteristics characteristic of bottom-up methods, researchers are able to build complex nanostructures with improved performance. This synergy enables control over nanoscale properties critical to catalysis, drug delivery, and tissue engineering applications. For example, top-down strategies can be used for constructing scaffolds of specific architectures that direct biomolecules self-assembly and therefore improve the therapeutic efficacy of nanocarriers in targeted therapy. Moreover, by the integration of these approaches, there is a more versatile design platform, since multifunctionality in nanomaterials is made possible.

As the field advances, the creation of top-down approaches will be central to the evolution of the synthesis and construction of novel nanomaterials and the capacity to compete with future innovations in technology.

4.1.1. Mechanical Milling

Mechanical milling is a vital process in nanomaterial construction and synthesis through reducing particle size to the nanometer range. The process involves the repetitive fracturing and cold welding of material powders in a milling device, usually a high-energy ball mill. Mechanical energy imparted to the particles facilitates extreme microstructural changes in the particles, resulting in enhanced properties like strength and reactivity. Mechanical milling is also a very efficient means of alloying, facilitating the manufacture of complex nanocomposite materials with unique functionalities. With this process, heterogeneous mixtures can be transformed into homogeneously dispersed nanostructures, a critical condition for many applications in electronics and catalysis fields, as described in the present developments of nanomaterials science. As nanomaterials knowledge becomes increasingly deeper, the significance of mechanical milling is increasingly evident in the design of ordered porous architectures. The process is not just significant in the reduction of size but also to achieving the nanostructure morphology as specified in specific applications. A case is the production of multi-component complex materials through mechanical milling that can involve the mixing of various elements which otherwise would be difficult to merge at the nanolevel. The transformation of conventional materials to nanostructured form demands optimum milling conditions to achieve uniformity and functionality. Technological advancements in milling technology, such as acclimatization of milling media and operation parameters, are directly affiliated with enhanced synthesis of nanoporous structure underpinning advanced technological requirements in energy and electronics (Nonnenmann et al., 2019). Synthesis of nanocomposites can also be achieved through mechanical milling, particularly in developing material like Al-Mn-Ce. By this technique, researchers are able to tailor the mechanical and physical properties of materials in order to meet the strict needs of many industries. By decreasing the particle size, the mechanical strength increases and thermal and electrical conductivity is improved compared to their bulk equivalents, which is a phenomenon of fundamental significance in applications in catalysis and structural components. Mechanical milling flexibility allows for composition and properties tailoring of the materials, which provides possibilities for new nanomaterials. As the synthesis of nanocomposites discussion shows, the process not only serves scientific goals but also meets the shifting needs of modern technology (Rai et al., 2008) (Figure 4.3).

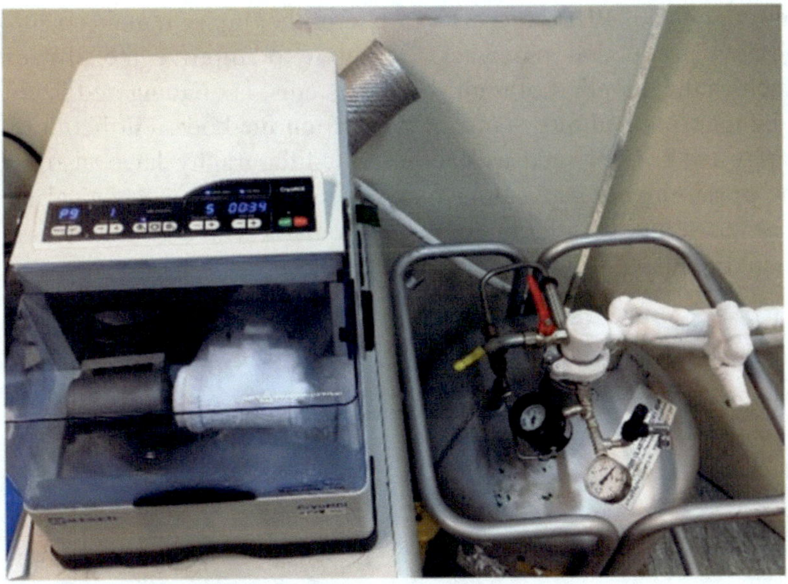

Figure 4.3. Mechanical milling of nanomaterials.

Source: https://www.mdpi.com/nanomaterials/nanomaterials-11-02484/article_deploy/html/images/nanomaterials-11-02484-g035-550.jpg.

4.1.2. Lithography Techniques

The nanotechnology progress enabled various lithography techniques to be achieved, which are important for the nano-scale patterning essential in nanomaterial synthesis and fabrication. Of the most advanced techniques is photolithography, where light is used to transfer a geometric pattern from a photomask into a light-sensitive chemical photoresist material on the substrate. It enables the production of intricate nanoscale features, which are essential in applications ranging from semiconductor devices to biomaterials. The level of resolution achievable through photolithography is largely determined by the wavelength of light used and the nature of the photoresist and as such there remains a need to continue searching for other techniques with the capability of achieving smaller feature sizes since device miniaturization increasingly becomes necessary. In contrast to photolithography, electron-beam lithography (EBL) involves the use of a focused beam of electrons to create high-resolution nanostructures. This technique offers improved precision, allowing the fabrication of features to the tens of nanometer order, now becoming increasingly important for emerging nanomaterials applications. However, the lower throughput of EBL compared to photolithography remains a limitation to large-area production processes. Research to improve the efficiency of EBL, such as the adoption of multi-beam systems, is ongoing and is critical if it is to be incorporated into standard fabrication methods. Furthermore, other lithography techniques such as nanoimprint lithography have also emerged with the possibility of replicating nanostructures using mechanical stress to produce a cost-efficient solution for pattern transfer with high fidelity and high-throughput manufacturing.

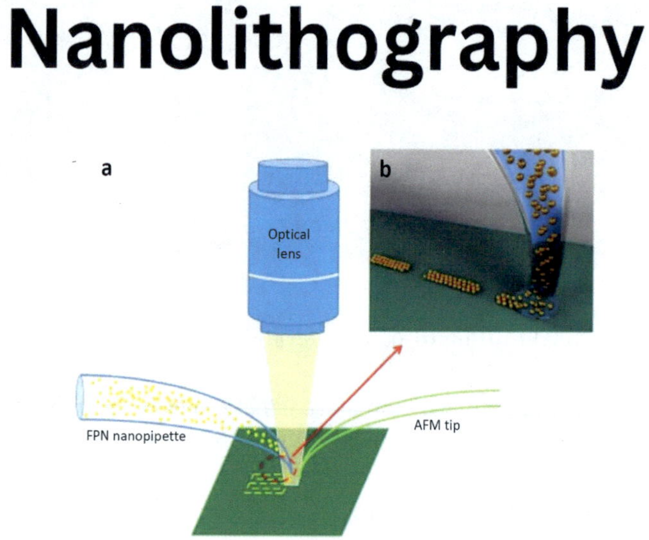

Figure 4.4. Nanolithography.

Source: https://scienceinfo.com/wp-content/uploads/2023/05/Nanolithography.jpg.

By hybridizing alternative types of lithography with other novel nanofabrication techniques, innovative solutions for nanomaterial development can be achieved. For instance, recent studies have explored the combination of chemical vapor deposition and lithography to create hybrid nanostructures that enhance the functional properties of the materials. As shown by the chemical vapor deposition of Cadmium Sulfide nanoparticles-decorated CuO nanowires following lithographic patterning, this process not only encompasses a way to fabricate intricate geometries but also the potential to engineer material properties in accordance with application needs, such as sensors and electronics (Mallick et al., 2015). In addition, lithographic techniques coupled with strain engineering have made it possible to explore the possibilities of flexible device formation allowing them to change and adjust to various surfaces, expanding even further the scope of nanomaterials application, such as in the fabrication of pressure sensing devices using graphene and nano-spheres (Dahiya et al., 2017) (Figure 4.4).

4.1.3. Etching Processes

Etching processes are important techniques involved in the manufacturing and synthesis of nanomaterials where it is feasible to etch and strip away material with precise accuracy at the nanoscale. Etching processes can be broadly classified into wet etching and dry etching. Wet etching employs liquid chemicals to dissolve targeted material layers, while dry etching applies gaseous reagents in vacuum chambers for similar applications. The advantage of dry etching is that it can potentially create higher fidelity and resolution features, which is particularly suitable to be applied in microelectronics and nanotechnology. Moreover, the control offered by dry etching also creates opportunities for novel combinations and configurations of materials, eventually resulting in enhanced device performance and functionality. The unique strengths of etching processes are significant in the broader field of nanoscale manufacturing by enabling tightly controlled pattern development. Advancements in plasma etching technology are only one demonstration of the capabilities needed for intricate design and accuracy in building nanoscale devices. At micrometer and nanometer confinement of plasma, emergent phenomena are possible to observe, which can also influence material synthesis with improved properties. It is critical to control and organize precursor materials at such quantities because it allows scientists to effectively control energy and matter, paving the way for new applications in semiconducting materials and self-organized nanostructures, which are critical for enhancing the performance of technological devices and their functionality (Meyyappan et al., 2013). Furthermore, the etching techniques are a crucial part of the synthesis of novel materials such as three-dimensional graphene networks, which have been of interest due to their multi-functionality in being energy storage and bioelectronic devices. The outstanding properties of the materials, including electrical conductivity and mechanical strength, are largely dependent on the microstructure development under controlled conditions. Techniques such as materials templating and dry etching not only improve precision but also enable the growth of ordered structures needed for optimization of functionalities of these networks. With further research in this area, highly controllable processes will be required to accomplish in order to design architectures that not only possess desirable material properties but also fulfill practical applications in various technologies (Mattevi et al., 2016).

4.2. BOTTOM-UP APPROACHES

Bottom-up synthesis methods of nanomaterials involve constructing material from atomic or molecular scale in an attempt to form structures with certain desired properties. This is contrasted with top-down synthesis methods, which make use of aggressive bulk material machining. Bottom-up methods allow for greater control over the material properties because they can accommodate the precise positioning of molecules and atoms. For example, use of chemical vapor deposition (CVD) or sol-gel chemistry allows synthesis of nanoparticles, thin films, or nanostructured materials for particular applications. Reproducibility and uniformity of such processes can lead to improved performance in fields from electronics to biomedicine, rendering these synthetic approaches essential to the development of nanotechnology. One of the primary advantages of bottom-up approaches is their inherent ability to create complex nanostructures with multi-functionalities.

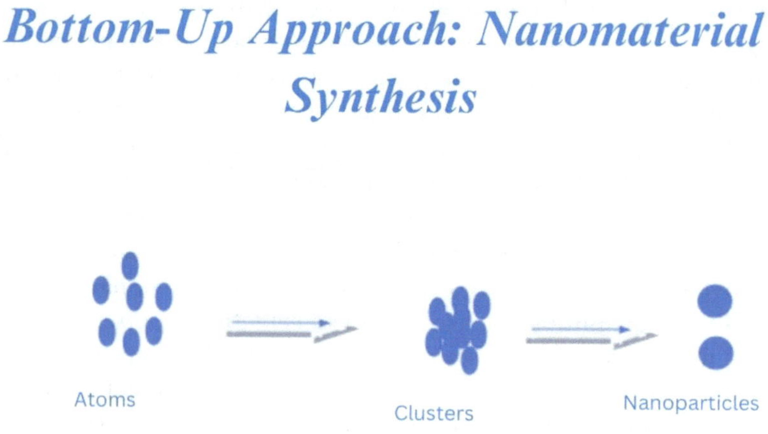

Figure 4.5. The bottom-up approaches in nanomaterial synthesis.

Source: https://scienceinfo.com/wp-content/uploads/2023/09/Bottom-up-approach.jpg.

This is best exemplified in the case of nanocomposites and hybrid materials, where different types of nanomaterials are blended in a bid to harness their synergies. By the blending of metallic, carbonaceous, and polymeric nanomaterials, researchers can synthesize sensors and devices with enhanced sensitivity and specificity. There has been proof shown in recent research that the integration of various nanomaterials into sensor platforms has allowed for the quantification of numerous various clinically relevant molecules, such as nucleic acids and cancer antigens, that has transformed healthcare diagnostics (Altintas et al., 2019). The diversity of bottom-up synthesis enables new solutions to be developed that can match technological and medical evolving demands. In addition, the tunability of nanomaterial properties synthesized using bottom-up approaches opens up new avenues for them to be incorporated in photonic devices. With the controlled composition and architecture of nanomaterials, researchers can achieve well-designed bandgaps that result in enhanced light emission and absorption. This aspect is particularly relevant to the design of LEDs and solar cells, where their performance could be significantly improved by adjusting nanomaterial properties. The ability to alloy various semiconductor materials at the nanoscale leads to the formation of new materials with fewer defects and higher efficiency, which is a step in the field of nanotechnology research (Dou et al., 2017). Bottom-up approaches are thus critical in bridging the gap between scientific theory and practical application and are hence crucial in the synthesis and formation of advanced nanomaterials (Figure 4.5).

4.2.1. Chemical Vapor Deposition (CVD)

CVD is a major technique of nanomaterial synthesis, renowned for the synthesis of thin films and functional nanostructures with high precision. CVD is a conversion process of gaseous precursors into a solid material that deposits on a substrate. The inherent flexibility of CVD allows the deposition of a large variety of materials, including metals, oxides, and carbides, for applications in electronics, energy storage, and catalysis. Highlighting the critical need for controlled material properties, CVD offers precise control over parameters such as temperature, pressure, and gas flow rates that have direct influences on films morphology and crystallinity. This adaptability has led to more sophisticated manufacturing methods, which are geared to the specific applications of different areas of technology and hence make CVD even more relevant to contemporary nanotechnology. One of the best examples of the effectiveness of CVD is its application in the synthesis of iron oxide-based nanocomposites, i.e., $\backslash u_3b_5$-Fe_2O_3 nanorod arrays. By using a step-by-step CVD and RF-sputtering synthesis, researchers have successfully fabricated such nanostructures, which are subsequently functionalized with gold and copper oxide nanoparticles. Such a novel synthesis technique enables a thorough investigation of the material's structural as well as magnetic properties with special emphasis on how the chemical composition as well as spatial distribution of nanoparticles can be judiciously controlled. The synergy between CVD and RF-sputtering showcases the potential for functional enhancements of nanomaterials to cater to evolving needs in various technological applications. These advancements expand the horizons for nanotechnology, showing how CVD can be used to enable the production of tailored materials for functional specificities (Barreca et al., 2018). Furthermore, CVD is also not restricted to traditional gaseous precursors; innovative techniques are also being explored, such as laser-plasma CVD methods that can be more efficient and clean in nanomaterial synthesis. This new technique uses laser-induced plasmas for material exfoliation, thereby making high-quality graphene of controlled layer counts possible. Research findings suggest that rapid laser activation is able to produce monolayers that then form multilayers with very low contaminants, thus preserving the quality of the graphene systems. The resultant materials show excellent potential for electronics and energy devices depending on their singular electrical behavior. As scientists keep on studying and enhancing these advanced CVD methods, the potential of adding numerous nanomaterials to viable applications will enhance, resulting in more innovation in efficient and sustainable nanomaterial production (Antici et al., 2017) (Figure 4.6).

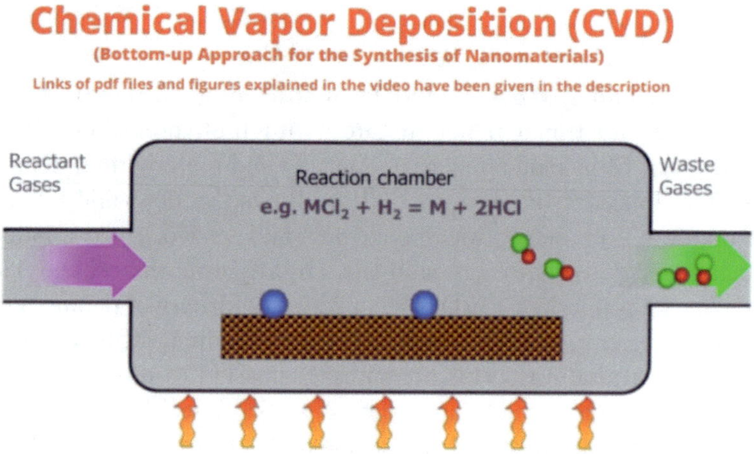

Figure 4.6. Chemical vapor deposition (CVD).

Source: https://external-content.duckduckgo.com/iu/?u=https%3A%2F%2Fi.ytimg.com%2Fvi%2FWXnCGcm5Ux8%2Fmaxresdefault.jpg&f=1&nofb=1&ipt=f723d11d2cd84176555fc5ba3bc753723d0170107b7fad5777184d46b26306210.

4.2.2. Sol-Gel Synthesis

Sol-gel synthesis process has been an essential process in nanomaterial preparation, with the reputation of having versatility and the capability of producing high-purity and homogeneous material. At its core, it is a transformation process of a solution (sol) into a network gel that, subsequently, continues to further processing towards the production of solid material. The sol-gel route allows for the introduction of several dopants and additives and provides for the tailoring of material properties to meet specific requirements. This aspect is particularly useful in nanomaterials development for various applications, ranging from catalysts to sensors. Further, compared to the majority of traditional synthesis methods that require severe conditions and dangerous chemicals, sol-gel can be performed under more acceptable conditions, thereby translating into more environmentally friendly nanotechnology fabrication processes (Drummer et al., 2021).

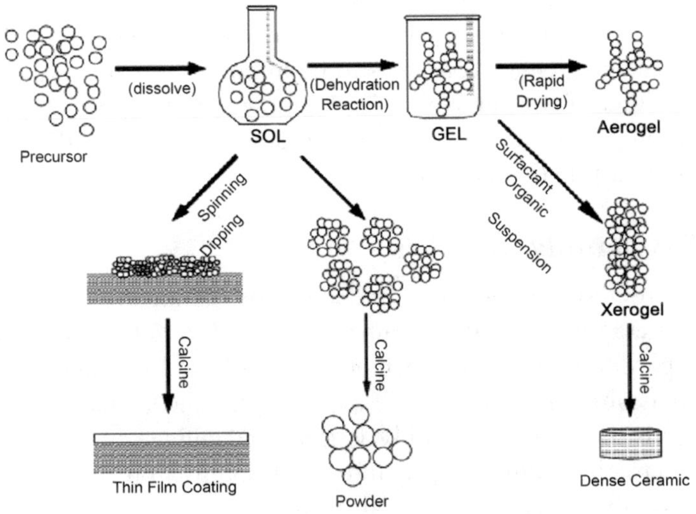

Figure 4.7. Sol-gel synthesis.

Source: https://external-content.duckduckgo.com/iu/?u=https%3A%2F% 2Fimage1.slideserve.com%2F2280695%2Fsol-gel-synthesis-l.jpg&f=1&nofb=1&ipt=4 fd209e719b854b0ca70e88a1a70bb6ac4d8013dafa18a5961bb531bf6130ff0.

Apart from its environmental benefit, sol-gel synthesis is key to the formation of functional nanomaterials, which are oxides and composites with unique optical, electrical, and magnetic characteristics. Through the modification of factors such as precursor concentration, temperature, and drying time, researchers can regulate the morphological features of resulting nanoparticles, hence their reactivity and surface area. The simplicity with which the method can enable the production of nanostructured materials such as nanoparticles, nanofibers, and films also underscores its applicability in various fields such as electronics, medicine, and environmental cleanup. Then, advancements in sol-gel synthesis techniques not only expand the scope of nanomaterial utilization but also come to provide the foundation for fundamental studies aiming at detailed exploration of intricate relationships between processing conditions and material properties (Naresh et al., 2021). The process of sol-gel synthesis, although useful, is not without difficulties and primarily concerns scalability and replicability. With researchers moving to scale up, it may become difficult to maintain the integrity of the material properties. Environmental changes, i.e., temperature and humidity during synthesis, can lead to variation and distribution in particle size. Secondarily, though the ability to dope with a variety of dopants is an advantage, it necessitates tight control over the chemistry involved because unwanted or poorly mixed reagents can adversely impact the final product. Addressing these challenges continues to be important for the development of useful applications of sol-gel synthesized nanomaterials, particularly in fields such as biosensing and catalysis, where accuracy and dependability are critical (Drummer et al., 2021) (Figure 4.7).

4.2.3. Self-Assembly Methods

Self-assembly methods are a core approach in the nanomaterial synthesis and fabrication, leveraging naturally occurring molecular and atomic interactions to form complex structures. These methods are based on random assembly of parts into desired forms without explicit guidance, enabling efficiency and cost-effectiveness. Self-assembly is a key advantage in the sense that it is capable of creating complex structures and textures on the nanoscale that may be impossible using traditional top-down approaches. The process applies to numerous applications, such as in the production of nanostructured materials with particular optical, electronic, and mechanical properties. With research focused on new self-assembly systems, the potential for preparing multifunctional nanomaterials expands, thereby expanding their usability in advanced applications such as drug delivery, biosensing, and energy storage. Integrating self-assembly methods into fabrication of nanomaterials has also encouraged innovation in biosensing devices, as evidenced by recent progress towards utilizing nanoparticle-based systems. For instance, carbon allotrope-based nanomaterials have been found to hold revolutionary potential in healthcare diagnosis on the basis of their inherent surface tunings and their tailor-made functionalities. The modularity of self-assembled nanocomposites allows various materials to be assembled to leverage synergies, whose impact can be significantly amplified on sensor performance. By adopting the principle of self-assembly, researchers can develop platforms capable of detecting an array of clinically relevant biomarkers, pathogens, and toxins, thereby facilitating early diagnosis and treatment. These advances underscore the necessity to include self-assembly strategies under the general framework of nanomaterial synthesis and the ensuing applications in biosensing, as observed in reviews of exhaustive biomedical research projects (Altintas et al., 2019). Furthermore, the prospect of self-assembly strategies is also justified through the inclusion of nanomaterials like graphene quantum dots (GQDs), as they illustrate how such a strategy can reshape sensor development. GQDs, with their unique optical and electrochemical characteristics, can be the center of attraction for biosensing purposes, and therefore may be ultra-fast and sensitive ultra-candidates for detecting a wide range of biological analytes. Through self-assembly, GQDs can be arranged as ordered structures for enhancing sensor performance in terms of sensitivity and specificity. This ability to maximize the assembly process ensures stable detection ability, a paramount requirement for the detection of low-concentration biomarkers in complex matrices of biological matter. With advancing science, continued research on self-assembly is not only promising for new sensor technologies but also accompanies an ongoing effort to ease nanomaterial synthesis for a wide range of biomedical applications, exemplified by recent reviews on GQDs and synthesis (Altintas et al., 2020) (Figure 4.8).

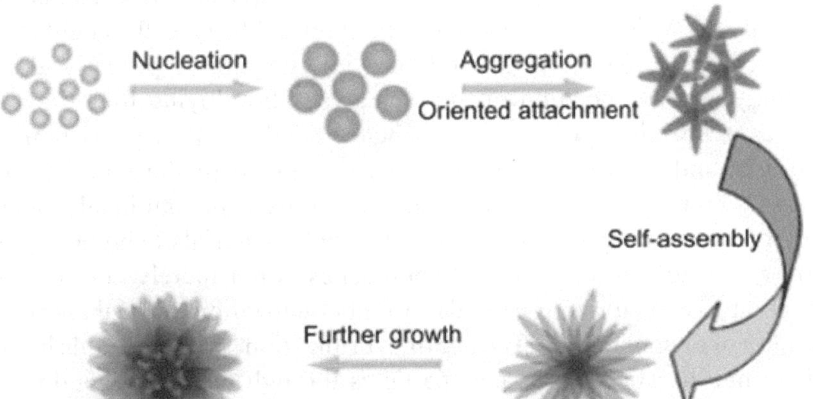

Figure 4.8. Self-assembly methods.

Source: https://external-content.duckduckgo.com/iu/?u=https%3A%2F%2Fchemistnotes.com%2Fwp-content%2Fuploads%2F2023%2F02%2Fimage-16.png&f=1&nofb=1&ipt=01682c134ed2cc93bf40c331a34d0bdde77cf14186c5d4a3dab67c740d145836.

4.3. HYBRID APPROACHES

For nanomaterial synthesis, hybrid methods have emerged as a breakthrough strategy that converges different techniques to enhance the performance and properties of nanostructures. By integrating techniques such as self-assembly, solution processing, and template-assisted synthesis, researchers are able to achieve new results that are unattainable with conventional methods. The versatility of hybrid approaches allows for the controlled manipulation of nanoparticles to create complex structures with more optical, electronic, and catalytic activity than their discrete counterparts. This collaboration can lead to the realization of materials with desired functionalities, particularly in drug delivery, sensing, and energy conversion. By way of illustration, studies on the self-assembly of polymer-capped nanoparticles have shown the vast potential to control nanoparticle ordering, creating ensembles with adjustable architectures that optimize their functional behavior (Zhang et al., 2022). Synthesis of hybrid nanomaterials also presents the potential to explore new functionalities by mixing dissimilar materials. For instance, the combining of conductive and semiconductive nanoparticles can produce materials with electrical as well as optical properties for application in photovoltaic devices and sensors. An example that is applicable is the application of hybrid catalysts for electrochemical processes, particularly hydrogen evolution reactions (HER). These catalysts, which may be composed of non-noble as well as noble metals, facilitate efficient hydrogen generation with the aid of properly designed morphology and composition of these catalysts. Employing hybrid synthesis techniques, researchers are eager to enhance catalytic activity by optimizing surface area and active site density. This fine grasp of the interdependence between structural parameters and catalytic activity is pivotal in advancing the efficacy of hydrogen generation from renewable materials (Zhu et al., 2019). Moreover, the utilization of hybrid approaches is not merely confined to the construction of structures; it extends to functionalization and enhancement of the stability of nanomaterials. The use of hybridization techniques aids in surface chemistry modification of the nanoparticles through polymer ligand grafting. The technique not only imparts new functionalities to the nanomaterials but also aids in dispersion in the matrices, which comes in handy in electronics and photonics applications. Moreover, controlled self-assembly of functionalized nanoparticles can allow the formation of hierarchically ordered nanostructures and thus improve advanced features such as chiroptical effects and enhanced reactivity. While research continues to broaden the frontiers of hybrid synthesis strategies, breakthroughs in the future can achieve record-breaker nanomaterials that are functional, green, and tailored to meet emerging technological requirements (Zhang et al., 2022).

4.3.1. Combination of Top-Down and Bottom-Up Techniques

The synthesis of top-down and bottom-up approaches is a revolutionary step in the synthesis and production of nanomaterials that takes the best of both worlds. Top-down approaches have traditionally been the reduction of macroscopic structures to small sizes by lithography and etching processes to achieve nanoscale accuracy. The technique is aided, however, by the intrinsic characteristics of the materials and can potentially leave defects or size and shape variations. Bottom-up methods, by contrast, build nanomaterials atom by atom or molecule by molecule and, as such, provide better control of the ultimate structure and the potential to access properties that emerge on the nanoscale. Through a combination of approaches, scientists can create precisely ordered nanostructures with the efficiency and versatility of bottom-up synthesis, with valuable applications to electronics and biomedicine, among others. Further, the hybrid approach has been of tremendous potential to enhance the functionality and usability of nanomaterials across various sectors. For instance, in biosensing devices, the combination of top-down and bottom-up approaches enables one to control nanostructures such that surface area is maximized and interaction with analytes of interest is enhanced.

It has been shown in more recent studies that combining various nanomaterials can harness their individual properties and allow them to generate synergistic effects that augment detection abilities. This is best illustrated in the creation of multifunctional sensing platforms that are capable of detecting a wide variety of clinically important molecules. Based on the comprehensive review of research studies, the integration of approaches is capable of making sensors feasible to identify advanced biological markers, thus presenting nanomaterials as vital devices in healthcare diagnostics and point-of-care testing (Altintas et al., 2019). In addition, hybrid fabrication methods unveil the interaction between plasma non-equilibrium processes and self-organization of nanoscale structures. For instance, local plasma character facilitates unprecedented control of nanomaterial self-assembly with the result being new structures with novel physical properties. Coupling of precursor species generation with transport is key to interpreting the operation of such hybrid approaches at the nanolevel. Here, a set of plasma-specific phenomena enables highly aberrant nano-solids to form, which can then enhance material properties vital for uses ranging from energy storage to catalysts. Blending top-down precision with bottom-up craftsmanship, combining these techniques not only opens the window of nanomaterial applications but also indicates the cross-disciplinary character of advanced material science (Meyyappan et al., 2013) (Figure 4.9).

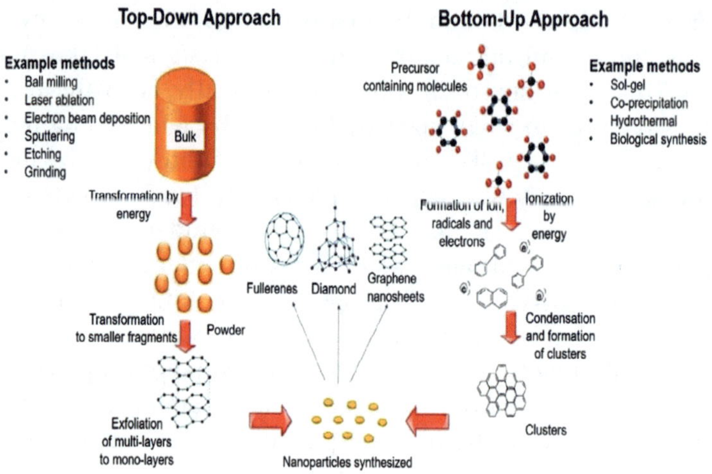

Figure 4.9. Top-down and bottom-up approaches.

Source: https://external-content.duckduckgo.com/iu/?u=https%3A%2F%2Fshop.nanografi.com%2Fproduct_images%2Fuploaded_images%2F1e550cbff272aee8ecf9a61f2bac9e59.png&f=1&nofb=1&ipt=022adecbd7fac9f10237c3552fc4d36f1c7267cf709c633240012ef2cff17711.

4.3.2. Advantages of Hybrid Methods

The advantages of hybrid approaches in nanomaterial fabrication and synthesis are becoming increasingly obvious, especially with researchers striving to overcome the shortcomings of conventional approaches. Hybrid approaches combine the strength of different material classes to create nanocomposites with enhanced properties than the sole material constituents. For example, the integration of conducting organic compounds and inorganic nanoparticles can enhance thermoelectric functionality, vital for applications such as efficient energy conversion in thermoelectric generators. Such emerging hybrid materials harness the unique properties of single constituents to enhance conductivity, stability, and overall functionality, as seen in contemporary reviews on trends in thermoelectric nanocomposites (Bisht et al., 2021). Lastly, hybrid approaches allow for the synthesis of materials that can be designed to specific application requirements and facilitate sustainability through effective use of available resources. Additionally, the tailorability of hybrid approaches provides ease in obtaining a high degree of flexibility in the synthesis of nanomaterials. By varying the quantity and types of materials used, scientists can optimize the physical, chemical, and biological properties of the resulting nanocomposites. The capacity to predict how different combinations impact performance is critical, in light of the rapidly evolving nature of nanotechnology. Such flexibility is especially crucial for applications where nanomaterials' performance is greatly influenced by their composition, such as drug delivery mechanisms and electronics. Besides, the hybridization method is also able to reduce production costs and environmental impacts in comparison to traditional techniques, since it has a tendency to discourage the need for excess amounts of post-synthesis processing (Jeevanandam et al., 2018). Thus, hybrid methods are not just enhancing functionality but also promoting responsible research and development in nanomaterials. Hybrid methods also contribute importantly to addressing a few of the pressing issues concerning nanomaterials, particularly toxicities and environmental friendliness. With the expanding use of nanomaterials in industries, their potential harmful effects also grow. Hybrid composites may be synthesized to offset these effects by employing safer and more environmentally friendly building units to guarantee safer use of nanotechnology. In addition, synthesis of hybrid nanocomposites can leverage naturally occurring nanostructures to produce materials with reduced toxicity profiles compared to what would be normally developed in completely synthetic states. This approach is supplementary to the growing demand for sustainable materials for technology, and therefore constitutes scientific advancement that not only advances science but also reduces public and environmental damage. Therefore, the advantages of hybrid approaches extend beyond improved material properties, making them more relevant to the recent argument surrounding nanomaterial safety and sustainability (Jeevanandam et al., 2018).

4.3.3. Examples of Hybrid Nanomaterial Applications

Hybrid nanomaterials have been largely utilized in the field of biosensing, particularly in healthcare diagnostics. The combinatory nature of the nanomaterials allows for the formation of multiple sensing platforms capable of detecting a range of clinically relevant biomolecules. Research indicates that a variety of nanomaterials, ranging from carbon allotropes to inorganic and organic molecules, possess unique properties that enhance detection mechanisms. For instance, compositing different nanomaterials can be capable of developing sensors not only with improved sensitivity but also specificity for the detection of pathogens, cancer antigens, or toxins in complex biological matrices. Such technological advancements in hybrid nanomaterial application can potentially revolutionize early disease diagnosis and point-of-care diagnostics and ultimately facilitate more effective healthcare interventions (Altintas et al.,

2019). Aside from biosensing, hybrid nanomaterials have proven themselves useful in semiconductor technology as well. Recent fabrication techniques, such as a new chemical patterning method, have enabled the assembly of nanostructures with great accuracy onto surfaces with selection of surface chemistry properties.

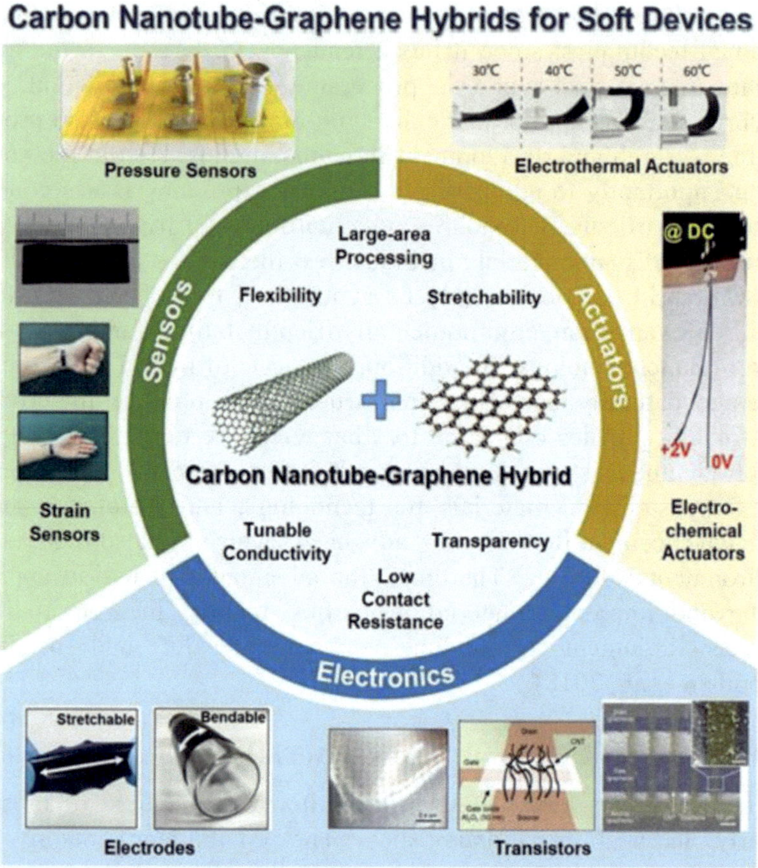

Figure 4.10. Examples of hybrid nanomaterial applications.

Source: https://external-content.duckduckgo.com/iu/?u=https%3A%2F%2Ftse1.mm.bing.net%2Fth%2Fid%2FOIP.KDCKpfFL3qHqLqXOZ0xo1QHa-Iq%3Fpid%3DApi&f=1&ipt=845aa1ef356ae009b9a8583d0e4db318908ab1683c59edf759c351fc1842f194.

The method relies on the combination of aminosilane molecular layers, which serve as sites for selective assembly of gold nanoparticles and DNA molecules, with poly(ethylene glycol) coatings, which are applied to avoid nonspecific adsorption. By application of this process, researchers have been able to create intricate multilayered structures that enhance the functionality and performance of semiconductor devices. These advancements in hybrid nanomaterial applications show the potential of nanostructure diversity integration with electronic devices, bringing about nanofabrication innovation and microelectronics enhancement (Alivisatos et al., 2005). Besides, hybrid nanomaterials find uses in the environment as effective agents of pollutant remediation and detection. Incorporation of nanomaterials into the structure of hybrid composites facilitates better adsorption as well as catalytic decomposition of toxic chemicals from water or air. For instance, hybrid nanomaterials can be engineered to focus on specific contaminants to detect, remove, or transform pollutants into harmless byproducts in real time. This capacity not only promotes environmental welfare but also aids in ecosystem sustainability. The synergy of hybridizing disparate nanomaterials creates specialty solutions for unique environmental issues, thus proving their new function in managing pollution and sustainable strategies in nanotechnology. Ongoing research revealing novel capabilities allows for remarkable potential for both environmental preservation and the development of nanomaterials (Altintas et al., 2019) (Figure 4.10).

4.4. CHARACTERIZATION TECHNIQUES

Characterization methods are crucial in nanomaterials because they enable scientists to assess the physical, chemical, and structural properties of material synthesized. Among many methods applied, microscopy methods such as Scanning Electron Microscopy (SEM) and Transmission Electron Microscopy (TEM) provide essential information about the shape and size of nanostructures. These methods allow for the examination of surface properties and interior structures at the nanoscale, which is crucial in the determination of how the properties of nanomaterials can be engineered for particular purposes. Additionally, X-ray diffraction (XRD) and Raman spectroscopy give insight into the crystallographic structure and phase composition of nanomaterials, allowing researchers to correlate physical properties with their synthesis parameters and fabrication processes. The effectiveness of characterization techniques was exemplified through the development of supported nanocomposites using the recently identified δ-iron(III) oxide polymorph, where a process of sequential radio frequency (RF)-sputtering and chemical vapor deposition (CVD) was employed. δ-Fe_2O_3 nanorod arrays were highly achieved through the CVD process, followed by the RF-sputtering that Au-functionalized and copper oxide nanoparticle-functionalized the nanorods. Characterization of the composites involved multi-technique approaches that explored material structure, morphology, and chemical composition. Experimental findings thereby obtained emphasized how modifications in chemical composition and spatial distribution immediately affect the magnetic properties of the nanocomposites, underlining the pivotal role of characterization in engineering nanomaterials for diverse technological applications (Barreca et al., 2018). Another new method of nanomaterial synthesis is laser-plasma driven techniques, particularly for the production of carbon-based nanomaterials such as graphene. Laser-plasma exfoliation of liquid phase suspended amorphous graphite leads to the production of bi- and few-layer graphene, and the process of layering is traced using Raman spectroscopy and X-ray Photoelectron Spectroscopy (XPS). From the observations, progressive multilayer development of graphene was uncovered and demonstrated the increase in the number of layers as a function of time under laser irradiation. Notably, the process produced high-purity colloidals, free from any impurities or graphene oxides, required in applications that call for clean and uniform nanomaterials. The ability to characterize these materials best indicates the need to use appropriate techniques in order to determine the quality and performance of nanomaterials (Antici et al., 2017) (Figure 4.11).

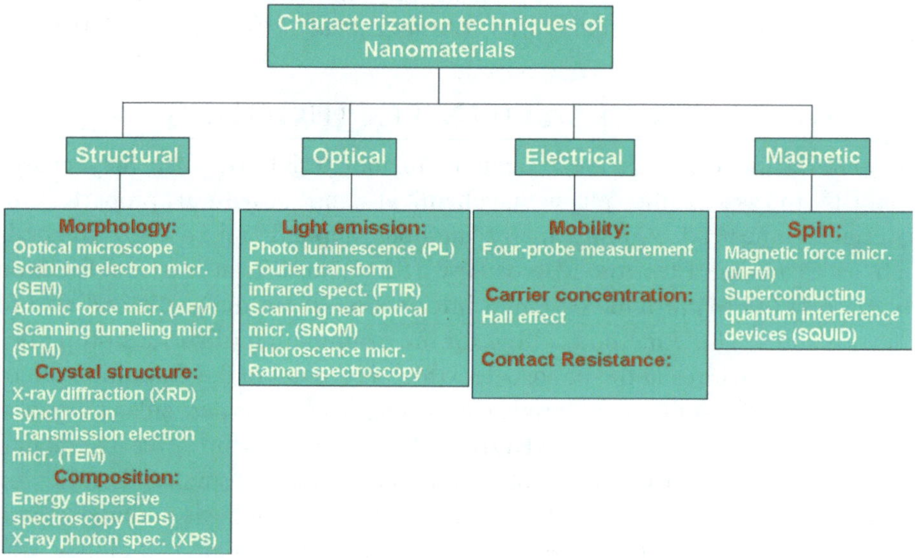

Figure 4.11. Characterization techniques.

Source: https://external-content.duckduckgo.com/iu/?u=https%3A%2F%2Flh6.googleusercontent.com%2F-cgwk2f4P0II%2FTXktfRlTgcI%2FAAAAAAAAB7A%2FCFhrJKjHKCs%2Fs1600%2Fcharacterization%2Btechniques.bmp&f=1&nofb=1&ipt=0445cb824d5d499458346455048af5482326c682517498756cc2b40b9343cdef.

4.4.1. Scanning Electron Microscopy (SEM)

SEM is an important analytical tool in nanomaterials as it allows researchers to visualize and characterize nanostructures. SEM works by scanning a focused beam of electrons over a sample surface, producing high-resolution images that reveal topography and composition data of the material to be analyzed. The technique is particularly useful in observing the unique features of nanostructures, where conventional microscopy would not work due to resolution limitations. While scientists synthesize the nanomaterials through various mechanisms, such as chemical vapor transport, the role of SEM in interpreting how the synthesis processes influence the morphological characteristics and structural integrity of the resulting materials cannot be overstated. The significance of SEM to the synthesis process is further underscored by its ability to provide real-time feedback on the growth dynamics of the nanostructures. For example, in the research on growth of titanium disulfide (TiS_2) and titanium trisulfide (TiS_3) using Surface Assisted Chemical Vapor Transport (SACVT), SEM played an essential role to correlate the structural development with changing ratios of sulfur-to-titanium. Such research yields minute details about how the morphology, such as flower-like structural elements, evolves as the synthesis is being carried out. SEM images also help unveil the relationship between synthesis parameters and the resulting nano-architectures, ultimately paving the way to optimized processing techniques that enhance the nanomaterials' properties for desired applications, for instance, electronics or energy storage. SEM's application further extends beyond the imaging function; quantitative analysis is provided using techniques such as energy-dispersive X-ray spectroscopy (EDX), capable of providing elemental composition data of the samples. This capability is particularly important in the evaluation of novel nanomaterials synthesized for technological applications. For instance, as observed in nanostructured germanium (Ge) composite electrode studies, SEM plays a critical role in confirming that the resulting products from synthesis have desirable electrochemical conduct. Coupling SEM imaging and analysis with synthesis protocols enables the complete comprehension of the performance properties of nanoscale materials. This multi-disciplinary approach ultimately enables nanotechnology developments by ensuring the synthesized materials meet the demanding specifications required in various industrial applications, particularly those related to energy.

4.4.2. Transmission Electron Microscopy (TEM)

Transmission Electron Microscopy is a key analytical technique of immense utility in nanomaterial characterization and is also vital in the understanding of their structural and composition-based properties. Using a concentrated beam of electrons, TEM enables researchers to image at the atomic level, illuminating the morphology, crystal structure, and defects in nanomaterials. High-resolution imaging is particularly valuable for studies of complex structures, e.g., bimetallic nanoparticles, where analysis of interfacial features and atomic configurations is critical. Various synthesis routes, chemical or physical synthesis, yield nanoparticles that can be investigated effectively by TEM. Consequently, this technique not only helps confirm the success of the synthesis processes but also aids in creating the total knowledge of the nanoparticle properties, which are essentially critical to their applications in catalysis, electronics, and medicine (Dlamini et al., 2023). Not only because of its imaging capability, TEM's capacity to perform advanced compositional analysis also aids its significance in nanomaterial research. Techniques such as energy-dispersive X-ray spectroscopy (EDX) can be integrated into TEM procedures for on-the-spot elemental analysis of nanostructures in their original form. This feature is particularly useful for material characterization such as non-noble-metal-based electrocatalysts, where the elemental distribution can be determined to be useful for designing high-performance catalysts for energy. The in-depth data obtained from TEM and EDX analysis plays a role in the clarification of the intricate behaviors of nanomaterials in different conditions, thereby guiding the synthesis processes toward higher functionality. The precision offered through these techniques is also helping to solve problems concerning reproducibility and scalability in nanomaterial fabrication (Zhi-Wu et al., 2020). The compatibility between TEM and the changing environment of nanomaterial synthesis points to the importance of comprehensive characterization methodology in creating future technology. As new technologies for synthesizing the material are being designed, TEM is set to change, providing critical information that supports the optimization of nanomaterial properties. A better insight from TEM studies leads to the identification of some synthesis parameters that have significant contributions to the functionality of nanomaterials and therefore the future of tailored applications. Such a synthesis, characterization, and performance evaluation cycle is imperative in the development of improved nanomaterials with enhanced functionalities to be used in future technologies across various industries. With advancements in nanotechnology, the pivotal role of TEM in elucidating the behavior and properties of materials remains to be the foundation on which consistent progress in the syntheses and fabrication techniques depends.

4.4.3. X-Ray Diffraction (XRD)

XRD is an important analytical instrument in the examination of nanomaterials, generating vital information concerning crystallographic structure, phase identification, and material properties. From the diffraction of X-rays off the periodic planes of atoms of a crystalline material, this method provides information that is quantifiable in terms of calculating interplanar spacing and lattice sizes. Of especial advantage to nanostructured materials are XRD and its ability to show how size and shape influence the crystal structure and the extent of defects, which are critical for tailoring material properties to an application. As a result, awareness of "nanomaterial development" enables researchers to optimize synthesis protocols to create nanomaterials with desired functionalities and hence improve their catalytic, electronic, and medical performances. Combination of XRD with recent synthesis techniques further elucidates the complexities existing in nanomaterials.

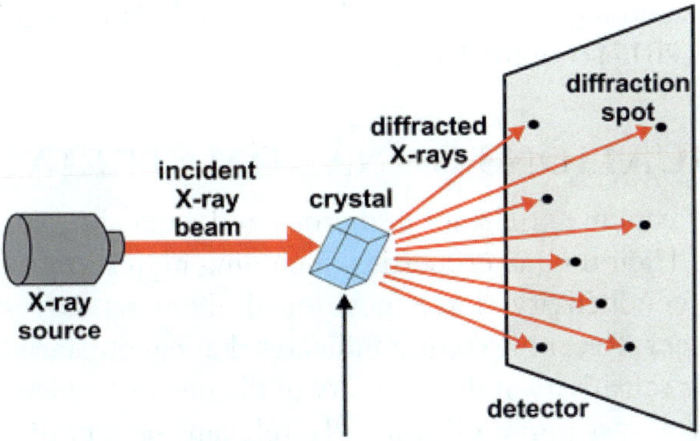

Figure 4.12. Understand X-Ray diffraction (XRD) in nanotechnology.

Source: https://external-content.duckduckgo.com/iu/?u=https%3A%2F%2Fwww.nanowerk.com%2Fnanotechnology-glossary%2Fimages%2Fx-ray-diffraction.png&f=1&nofb=1&ipt=0e610058acbe98b72b86030f42fbeff2fa08c3ce925907e2f3981562946f07ad.

For example, surface-assisted chemical vapor transport (SACVT) methods have been successfully utilized to synthesize titanium disulfide (TiS_2) and titanium trisulfide (TiS_3) nanostructures, for which XRD analyses are crucial in monitoring the development of such nanomaterials. As demonstrated in recent studies, variations in the ratio of titanium to sulfur will proportionally affect the ensuing morphology of the nanostructures that can be identified by distinct XRD patterns corresponding to different crystalline phases. XRD, complemented by auxiliary techniques like electron microscopy and Raman spectroscopy, can reliably track composition and structural evolution, thus enabling researchers to continue refining their methodology and enhancing nanomaterial performance for intended applications like energy storage and magnetism, as seen in the results for TiS_2 and TiS_3 structures (Denholme et al., 2012). Apart from structural studies, XRD is a foundation within the big picture of nanomaterial uses, namely for noble metal nanoparticles like gold and silver. Such compounds possess unique attributes that are greatly size- and shape-dependent, which necessitates precise control during synthesis to arrive at wanted properties. Recent literature is focusing on the application of using XRD to characterize such nanoparticles since it allows for the determination of size distribution and crystal growth orientation. The ability to synthesize stable nanoparticles with well-defined morphological properties creates new avenues in far-reaching disciplines such as nanomedicine and sensor technology, where the integrity of the structure of the material is a determinant of the effectiveness of the material. Thus, the use of XRD in the synthesis process increases not only our understanding of nanomaterials but also fuels innovation in their use (Fratoddi et al., 2017) (Figure 4.12).

4.5. APPLICATIONS OF NANOMATERIALS

Applications of nanomaterials are numerous and extensive, particularly in the health industry. Their unique properties, including high surface area, reactivity, and electrical conductivity, have positioned them as key components in biosensing devices. Recent literature indicates that nanomaterials are amenable to enhancing the sensitivity and specificity of diagnostic tests to the point where they can detect a vast range of clinically relevant targets like nucleic acids, viruses, and cancer biomarkers. The flexibility of such materials makes them adaptable to various sensing platforms, which can display a combination of several forms of detection mechanisms. Such synergy arising from different types of nanomaterials makes their functionalities more effective by enabling novel means for next-generation diagnostics to be created that not only help in the early detection of disease but also enable point-of-care testing platforms, enhancing patient outcomes and the effectiveness of healthcare (Altintas et al., 2019). Moreover, the synthesis and fabrication techniques employed to create nanomaterials contribute significantly toward their suitability for use in various industries. For instance, techniques such as CVD and RF sputtering have made it possible to create supported nanocomposites with certain properties. Using these techniques, researchers have fabricated new structures like \u0$_3$b$_5$-Fe_2O_3 nanorod arrays, which can further be functionalized with noble metal nanoparticles like gold or copper oxide. These techniques not only decide the resulting nanomaterials' chemical composition and morphology but also enhance their magnetic properties, thereby expanding their applications from electronics to biomedicine. These advances in synthesis techniques are necessary in order to meet the ever-growing demands for expert material produced for distinctive technological application and therefore highlight the need for ongoing research and innovation in this area (Barreca et al., 2018). Beyond electronics and medicine, nanomaterials are being found to be widely used in environmental applications, namely remediation and pollution control. Their ability to adsorb contaminants and catalyze chemical reactions at the nanolevel makes them capable of targeting and neutralizing hazardous chemicals in various environments, including water and soil. For instance, nanomaterials can purify heavy metals, pesticides, and pharmaceutical residues, and this provides an effective solution to environmental remediation processes. Furthermore, their specific photonic and catalytic properties can potentially increase the efficiency of energy transformation devices, such as solar cells and fuel cells. With rising environmental sustainability issues, nanomaterials' ability to reduce such issues is a major application area, which reflects the ability of nanotechnology to act positively in terms of ecological well-being and sustainability measures (Figure 4.13).

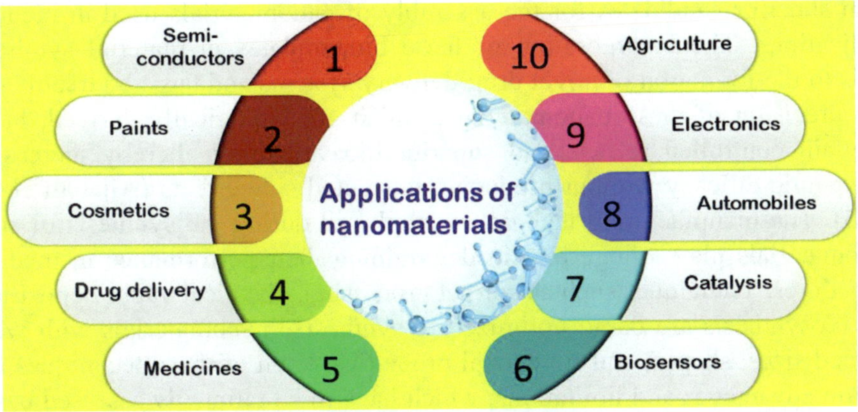

Figure 4.13. Applications of nanomaterials.

Source: https://external-content.duckduckgo.com/iu/?u=https%3A%2F%2Fthf.bing.com%2Fth%2Fid%2FOIP.LaQ0M7eV3-i24GETFF8nzwHaDl%3Fr%3D0%26cb%3Dthfc1%26pid%3DApi&f=1&ipt=2b9ba4d755109497a541d38d8ec3f-5faae17b0be1c46f8498ca824b5754672a5.

4.5.1. Nanomaterials in Medicine

Nanomaterials have been central in advancing medical science, particularly diagnostics and therapeutics. The unique characteristics of nanomaterials, owing to their nanoscale dimensions, enable more effective interaction with biological systems and therefore result in better detection and treatment strategies. The compositional diversity of nanomaterials allows for the development of highly sensitive biosensing platforms that can detect a wide variety of clinically relevant molecules, such as nucleic acids, bacteria, and cancer antigens. Recent studies suggest that it may be feasible to combine different nanomaterials and take advantage of their synergistic collective properties for novel sensor applications, resulting in the acceleration of healthcare diagnostics exponentially. The ability to embed these materials in complex biological environments raises sensitivity and specificity levels of detection methods, which paves the way for more effective early disease screening and point-of-care testing (Altintas et al., 2019). Furthermore, the structural integrity and safety profiles of certain nanomaterials render them uniquely applicable for application in biomedical uses. A good example includes plant virus capsids that illustrate the superiority of biological templates in synthesis. The capsids possess strongly monodisperse three-dimensional structures, are amenable to mass production scaling, and are not infectious towards mammalian cells. Their intrinsic safety profile renders them superior candidates for the assembly of nanomaterials used in medical applications. The incorporation of these biotemplates in material synthesis leads to the formation of novel drug-delivery systems and targeted treatments. The precision of these nanomaterials, which are biologically derived, helps to attain controlled release and superior bioavailability, thereby increasing therapeutic efficacy without producing unwanted side effects (Aljabali et al., 2015). The manufacturing techniques employed during the synthesis of such nanomaterials play a huge role in determining their performance in medical procedures. Techniques such as sol-gel processing, chemical vapor deposition, and biosynthesis are being optimized to synthesize nanomaterials with well-defined sizes, shapes, and functional properties. Each of these techniques has certain advantages and limitations, which have to be rationally assessed based on the desired medical application. For instance, the choice of the technique influences the surface functionalization and stability of nanomaterials, which are prerequisites for integrating nanomaterials into biological systems. Present research directed toward the optimization of such fabrication techniques is paramount to enhance the scope of nanomaterial utilization in medicine to make them reliable enough to address the new requirements for sophisticated diagnostics and therapeutics in diverse healthcare settings (Altintas et al., 2019).

4.5.2. Nanomaterials in Electronics

Nanomaterials have revolutionized electronics with their applications varying from high-end microprocessors to clean energy applications. These materials, whose dimensions are in the order of nanoscale, show unique properties differing significantly from bulk materials. Nanomaterials, for instance, have higher electrical conductivity, higher surface areas, and lower power consumption, and are thus highly desirable to utilize in the development of faster, smaller, and more efficient electronic devices. Their synthesis is typically with cutting-edge fabrication techniques, which can be electrospinning and chemical vapor deposition, wherein they have stringent control over composition and morphology. Based on these properties, researchers and engineers can push the boundaries of electronic performance, enabling new technologies such as very sensitive sensors and flexible electronics. Furthermore, the introduction of nanomaterials into electrical systems has far-reaching implications for healthcare technology, particularly in biosensor and implantable device engineering. Recent advances have highlighted their capacity to create fully embedded and non-invasive systems necessary for continuous health monitoring. For example, nanomaterials enable the fabrication of microelectronic devices capable of sensing multiple biological markers with biocompatibility and robustness. Here, the production techniques must ensure not only precision in nanomaterial fabrication but also production of faultless interfaces with biological systems to provide security, privacy, and long-term functionality. With the healthcare sector heading towards individualized medicine, the role of nanomaterials remains pivotal in transforming how real-time health information is collected and processed (Baj-Rossi et al., 2012). In addition, the versatility of nanomaterials extends to semiconductor application, where their nanoscale characteristics enable new functionality across a broad spectrum of electronic devices. Recent developments have indicated radical progress toward the development of nanoscale semiconductor materials with tunable bandgaps via alloying between various semiconductors. Control of such materials to process them without defects typical of bulk materials allows for maximum device performance in lasers, solar cells, and artificial photosynthetic devices. The invention is founded on processing techniques allowing it to tightly control lattice mismatches, thus making it possible for a range of applications in photonic technologies as well as energy conversion. With the growing need for sustainable and efficient electronic components, nanomaterial discovery and development will definitely shape the future of the electronics industry (Dou et al., 2017).

4.5.3. Nanomaterials in Environmental Science

Application of nanomaterials in environmental science is a breakthrough approach to addressing various ecological challenges. Due to their unique attributes such as large surface area, enhanced reactivity, and tunable properties, nanomaterials are being investigated for application in the sensing of pollution, cleanup of waste, and environmental monitoring. For instance, nanoparticles have the potential to act as efficient catalysts in degrading pollutants in water and hence reducing the impact of industrial effluents. The ability of these materials to communicate with the pollutants at the molecular level increases their effectiveness, and it is a promising solution for detoxifying toxic wastes and improving water quality. Furthermore, the use of nanostructures in biosensing technology is at the core of real-time environmental contaminant monitoring. These advances are a reflection of the need for constant research on the synthesis and fabrication techniques of nanomaterials, maximizing their potential for use in the environment (Naresh et al., 2021). Other than remediation, nanomaterials also play a crucial role in developing advanced sensing devices that detect trace environmental pollutants. Miniaturization and enhancement of sensor technology with the aid of nanostructured materials lead to sophisticated biosensors that have unprecedented sensitivity and specificity. With the aid of the unique properties of nanomaterials, such as carbon nanotubes and metal nanoparticles, it is feasible to design sensors that monitor many environmental variables, ranging from the quality of water to airborne pollution. The sensors can be engineered to exhibit quick response times, thus facilitating effective monitoring of environmental health. As these technologies evolve, fabrication issues related to them, such as scalability and reproducibility, must be addressed. Long-term enhancement of nanomaterial synthesis will be essential in developing stable environmental sensing solutions, which will make an impact on public health as well as environmental stewardship (Zhu et al., 2019). The application of nanomaterials also extends beyond sensing and cleaning up, such as with new approaches toward production of sustainable energy. For example, nanostructured materials are increasingly being utilized in hydrogen-producing reactions for water electrolysis, which presents a clean alternative to fossil fuels. Electrocatalytic performance of these materials is the most important, since the efficiency of the hydrogen evolution reaction depends on the potential for hydrogen as a renewable energy resource: Research hopes to maximize the morphology, composition, and synthesis methods of these catalysts to enhance their performance, hence enabling the production of hydrogen at a reasonable cost. The search for effective nanocatalysts is part of the broader agenda for environmental sustainability and energy management, emphasizing the pivotal role to be taken by synthesized nanomaterials in shaping the future of energy solutions (Zhu et al., 2019).

4.6. CHALLENGES IN SYNTHESIS AND FABRICATION

Synthesis and fabrication of nanomaterials present a variety of challenges that stem from their unique properties and dimensions. Nanoscale dimensions confer distinct physical and chemical characteristics, which contribute to issues with uniformity and reproducibility of manufacturing. Traditional methods may not be sufficient, as they fail to meet the required accuracy in size and shape. Additionally, ensuring the homogeneity of material makeup is the biggest issue; even slight deviations may lead to drastic variations in performance outcomes. Such complexities necessitate the creation of new synthesis techniques that retain the integrity of such materials and allow for large-scale production. In trying to achieve consistency, scientists must walk the tightrope between getting desired properties and maintaining efficiency in synthesis routes. Scalability poses the biggest challenge in nanomaterial synthesis. While lab-scale processes can provide amazing results, it has been wearisome to scale up such processes to large scales of manufacture. Such a shift usually has additional technical challenges that include specialized equipment and controlled environments that preserve the unusual properties of the materials. Furthermore, scale differences in manufacture could lead to material quality variations that impact their applicability to actual use.

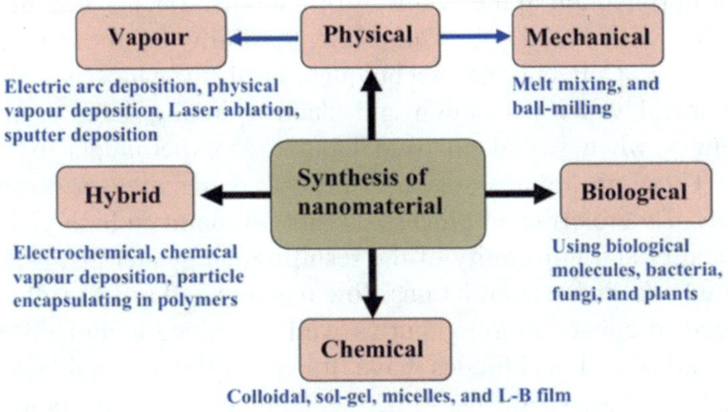

Figure 4.14. Synthesis of nanomaterials.

Source: https://external-content.duckduckgo.com/iu/?u=https%3A%2F%2Fwww.lifeasible.com%2Fwp-content%2Fthemes%2Fv1%2Fimages%2FDesign-and-Synthesis-of-Nanomaterials-1.jpg&f=1&nofb=1&ipt=c246f7ea7563e91497cf0e39c53a93bdb9a845d23d90df14e910828a3172f275.

The development of methods such as roll-to-roll processing or continuous flow processes has shown promise for addressing these issues; however, scientists still need to resolve issues of cost-effectiveness and sustainability. Without forefront solutions to these issues, the entire gamut of possibilities of nanomaterials, particularly in biomedical technology, may never be realized. Another fundamental challenge of synthesizing and fabricating nanomaterials revolves around the environmental and health impacts of certain synthesis techniques. The majority of the traditional techniques are based on toxic or hazardous chemicals that do not only do harm to the environment but to the individuals handling the production processes as well. As there is heightened awareness of such safety concerns, there is greater need for greener and more sustainable synthesis procedures. Researchers are exploring new materials and processes that inflict little environmental damage but are as efficient as their traditional counterparts. For instance, the use of biological systems in green synthesis has been a new trend, with safety and efficiency ensured. Moreover, solving these environmental issues can also allow public acceptance and approval by regulatory bodies for the use of nanomaterials in various industries, thereby fast-tracking their development in industries such as healthcare and electronics (Figure 4.14).

4.6.1. Scalability of Production

Scalability of production in the synthesis of nanomaterials is also an important aspect, as this will directly determine the feasibility of these materials for commercial use. Conventional techniques used in nanoscale fabrication, such as chemical vapor deposition and electrospinning, have a tendency to face challenges when scaled up from laboratory experiments to large-scale production. These procedures are time and cost-demanding, necessitating the development of more efficient processes that will maintain high yields as well as ensure quality and uniformity of the resulting nanomaterials. New scalable synthesis methods such as continuous flow reactors and roll-to-roll processing have emerged to enhance throughput as well as reduce manufacturing costs. Also, these advanced techniques have the potential to enable scaling up nanomaterials to large-scale use, making them even more competitive for other sectors like electronics, medicine, and energy. Scalability of production is critical in nanomaterial applications to facilitate the widespread adoption of new technology. An example is the application of nanomaterials in light aerospace structures where it demands scalable fabrication processes without compromising the integrity and performance of the materials. Nanomaterials such as multi-walled carbon nanotubes (MWCNTs) are of particular interest because of their mechanical performance and customized functionality in structural applications. Their integration requires cost-effective scaling processes with low changes in the material behavior. Furthermore, while lightweight design techniques are perfected, the production of these materials on a large scale is part of the process for achieving sustainability goals in high-performance markets. The efficient synthesis and production techniques that enable scalability can ultimately drive aerospace development through enhanced fuel efficiency, greater loading capacity, and structural integrity (Loh et al., 2017). In a bid to respond to the scalability requirement, scientists keep studying simpler and more agile nanomaterial synthesis techniques. For instance, techniques such as hot lamination and drop casting have been combined to produce effective sensing devices that respond to minimal forces, demonstrating the potential for large-scale fabrication for flexible applications. The flexibility of such techniques enables one to employ effective fabrication processes that can easily be modified for other applications, thus facilitating broader dissemination of nanomaterial technologies. As such synthesis techniques improve, they are able to provide a model for the production of nanomaterials that not only function but that are economically viable for use on mass markets. Such balance between manufacturing scalability and effective synthesis techniques is a key turning point toward reaching the full potential of nanomaterials in wide-ranging industries, from robotic e-skin and future sensing technologies (Dahiya et al., 2017).

4.6.2. Quality Control and Reproducibility

Reproducibility and quality control are absolutely critical when it comes to synthesizing and producing nanomaterials. With researchers seeking new breakthroughs, reproducible and repeatable results cannot be stressed enough. The nanoscale has the tendency to introduce uncontrollable variables due to the unique properties of the material, such as quantum effects and high surface area-to-volume ratios. To minimize such challenges, rigorous quality control must be employed, including statistical consideration and observation of the process of synthesis. For instance, keeping the parameters used in the fabrication processes constant relative to batches allows for accurate comparisons and determination. Utilization of nanomaterials in numerous applications justifies the necessity of these controls; deviations from expected performance have the potential to lead to catastrophic failures of devices like biosensors or catalytic converters, where function fidelity is essential to correct operation, based on recent summaries of the literature. Synthesis of nanomaterials often involves complex methods that necessitate reproducibility not only for academic but also for technological application. Techniques like sol-gel processes, chemical vapor deposition, and electrospinning all possess inherent difficulties which can influence product consistency. To attain the needed reproducibility, scientists are increasingly depending on standardizing methods and protocols in the course of these techniques. Advanced characterization techniques, such as electron microscopy and spectroscopy, are extremely important in establishing the quality of synthesized nanomaterials. By using these techniques, scientists are able to ensure particle size, shape, and composition meet specified requirements. Control like this is crucial in applications such as biosensors when detection system functionality relies on reproducible response to stimulation, highlighting the significant interface between quality control and functional application in novel technologies (Naresh et al., 2021). Further, advancement of nanomaterials in practical applications depends on efficient coupling between processes of quality control and scalability of synthesis methods. With industries adopting nanotechnology in large scale, the conservation of nanomaterial inherent properties while providing consistency and reliability becomes increasingly challenging. Innovation in catalyst development, particularly for hydrogen evolution and energy conversion applications, is an example of such requirement. It has been established through studies that enhancing the structural and compositional integrity of catalysts directly affects their performance and lifespan (Zhu et al., 2019). Future endeavors should, in this case, be devoted to the coordination of high-performance synthesis processes with strict quality control. By synchronizing these activities, researchers can definitively identify reproducible fabrication parameters, both with scientific research and practical applicability, and also increase the realization of a sustainable and efficient utilization of nanomaterials in various industries.

4.6.3. Environmental and Health Concerns

The synthesis and preparation of nanomaterials entail severe environmental and health concerns that must be seriously addressed. Since these materials are increasingly being integrated into products—ranging from electronics to biomedical equipment—issues regarding their environmental impact become more critical. For instance, processes used to create nanomaterials may involve hazardous chemicals and waste streams that become toxic to the environment locally. The solution to these environmental problems must be addressed on two fronts: with environmentally friendly manufacturing processes that minimize ecological damage and facilitate efficient resource consumption. Scientists have highlighted the importance of developing non-toxic material and enforcing stringent regulatory protocols to ensure that the benefits of nanotechnology must not be in violation of environmental integrity (Luo et al., 2023). Without this concerted action, the extensive applications of nanomaterials can have unintended implications, threatening both biodiversity and human health. Medical implications of exposure to nanomaterials are also a matter of pressing research. Although the unique properties of nanomaterials, i.e., increased reactivity and antimicrobial action, make them of particular significance for medical uses, they are also very harmful to human health. Silver nanoparticles (AgNPs), for instance, are universally recognized to be antimicrobial in nature and have proven to be very effective against a chain of pathogens, including resistant bacterial strains (Bruna et al., 2021). But their potential toxicity, particularly in cellular interaction and bioaccumulation, raises concern during safety assessments. Exposure of nanomaterials applied as products by inhalation or dermal contact can lead to adverse health effects that require intense toxicological studies. It is essential to understand the balance between their beneficial applications and the accompanying health risks in order to offer safer deployment across various industries. To check the environmental and health challenges relating to nanomaterials, the multi-faceted strategy of policy, education, and technology development is required. The regulatory agencies must establish comprehensive frameworks for supporting the safe development and use of nanomaterials and also improve transparency and public awareness. Education is further vital in empowering the scientists, manufacturers, and users with adequate information to make appropriate judgments regarding nanotechnology. Moreover, green chemistry investment and alternative synthesis techniques can reduce hazardous impacts characteristic of traditional fabrication techniques. Such collaborative strategy—spanning scientific discovery, policy creativity, and public engagement—aims to develop the science of nanomaterials, not at the expense of environmental health and safety (Luo et al., 2023). By solving these interdependent challenges, we can use the potential of nanomaterials while safeguarding humans and the planet.

SUMMARY

Its finding of methods of synthesis and fabrication of nanomaterials, one can appreciate that these processes play an essential role in material science improvement. The different processes yield the materials with specific physical, chemical, and electrical properties that enhance their application in various sectors. Techniques such as chemical vapor deposition, sol-gel processing, and electrospinning play a crucial role in fabricating nanostructured materials that exhibit enhanced performance in electronics and renewable energy fields. Besides, growing demand for multifunctional materials creates room for subsequent innovation and optimization by the flexibility of synthesis routes. This dynamic situation fosters the demand for sustained research and development in the field so that nanomaterials can take up society's evolving technological needs. Characterization methods employed to assess nanomaterials are also a component of demonstrating the competence of synthesis procedures. Non-destructive techniques such as Raman spectroscopy make it possible to closely observe the structural and electronic properties of such materials and subsequently inform their possible applications in fields such as optoelectronics. The use of an economically priced, home-built Raman system, as reported in the literature, indicates the dual value of affordability and precision in research. By democratizing access to these advanced methodologies, scientists are able to explore the properties of many different types of nanostructures, including Ge nanoparticles and carbon nanotubes, opening up innovations that may not be possible with conventional, more expensive equipment (Singha et al., 2004). This availability can fuel significant advancements in the field, ultimately serving a broader range of technological applications. Finally, consideration must be given to the prospects in the future for the fabrication technique and properties of nanomaterials. Progress toward low-temperature chemical synthesis to create metallic magnetic nanowires indicates a trend toward development of materials with high coercivity that offer decent alternatives to traditional magnets like AlNiCo or SmCo. Their application in high-density magnetic recording is particularly noteworthy, with the ability to revolutionize storage technologies by bringing devices down in size and improving efficiency (Craik et al., 2007). With ongoing improvements in these techniques and studies on the properties of nanomaterials, there lies a huge potential to revolutionize industries. Therefore, concluding thoughts are in line with the importance of investing long-term into research, with a view to introducing advancements that are aligned with the changing requirements of our technology-driven society.

In research on nanomaterials, some synthesis techniques have drawn considerable interest due to their potential to tailor the final material's properties and functionalities. Of particular interest is the fabrication of strained Germanium (Ge) nanomembranes. These nanometer-thick single-crystal sheets are sufficiently strong to endure significant tensile strain, which creates new properties such as enhancing photoluminescence and the creation of direct-bandgap Ge. This is especially relevant to breaking problems inherent in standard Group-IV semiconductors that possess indirect bandgaps being unfavorable in light-emitting applications. With the incorporation of tensile strain, researchers have promoted an approach that not only offers channels for integration of electronic and photonic functions on silicon chips but also positions strained Ge as a potential prospect for CMOS-compatible laser technology, thus enhancing device potential and performance (Lagally et al., 2018). Another essential synthesis method focusing the nanomaterials' topology is controlled vacuum filtration process, particularly for Carbon Nanotubes (CNTs). The one-dimensional nanomaterial with its excellent electronic, thermal, and optical performances is limited in macroscopic application by the challenges of uniformity and scalability. With the development of utilizing controlled vacuum filtration, films on a wafer scale can now be made crystalline and rich in chirality. These advancements not only facilitate extensive study in elementary physics but also advance the area of optoelectronic device application. The potential for controlling and preparing the film of CNTs opens the gates to the future in numerous areas, continuing to maintain authority in chemistry, material science, and engineering (Gao et al., 2018). Where these synthesis and fabrication techniques intersect is a revolutionary era in nanotechnology. As evident in the enhancement of both Ge nanomembranes and CNT films, material nanoscale engineering plays a key role in fulfilling current needs in optoelectronics and electronics. With the exploitation of the properties of strained Ge, outstanding breakthroughs have been experienced in overcoming the inherent barriers posed by indirect bandgaps, thereby opening new opportunities for integrated photonic devices. Conversely, the improvements in synthesizing high-quality films of CNTs through novel methods confirm the indispensable role of control techniques in the enhancement of material properties for engineering applications. The ongoing exploration of the fields demonstrates not only the importance of synthesis techniques but also the profound implications of future technologies on the basis of the unique properties of nanomaterials.

REFERENCES

1. Alivisatos, P., Majumdar, A., Kannan, B., Halperin, W., Hoff, M., Kannan, R., Castelino, K., et al. (2005). Chemical patterning for the highly specific and programmed assembly of nanostructures. https://core.ac.uk/download/4873165.pdf (accessed on 25th September 2025).
2. Aljabali, A. A. A., & Evans, D. (2015). Bionanomaterials from plant viruses. https://core.ac.uk/download/151161135.pdf (accessed on 25th September 2025).
3. Altintas, Z., & Mansuriya, B. D. (2020). Applications of graphene quantum dots in biomedical sensors. https://core.ac.uk/download/322792122.pdf (accessed on 25th September 2025).
4. Altintas, Z., & Pirzada, M. (2019). Nanomaterials for healthcare biosensing applications. https://core.ac.uk/download/287885382.pdf (accessed on 25th September 2025).
5. Antici, P., & Barberio, M. (2017). Laser-plasma driven synthesis of carbon-based nanomaterials. https://core.ac.uk/download/154948443.pdf (accessed on 25th September 2025).
6. Baj-Rossi, C., Burleson, W., Carrara, S., Cavallini, A., et al. (2012). Fully integrated biochip platforms for advanced healthcare. https://core.ac.uk/download/286729684.pdf (accessed on 25th September 2025).
7. Barreca, D., Carraro, G., Maccato, C., Peddis, D., et al. (2018). Magnetic properties of epsilon iron (III) oxide nanorod arrays functionalized with gold and copper (II) oxide. https://core.ac.uk/download/154329226.pdf (accessed on 25th September 2025).
8. Bartholomew, R., Thomas, S. A., Reddy, A., Reddy, M. R., & Cherusseri, J. (2023). Minireview on fluid manipulation techniques for the synthesis and energy applications of two-dimensional MXenes: Advances, challenges, and perspectives. Energy & Fuels, 37, 6999–7013. https://doi.org/10.1021/acs.energyfuels.3c00589 (accessed on 25th September 2025).
9. Bisht, N., More, P. V., Khanna, P. K., Abolhassani, R., Mishra, Y. K., & Madsen, M. (2021). Progress of hybrid nanocomposite materials for thermoelectric applications. Materials Advances, 2, 1927–1956. https://doi.org/10.1039/d0ma01030h (accessed on 25th September 2025).
10. Bruna, T., Maldonado-Bravo, F., Jara, P., & Caro, N. (2021). Silver nanoparticles and their antibacterial applications. International Journal of Molecular Sciences, 22, 7202–7202. https://doi.org/10.3390/ijms22137202 (accessed on 25th September 2025).

11. Chamanehpour, E., Thouti, S., Rubahn, H. G., Dolatshahi-Pirouz, A., & Mishra, Y. K. (2023). Smart nanofibers: Synthesis, properties, and scopes in future advanced technologies. Advanced Materials Technologies, 9. https://doi.org/10.1002/admt.202301392 (accessed on 25th September 2025).
12. Craik, D. J., Ott, F., Chaboussant, G., Viau, G., Piquemal, J. Y., Maurer, T., & Soumare, Y. (2007). Magnetic nanowires as permanent magnet materials. http://arxiv.org/abs/0711.1978 (accessed on 25th September 2025).
13. Cui, X., Liu, Y., & Chen, Y. (2024). Ultrafast micro/nano-manufacturing of metastable materials for energy. National Science Review, 11. https://doi.org/10.1093/nsr/nwae033 (accessed on 25th September 2025).
14. Dahiya, R. S., Shakthivel, D., & Yogeswaran, N. (2017). Towards graphene-based flexible force sensor.
15. Denholme, S. J., Dobson, P. S., Gregory, D. H., MacLaren, D., et al. (2012). Growth and characterization of titanium sulphide nanostructures by surface-assisted vapor transport methods; from trisulphide ribbons to disulphide nanosheets. https://core.ac.uk/download/9396199.pdf (accessed on 25th September 2025).
16. Dlamini, N. G., Basson, A. K., & Pullabhotla, V. S. R. (2023). Synthesis and characterization of various bimetallic nanoparticles and their application. Applied Nano, 4, 1–24. https://doi.org/10.3390/applnano4010001 (accessed on 25th September 2025).
17. Dou, L., Yang, P., & Yang, C. Z. N. (2017). Bandgap engineering in semiconductor alloy nanomaterials with widely tunable compositions. https://core.ac.uk/download/287622640.pdf (accessed on 25th September 2025).
18. Drummer, S., Madzimbamuto, T. N., & Chowdhury, M. (2021). Green synthesis of transition metals nanoparticle and their oxides: A review. https://doi.org/10.20944/preprints202102.0380.v1 (accessed on 25th September 2025).
19. Fratoddi, I. (2017). Hydrophobic and hydrophilic Au and Ag nanoparticles: Breakthroughs and perspectives. https://core.ac.uk/download/153366182.pdf (accessed on 25th September 2025).
20. Gao, W., & Kono, J. (2018). Science and applications of wafer-scale crystalline carbon nanotube films prepared through controlled vacuum filtration. http://arxiv.org/abs/1810.02928 (accessed on 25th September 2025).

21. Ismael, M., Sharma, A., & Kumar, N. (2024). An extensive catalytic potential of sustainable TiO_2-based materials fabricated via flame spray pyrolysis: A comprehensive review. Sustainable Materials and Technologies, 40, e00826–e00826. https://doi.org/10.1016/j.susmat.2024.e00826 (accessed on 25th September 2025).

22. Jeevanandam, J., Barhoum, A., Chan, Y. S., Dufresne, A., & Danquah, M. K. (2018). Review on nanoparticles and nanostructured materials: History, sources, toxicity and regulations. Beilstein Journal of Nanotechnology, 9, 1050–1074. https://doi.org/10.3762/bjnano.9.98 (accessed on 25th September 2025).

23. Jose, J., Thomas, V., Vinod, V. P., Abraham, R., & Abraham, S. C. (2019). Nanocellulose based functional materials for supercapacitor applications. Journal of Science: Advanced Materials and Devices, 4, 333–340. https://doi.org/10.1016/j.jsamd.2019.06.003 (accessed on 25th September 2025).

24. Lagally, M. G., & Paiella, R. (2018). Optical properties of tensilely strained Ge nanomembranes. https://core.ac.uk/download/286139374.pdf (accessed on 25th September 2025).

25. Li, Q., Schenning, A. P. H. J., & Bunning, T. J. (2019). Light-responsive smart soft matter technologies. Advanced Optical Materials, 7. https://doi.org/10.1002/adom.201901160 (accessed on 25th September 2025).

26. Loh, K. J., Schagerl, M., Viechtbauer, C., Zhao, H., et al. (2017). Characterizing the conductivity and enhancing the piezo resistivity of carbon nanotube-polymeric thin films. https://core.ac.uk/download/323064131.pdf (accessed on 25th September 2025).

27. Luo, Y., Abidian, M. R., Ahn, J. H., Akinwande, D., Andrews, A. M., Antonietti, M., Bao, Z., et al. (2023). Technology roadmap for flexible sensors. ACS Nano, 17, 5211–5295. https://doi.org/10.1021/acsnano.2c12606 (accessed on 25th September 2025).

28. Mallick, P. (2015). Synthesis and characterization of CuO–CdS heterostructure. https://core.ac.uk/download/80148121.pdf (accessed on 25th September 2025).

29. Mattevi, C., & Sherrell, P. C. (2016). Mesoscale design of multifunctional 3D graphene networks. https://core.ac.uk/download/77008447.pdf (accessed on 25th September 2025).

30. Meyyappan, M., Neyts, E. C., & Ostrikov, K. (2013). Plasma nanoscience: From nano-solids in plasmas to nano-plasmas in solids. http://arxiv.org/abs/1306.6711 (accessed on 25th September 2025).

31. Naresh, V., & Lee, N. (2021). A review on biosensors and recent development of nanostructured materials-enabled biosensors. Sensors,

21, 1109–1109. https://doi.org/10.3390/s21041109 (accessed on 25th September 2025).

32. Nonnenmann, S. S., & Zhou, Z. (2019). Progress in nanoporous templates: Beyond anodic aluminum oxide and towards functional complex materials. https://core.ac.uk/download/287624591.pdf (accessed on 25th September 2025).

33. Rai, S., & Roy, D. (2008). Sintering characteristics and characterization of Al–Mn–Ce nanostructured material prepared by mechanical alloying. https://core.ac.uk/download/53186928.pdf (accessed on 25th September 2025).

34. Seo, M. H. (2011). Ge nanowires anode sheathed with amorphous carbon for rechargeable lithium batteries. https://core.ac.uk/download/79702833.pdf (accessed on 25th September 2025).

35. Singha, A., Roy, A., Grethlein, C. E., Lamb, H., Dhar, P., & Rytov, S. M. (2004). A non-destructive analytic tool for nanostructured materials: Raman and photoluminescence spectroscopy. http://arxiv.org/abs/cond-mat/0406096 (accessed on 25th September 2025).

36. Trache, D., Tarchoun, A. F., Derradji, M., Hamidon, T. S., Masruchin, N., Brosse, N., & Hussin, M. H. (2020). Nanocellulose: From fundamentals to advanced applications. Frontiers in Chemistry, 8. https://doi.org/10.3389/fchem.2020.00392 (accessed on 25th September 2025).

37. Wu, Z. P., Lu, X. F., Zang, S. Q., & Lou, X. W. (2020). Non-noble-metal-based electrocatalysts toward the oxygen evolution reaction. Advanced Functional Materials, 30. https://doi.org/10.1002/adfm.201910274 (accessed on 25th September 2025).

38. Zhang, N. N., Shen, X., Liu, K., Nie, Z., & Kumacheva, E. (2022). Polymer-tethered nanoparticles: From surface engineering to directional self-assembly. Accounts of Chemical Research, 55, 1503–1513. https://doi.org/10.1021/acs.accounts.2c00066 (accessed on 25th September 2025).

39. Zhu, J., Hu, L., Zhao, P., Lee, L. Y. S., & Wong, K. Y. (2019). Recent advances in electrocatalytic hydrogen evolution using nanoparticles. Chemical Reviews, 120, 851–918. https://doi.org/10.1021/acs.chemrev.9b00248 (accessed on 25th September 2025).

CHAPTER 5

NANOMATERIALS IN CEMENT AND CONCRETE

Contents

INTRODUCTION

Civil engineering has experienced revolutionary advances with the introduction of nanomaterials, particularly in the fields of cement and concrete technology. Conventional concrete, while widely applied, generally has its mechanical features and sustainability limited. Nanomaterials, by virtue of their remarkable properties at the nanoscale, can greatly enhance the features of concrete. Utilization of materials such as nanosilica and graphene oxide in concrete mixtures has the potential to improve strength along with sustainability against environmental forces. As sustainable and durable construction material pressure is increasing, finding nanomaterials is a decisive step towards employing advanced materials meeting functional along with environmental requirements. A study has indicated that specific approaches are successful in the incorporation of nanomaterials into concrete systems. For instance, the present study indicates that enhancement of the interface between cement paste and coarse aggregates through the addition of nanomaterials is capable of bringing about substantial enhancement in transport properties.

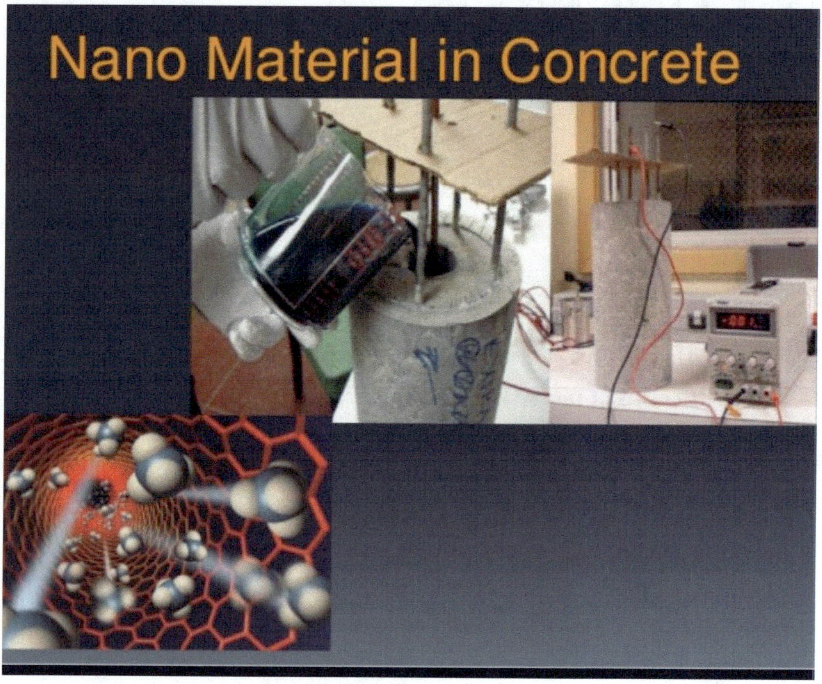

Figure 5.1. Nano materials in concrete.

Source: https://external-content.duckduckgo.com/iu/?u=https%3A%2F%2Fimage.slidesharecdn.com%2Fnanomaterialinconcrete-141017052018-conversion-gate02%2F95%2Fnano-materials-in-concrete-1-638.jpg%3Fcb%3D1413523294&f=1&nofb=1&ipt=d1029a73847cadbfc1697e44ed2526bdf74176f181ed3e85ffba073cdaf5aa49.

The use of nanomaterial-coated coarse aggregates in high-volume fly ash (HVFA) concrete has provided outcomes of reduced permeability, thereby enhancing durability. However, it needs to be noted that while such changes are productive in increasing the transport characteristics of concrete, increases in the compressive and splitting strengths are not always possible with regard to direct mixing methods. This nuance of how nanomaterials would react to the constituents of concrete is significant for optimizing performance and durability for building applications (Du et al., 2018). Other nanomaterial innovation beyond the enhancement of already established concrete properties is the development of new formulations. An example is the incorporation of graphene oxide (GO) in fly ash pervious concrete, showing the potential of the future for improving stormwater management in urban infrastructure. GO not only improves permeability but also provides removal of environmental pollutants from stormwater runoff. Evidence from studies demonstrates that GO-modified fly ash concrete is superior in resistance to freeze/thaw cycles and salt attack compared to traditional mixes. These advances point to the two-fold benefits of efficacy and sustainability, combining existing engineering practice with environmental stewardship. Overall, nanomaterial application to cement and concrete presents an appealing new field that brings together innovation and pragmatic engineering concerns (Shi et al., 2018) (Figure 5.1).

Definition of Nanomaterials

Nanomaterials are defined as having at least one dimension between 1 and 100 nanometers, which enables dramatic changes in their physical and chemical properties in comparison to their bulk counterparts. This size range allows nanomaterials to exhibit behaviors that are not only new but potentially useful in a wide range of applications, including the construction industry. Manipulation of material at the nanoscale can lead to improved strength, toughness, and resistance to environmental factors in cement and concrete materials. The insertion of nanomaterials in construction practice thus represents a great material science achievement with the potential to improve the performance of traditional construction materials through technology and innovation. A significant area of use for nanomaterials is improving the performance of cement and concrete. Nanoparticle inclusion, like graphene oxide or titanium dioxide, has been observed to enhance several characteristics of concrete, including its environmental sustainability and mechanical strength. For instance, the addition of titanium dioxide nanoparticles into rendering plastering mortars has been observed to enhance photocatalytic activity, thereby resulting in cleaner building facades and better thermal management properties. As discussed, the increased application of such materials by the construction sector is attributable to the fact that they can create buildings that are not only stronger but also more energy-efficient, effectively revolutionizing building procedures to meet the demands of contemporary sustainability (Carneiro et al., 2014). Furthermore, the use of nanomaterials in concrete mixes enables enhanced durability and resistance. For example, the development of graphene oxide-modified pervious concrete has displayed excellent performance in water infiltration capacity and contaminant removal in stormwater management systems. Test results indicate that such modification, by nanomaterials, can significantly enhance freeze/thaw resistance and salt scaling properties, which are extremely significant in most geographical positions. This new approach is a step towards a future where nanomaterials can be anticipated to contribute toward extending the life of concrete buildings, minimizing maintenance processes and life extension (Shi et al., 2018). Increased research and application of nanomaterials in cement and concrete are a start in civil engineering where high technology is effectively blended with traditional construction methods.

Importance of Cement and Concrete in Construction

Concrete and cement are the fundamental ingredients of modern construction of buildings due to their unparalleled hardness, strength, and versatility. These materials are the backbone of buildings, from residential homes to humongous bridges, enabling the formation of long-lasting architectural designs. Cement as a binding agent and concrete formed from aggregates and water are the essential attributes that enable engineers to construct buildings strong enough to withstand weather stresses and various loads for several years. With the projected increase in the world's population to more than 9.7 billion by 2050, the demand for long-lasting construction materials has never been more crucial. As a result, innovations such as nanomaterials added to cement and concrete are being researched increasingly to enhance their mechanics and life cycle, marking a revolutionary shift in material engineering in construction. Besides structural stability, cement and concrete's viability is also put into question with respect to environmental issues and climate change. The construction industry contributes almost two-thirds of human-caused greenhouse-gas emissions, with energy consumption projected to increase significantly by 2050 (Robert et al., 2017). With mounting urbanization, energy-efficient building materials are in high demand. Including nanomaterials in cement and concrete could improve their performance while reducing their environmental footprint. For instance, incorporating nano-silica can make concrete stronger and more durable, therefore reducing the need for excessive resource consumption. While the industry attempts to solve sustainability issues, understanding the role of advanced materials is important in reducing the environmental footprint of building. Furthermore, the entry of nanotechnology into the design of cement and concrete is a promising future in civil engineering, which is often a more evolutionary field (Robert et al., 2017). The application of advanced methods, such as pulsed laser deposition (PLD), has demonstrated capability to enhance considerably the quality of deposited thin films, which can be extended to influence the surface properties of cement-based materials (Robert et al., 2017). Such a degree of innovativeness not only addresses engineering challenges but also offers potential for the development of monumental jumps in the commercial realizability of innovative construction ideas. As civil engineering incorporates ever more high-tech solutions, including the utilization of nanomaterials, the sector can evolve to meet current requirements and future demands in infrastructure supply and energy performance, enabling the quest for a sustainable built environment (Figure 5.2).

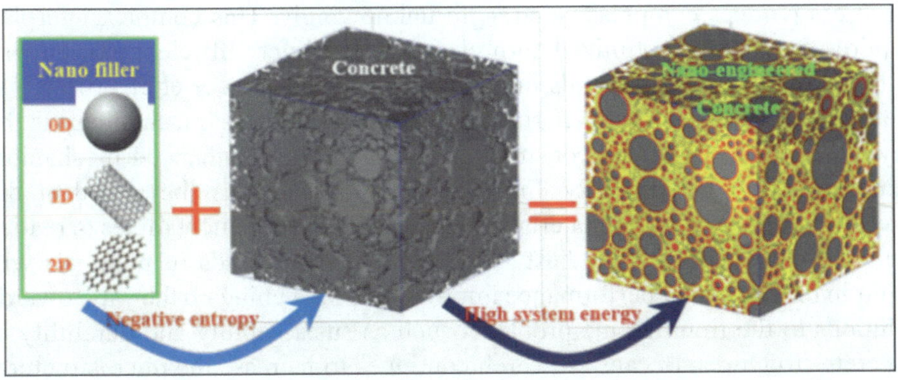

Figure 5.2. Nanomaterials in cement and concrete in construction.

Source: https://lupinepublishers.com/civil-engineering-journal/images/Lupine-publishers-openaccess-journals-Civil-engineering-Architechture.ID.000101.G004.png.

Overview of the Role of Nanomaterials in Enhancing Cement and Concrete Properties

The addition of nanomaterials in the construction industry is a giant step towards enhancing cement and concrete properties. Their application ranges from mechanical strength enhancement to durability enhancement, thus addressing the demand for high-performance concrete. Amongst all nanomaterials, graphene oxide (GO) is the most notable due to its exceptional properties. Research indicates that the addition of GO nanosheets to ultra-high-performance concrete (UHPC) results in improved compressive and flexural strengths, particularly at optimal concentration levels of 0.02% and 0.04% for normal and steam curing conditions, respectively. This enhancement is crucial in developing economically viable and sustainable infrastructural products, and it reflects that smart choice and precise dosage of nanomaterials can significantly influence the overall properties of cementitious composites (Lambert et al., 2020). Besides, the modification of the interfacial transition zone (ITZ)—the region between aggregate and matrix—is also critical to further enhance the properties of concrete. Microstructure properties in the ITZ can be improved through the application of various nanoparticles as surface coatings on aggregates. Studies comparing various nanoparticles like nano-$CaCO_3$ and nano-SiO_2 indicated that variation in coating thickness strongly affects the properties of the concrete, i.e., compressive strength and chloride penetration resistance. While coatings such as slag and nano-SiO_2 improve chloride penetration resistance, higher thickness reduces compressive strength unknowingly. This complex interplay requires maximally optimized formulations to deliver sufficient expressions of the benefit of nanomaterials in real applications (Schutter et al., 2019). On the whole, the use of nanomaterials in cement and concrete technology is the key to revolutionizing the construction sector through enhanced mechanical characteristics and durability. Further investigations must be aimed at the interaction of various nanomaterials with the traditional cement mixes to realize the maximum benefits. The best application of these kinds of materials will result in revolutionary performance improvements, opening up the way towards solutions to the troublesome problems such as sustainability and durability in concrete structures. Because research continues to increase the microstructural relevance of nanomaterials, their use in concrete structures will continue to increase, creating a new era for creative construction materials designed for modern-day engineering use (Lambert et al., 2020; Schutter et al., 2019).

5.1. TYPES OF NANOMATERIALS USED IN CEMENT AND CONCRETE

The incorporation of nanomaterials into cement and concrete has been a revolutionary means of enhancing the physical and mechanical strengths of these essential construction materials. The most widely noted type of nanomaterial incorporated is graphene oxide (GO) nanosheets, which possess an outstanding strength and high aspect ratio. The research has indicated that incorporation of GO to ultra-high-performance concrete (UHPC) can significantly augment compressive and flexural strength without reducing workability. For example, experimental work indicates that optimal addition of GO by about 0.02%–0.04% by weight of cement yields beneficial mini slump flow properties and enhanced durability. This innovation not only paves the way to greener options in infrastructure but also answers the pressing challenge of meeting current building standards (Lambert et al., 2020). The ongoing research on GO emphasizes the potential of nanotechnologies to revolutionize conventional concrete strategies. In addition to graphene oxide, nanosilica is another efficient nanomaterial for cement and concrete applications. The enhanced surface area and reactivity of nanosilica render it an ideal activator for pozzolanic reactions responsible for the strength and stability of cementitious composites. Nanosilica, when incorporated in high-volume fly ash (HVFA) concrete, helps improve microstructural properties, mainly through increasing the interfacial transition zone between the coarse aggregate and cement paste. Aggregates coated with nanosilica have been found to reduce permeability and, as a result, increase the transport properties of the concrete.

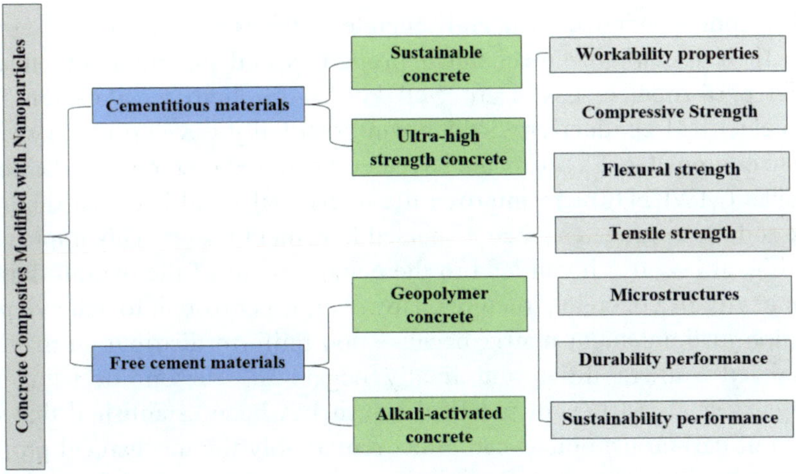

Figure 5.3. Types of nanomaterials used in cement and concrete.

Source: https://www.mdpi.com/jcs/jcs-07-00067/article_deploy/html/images/jcs-07-00067-g002.png.

However, scientific studies indicate that while this coating process increases the durability, it has very slight impacts on compressive and splitting strengths compared to the traditional mixing processes. The comparative study identifies the potential of nanosilica and requires large-scale studies into its varied uses and forecast in concrete technology (Du et al., 2018). Another category of recently-arrived nanomaterials in concrete are carbon nanotubes (CNTs), which are being touted to be the best concerning mechanical properties and electrical conductivity. Incorporation of CNTs can lead to tensile strength and impact resistance enhancement, which is crucial for structures to endure dynamic loading. But the hitch is dispersing the CNTs in the concrete matrix in a uniform manner, which is highly crucial to witness their maximum potential being realized. Techniques such as surface functionalization and the use of surfactants have been developed to maximize their integration and dispersion. These technologies are being optimized further and the synergistic effect of CNTs with other nanomaterials, for example, nanosilica or GO, is being considered in an effort to create composite materials with enhanced properties. This multi-disciplinary approach not only acknowledges a greater understanding of nanomaterials in the construction industry but also promotes innovative engineering solutions for green growth within the industry (Du et al., 2018; Lambert et al., 2020) (Figure 5.3).

5.1.1. Carbon Nanotubes (CNTs)

CNT incorporation into cement and concrete matrices has drawn much attention due to their remarkable mechanical properties and potential for enhancing material performance. CNTs are well known for high tensile strength and unique electrical conductivity, which can potentially develop next-generation concrete composites. Experiments have demonstrated that multi-walled carbon nanotubes (MWCNTs) can improve the interfacial bond between the cement matrix and the aggregates when dispersed in cement pastes. This improvement is significant because it can lead to the enhancement of the overall durability and longevity of concrete structures. However, it is difficult to achieve perfect dispersion in the cement matrix because non-uniform distribution may result in localized vulnerabilities and finally negate the reason for using CNTs (Rodríguez et al., 2017). In practical use, it has been established that minor addition of carbon nanotubes can affect remarkably the mechanical properties of cement-based composites. For instance, research has proved that an addition of 0.15% MWCNT is sufficient to change the strength characteristics and density of cement pastes. Introduction of surfactants into the mixing process is typically needed in order to gain a homogeneous distribution of CNTs, since this is necessary in utilizing their strengthening properties. Interestingly, while the addition may initially support some properties, it also needs to consider that surfactant use can in turn produce added porosity, in the best-case scenario lowering the general mechanical performance if not well controlled (Rodríguez et al., 2017). Thus, maintaining proper balance of CNT volume and effective dispersion strategies is key in maximizing gains in concrete application. Moreover, the addition of CNTs in concrete influences resistance to environmental abrasion, particularly in dynamic loading conditions such as water jet attack. Experimental tests have indicated that CNT/concrete composites exhibit greater endurance against erosive action compared to conventional concrete, demonstrating resistance against pulsating and continuous water jets. This resistance is linked to the enhanced mechanical strength provided by the CNTs, which enhance the materials' endurance against forces that would otherwise cause degradations. Such improvements in erosion resistance are particularly useful in environmental applications where structural endurance is the primary concern. These findings demonstrate the versatility of carbon nanotubes to not just enhance traditional mechanical properties but also to enhance the functional life of concrete structures in cases of adverse exposures.

5.1.2. Nanoclays

The application of nanoclays within cement and concrete mixtures has been an emerging trend in materials engineering, with specific focus placed on enhancing mechanical and physical properties. Nanoclays, due to their silicate layer structure, are contributing to high performance in most types of concrete, and they find applications ranging from high-strength buildings to eco-friendly construction materials. Their unique properties allow for increased interfacial bonding between aggregates and the cement matrix, which can result in reduced permeability and increased durability. On the heels of study by the European project SCAFFOLD, the use of nanomaterials, like nanoclays, has been found to possess high potential to increase the durability and lifespan of building materials, showing that its application may be a crucial factor in sustainable infrastructure development (Pintea et al., 2015). In high-volume fly ash (HVFA) concrete, nanoclays are regarded as influential agents of the transition zone between coarse aggregates and cement paste. Employing a focused approach—coating coarse aggregates with a nanomaterial-enriched cement paste such as nanoclays—a revolutionary impact can be observed on the mechanical behavior and transport properties of the final concrete. Studies indicate that the method can effectively improve the durability characteristics of HVFA concrete, such as water sorptivity and chloride migration. This is evidence that applying nanoclays at the interface is an evident advantage over other conventional methods involving direct mixing of the nanomaterials in the concrete matrix, demonstrating evidence on their mechanistic gains in concrete application (Du et al., 2018). In addition, the mechanical strength of concrete is also greatly enhanced by the specific use of nanoclays because they have the potential to reduce permeability and long-term degradation mechanisms. Small particles of nanoclays interweave within the cement paste to form a denser structure with higher strength. It is this reinforcing of the structure that can render the material more resistant to environmental factors, such as freezing and thawing cycles, thereby extending the service life of the concrete in various construction environments. As the construction industry increasingly uses nanotechnology, the experiences from such projects as SCAFFOLD highlight the necessity of providing value to innovative material solutions, in this instance nanoclays, for addressing the contemporary requirements of sustainable construction practice and ensuring the long-term durability of infrastructure (Pintea et al., 2015) (Figure 5.4).

Figure 5.4. Nanoclays used in cement and concrete.

Source: https://assets.newatlas.com/dims4/default/50e8d74/2147483647/strip/true/crop/3023x2015+0+0/resize/2880x1920!/format/webp/quality/90/?url=https%3A%2F%2Fnewatlas-brightspot.s3.amazonaws.com%2F78%2F4b%2Fabd796a94b829157a19bdb643b96%2Ffigure3.jpeg.

5.1.3. Nanosilica

The use of nanosilica in cement and concrete technology is a significant step towards enhancing the quality of building materials. Nanosilica possesses an extremely high surface area as well as reactivity of silica, which increases the mechanical properties and sustainability of concrete mixtures. When incorporated into concrete mixtures, nanosilica can enhance the microstructure through filling pore space between cement particles and enhancing the formation of more calcium silicate hydrate (C-S-H), the major binding phase in cement. This increases compressive strength and decreases permeability and thus enhances the durability and performance of concrete structures. Successful implementation of nanosilica also indicates a shift towards greener construction practice because its properties can reduce the overall content of cement required in concrete mixtures and consequently reduce the environmental impact associated with cement manufacturing. Moreover, the utilization of nanosilica also extends beyond enhancements in mechanical properties; it also bears significant effects on the transport behavior of concrete. A study on the performance of nanosilica-coated coarse aggregates in high-volume fly ash (HVFA) concrete supports this claim strongly. By increasing the transition zone interface between the aggregate and the cement paste, the study established that the use of nanosilica coatings was capable of properly reducing the permeability of the concrete while increasing its resistance to incursion by deleterious chemicals, such as chlorides and sulfates. The improvement of the transport properties was particularly important, as they contribute to the material's resistance against attack by the environment, including freeze-thaw and chemical attack. However, the transport properties were improved but no notable strength enhancement in compressive and splitting strengths was observed in the use of coated aggregates compared to directly inserting nanosilica into the concrete, indicating the need for more studies in maximizing the use of nanosilica in concrete. Nanosilica also aids in controlling shrinkage, a common defect inherent in concrete. Over the last few decades, industry developments have been geared towards producing cement finer to gain strength faster. This has inadvertently led to more concrete shrinkage after hardening. Adopting coarse cements with nanosilica seems to be a likely solution to overcome this problem. Applying coarse cements reduces shrinkage, while introducing nanoparticles enhances early strength gain, achieving a balance between rapid setting times and physical stability. Experimental work demonstrates that mixtures with coarse cement and nanosilica not only possess greater seven-day compressive strength than standard concrete mixtures but also much less shrinkage. The two-fold benefit positions nanosilica as an essential ingredient for producing durable, low-shrinkage concrete, particularly in regions where durable material performance matters (Du et al., 2018; Faltin et al., 2022) (Figure 5.5).

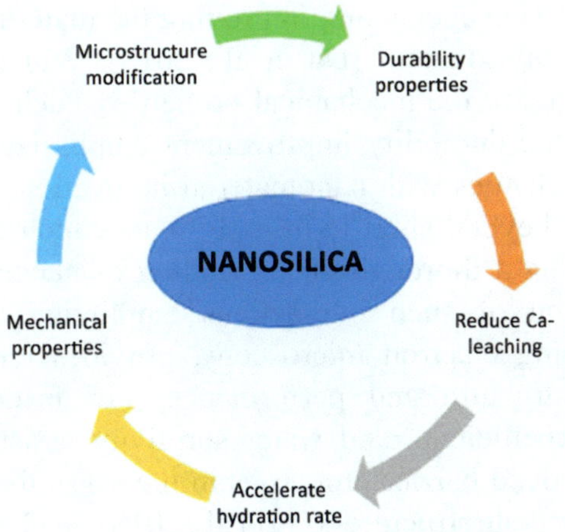

Figure 5.5. Nanosilica used in cement and concrete.

Source: https://ars.els-cdn.com/content/image/1-s2.0-S0920410520304691-fx1.jpg.

5.2. MECHANISMS OF ACTION

The addition of nanomaterials to cement and concrete involves advanced modes of action that enhance the composites overall performance. One of these key modes is the increased interface interaction between the coarse aggregate and the cement paste, which can be facilitated by adding nanomaterials. Through the modification of this interface, nanomaterials can minimize the issues related to bonding and, as such, enhance the overall matrix. This increased interfacial adhesion is of utmost significance to ensure maximum mechanical properties of concrete, such as tensile strength and durability, that are important in structures demanding utmost structural integrity (Rodriguez et al., 2018). In addition, understanding these mechanisms provides a foundation for the production of stronger concrete with extended lifespan since effective interfacial bonding minimizes recurring pathologies such as cracking and spalling, thereby extending infrastructure lifespan.

Most striking is the mechanism of action of graphene oxide and nanosilica when used to enhance high-volume fly ash (HVFA) concrete. Instead of mere blending of these nanomaterials within the matrix of concrete, recent evidence suggests a particular coating mechanism that enhances the interface transition zone optimally. Through coating of coarse aggregates with a cement paste layer containing these nanomaterials, scientists have observed enhancements in the transport characteristic of HVFA concrete. This treatment reduces permeability and keeps the aggressive agents out, improving the material durability against degradation in the environment (Du et al., 2018). While this treatment did not significantly increase the mechanical properties, such as compressive and splitting strengths, the durability improvement emphasizes the imperative to drive interfacial mechanics with nanomaterial innovation. The effects of such mechanisms extend beyond short-term performance metrics, emphasizing the importance of gaining a thorough understanding of nanomaterial interaction. Routine investigation of such interactions employing experimental tools, for instance, scanning electron microscopy, provides temporal information on material microstructure and performance. For instance, the study of chloride diffusion coefficients and water sorptivity assists in understanding nanomaterials' improved barrier properties in the sense that they can enhance the longevities of concrete structures (Du et al., 2018). Such research thus assists in the development of design guidelines and standards that embrace the use of nanomaterials, therefore ensuring the use of cutting-edge technologies in the construction industry. Through a systematic analysis of such mechanisms, the application of nanomaterials in concrete progresses toward the next generation of sustainable and durable infrastructure.

5.2.1. Particle Size and Surface Area Effects

The size of the particles and surface area of the materials used have a significant impact on the properties of cement-based materials, particularly when applying nanomaterials. In the case of nanotechnology for cement and concrete, smaller particle sizes provide a higher specific surface area, which enhances the materials' reactivity. This is mostly due to increased contact area with the surrounding cement paste, which enables better hydration and interparticle binding interactions. Consequently, nanoparticles can fill gaps between larger particles, forming a denser material with reduced microcracks and pores, ultimately resulting in an increased durability and mechanical resistance of the concrete mixture.

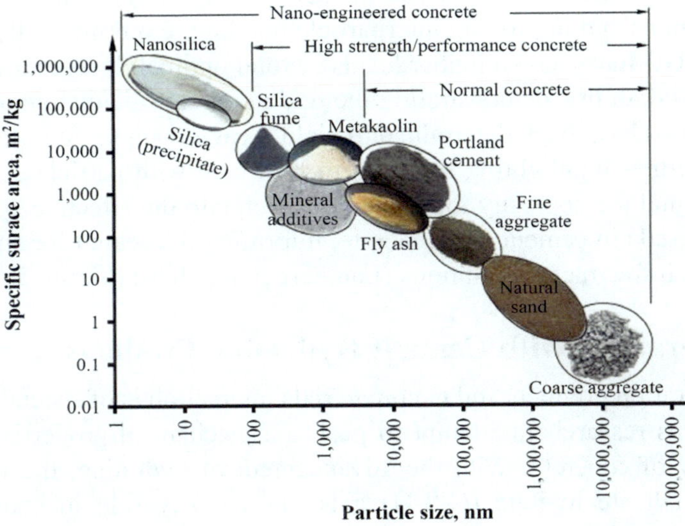

Figure 5.6. Particle size and surface area effects in cement and concrete.

Source: https://www.researchgate.net/profile/Prashant-Kumar-167/publication/303365610/figure/fig6/AS:613509054357510@1523283365198/Particle-size-and-specific-surface-area-of-concrete-materials-taken-from-Kumar-et-al.png.

Thus, it is necessary to grasp the fine balance of particle size to obtain the maximum out of nanomaterials' performance in concrete. When nanomaterial coatings are examined on aggregate surfaces, scientists found that variations in particle size and coating thickness can highly influence the compressive strength and durability properties of concrete. Nano-$CaCO_3$ and nano-SiO_2

coatings had varying impacts on interfacial transition zone (ITZ), which plays a critical role for the overall performance of concrete. While thicker coating improved resistance to chloride penetration, it negatively affected compressive strength, indicating the complex interrelation between particle size, surface area, and mechanical properties. These findings suggest proper control of these parameters can optimize ITZ microstructure for enhanced overall performance of concrete, particularly under dynamic loads. Balancing these effects is essential in manufacturing high-performance concrete with the ability to meet modern requirements of construction (Schutter et al., 2019). Furthermore, incorporation of nanomaterials such as graphene oxide (GO) has been found to play a significant part in enhancing the mechanical properties of ultra-high-performance concrete (UHPC). The unique two-dimensional nature of GO allows it to be dispersed evenly in the cement matrix, thereby maximizing surface area interactions and improving interparticle bonding. Experimentally, varying amounts of GO have shown higher levels beyond optimal dosages may lead to decreased workability, demonstrating the trade-offs in utilizing nanomaterials. Ideal GO ratios have been determined to yield greater compressive and flexural strengths, further highlighting the need to take care with particle size. These advances signal the necessity for further research into the effective application of nanomaterials in cement and concrete, informing the search for sustainable and resilient infrastructure solutions (Lambert et al., 2020) (Figure 5.6).

5.2.2. Interaction with Cement Hydration Products

Cement hydration products and nanomaterials interaction is of special concern, particularly as researchers attempt to push the mechanical properties as well as durability of concrete. With the advancement of hydrating, the formation of calcium silicate hydrate (C-S-H) gels plays a key role in cementitious materials' strength maturing. Nanomaterials such as graphene oxide (GO) nanosheets have been identified as potential additives that can modify this microstructure evolution. Research has shown that the inclusion of GO in ultra-high-performance concrete (UHPC) increases the flexural and compressive strength considerably. For example, by maintaining the water-cement ratio at a constant value and varying the amount of GO, one not only controls the mini slump flow but also optimizes the mechanical performance of the composite to maximize the interaction between the nanomaterial and hydration products. This strategy is an important step towards the achievement of cost-effective and sustainable infrastructure, as has been evident in results from studies (Lambert et al., 2020). Aside from GO, application of plant-derived nanomaterials, such as nanocellulose, presents another possibility for increased interaction with cement hydration products. Nanocellulose has attracted attention because it possesses superior mechanical properties and environmental friendliness and presents significant potential for use in cement-based composites. Plant fibers and their derivatives' reaction with the cement hydration products has the potential to change the microstructural characteristics of the composite, enhancing not only physical characteristics but also material durability. By overcoming the challenges associated with dispersion, fiber-matrix interfaces, and the long-term character of the composites, scientists can better understand the effects of natural fibers on cement hydrates. Such an understanding could lead to the development of high-performance and sustainable engineered cement-based composites (Ahmad et al., 2017). Lastly, the progress in nanomaterials introduces new aspects to the underlying processes in cement hydration and its subsequent interactions. Extending the knowledge about these interactions can guide the design of innovative cementitious materials that utilize both organic and inorganic nanomaterials. How these nanomaterials influence the properties of hydration products is important to know in understanding for developing more mechanically advanced concrete and weather degradation resistance. With research progressing, nanomaterials' integration into the cement matrix not only foretells improved forms of existing technology but also opens doors to new civil construction applications, which are greatly demanded in civil engineering today because of the sustainability imperatives.

5.2.3. Influence on Microstructure and Porosity

The incorporation of nanomaterials into cement and concrete plays a significant role in controlling the microstructure and porosity of the resulting composite, leading to its enhanced mechanical and durability properties. By modifying the interfacial transition zone (ITZ) between aggregates and cement matrix, the pore distribution and size can be precisely controlled. For instance, nanoparticle coatings like nano-SiO_2 or nano-$CaCO_3$ on coarse aggregates were reported to have altered the microstructural properties of concrete. Specifically, the nanoparticle coatings enhance the resistance to chloride ingress, an important factor towards longevity and durability, and it controls the pore density and distribution in this region. As noted by Schutter et al. (2019), it is possible to enhance chloride resistance through greater coating thickness, although this could potentially lower compressive strength, emphasizing the intricate compromise in optimizing concrete mixes. In addition, the specific application of nanomaterials affects the mechanical properties of high-volume fly ash (HVFA) concrete directly. With nanomaterials as surface coatings on coarse aggregates, the interfacial zone is improved by improved bonding and porosity reduction, leading to increased strength of the structure. Experimental data indicate that while addition of such nanomaterial enhances transport properties of concrete, i.e., permeability is reduced effectively, it does not necessarily translate into increased compressive or splitting strengths over conventional mixing practices of incorporating the nanomaterial directly into the concrete. As shown by Du et al. (2018), the enhanced transportation properties result in a material that is resistant to environmental aggressors, ultimately prolonging the life and performance of concrete structures. Continued research into the nanomaterial effects on microstructure and porosity further emphasizes the paramount significance of an extensive understanding of these materials interactions in concrete. The nanoscale alterations could control performance parameters of importance to construction applications, including freeze-thaw resistance and general chemical attack resistance. Advanced characterization tools such as scanning electron microscopy and energy-dispersive X-ray spectroscopy allow researchers to visualize and quantify such microstructural alterations and derive insights that drive innovation in concrete technology. As knowledge synthesis goes on, the application of nanomaterials' unique properties will definitely push the development of high-performance concrete to suit the strict demands of modern construction (Figure 5.7).

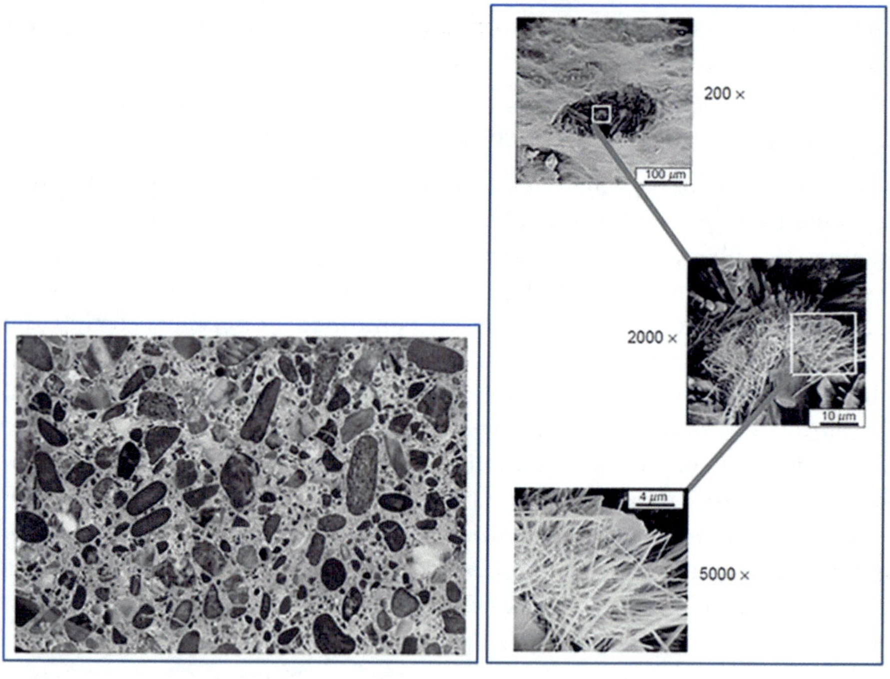

Figure 5.7. Microstructure of concrete.

Source: https://cdnintech.com/media/chapter/51720/1512345123/media/fig1.png.

5.3. BENEFITS OF INCORPORATING NANOMATERIALS

The incorporation of nanomaterials into cement and concrete mixes has brought a lot of advantages that not only enhance the mechanical properties and sustainability of construction materials but also make these materials more durable. One of the important benefits is the ability of nanomaterials to improve the interface between coarse aggregates and cement paste. Utilizing a method whereby coarse aggregates are enveloped with cement paste loaded with nanomaterials sees interface transition zone interaction seeing dramatic improvements. Though regular blending of nanomaterials into the new concrete is not necessarily achieving high compressive and splitting strengths, the process of applying coated aggregates can significantly reduce permeability in HVFA concrete, thereby enhancing its overall transport properties. This quality is important, as it potentially results in longer-lasting structures less susceptible to environmental forces, such as moisture intrusion and attendant degradation (Du et al., 2018). In addition, the incorporation of graphene oxide (GO) in concrete mixtures has been shown to be significantly beneficial, particularly in the application of pervious concrete. In a study analyzing GO-modified fly ash pervious concrete, performance in stormwater management improves as noted, including improved infiltration and pollutant removal. The chemical and mineral properties of fly ash pastes, with graphene oxide modification, distinguish them from traditional cement pastes, which exhibit enhanced freeze/thaw cycling and salt attack resistance. All are important enhancements in infrastructure subject to variable weather, since the durability improvements translate directly to reduced maintenance costs and increased lifetimes for concrete structures. The employment of advanced analytical tools, such as nuclear magnetic resonance and X-ray diffraction, also delineates the central role played by GO in making concrete more resistant to environmental stresses (Shi et al., 2018). Ultimately, the strategic application of nanomaterials in cement and concrete is an anticipatory solution to the future-looking needs of construction today. Due to growing sustainability and efficiency needs, the use of material like graphene oxide and nanosilica can reduce the environmental footprint of making concrete while simultaneously improving performance. These nanomaterials not only achieve the mechanical and durability requirements but also enable functional versatility of the mixes of concrete in order to accommodate specific engineering applications. With the advancement of the construction industry, the application of the unique properties of nanomaterials is a potential path to create and strengthen building materials in a manner that they can fulfill the demands of present-day architectural and infrastructural needs.

5.3.1. Improved Mechanical Properties

Incorporation of nanomaterials into concrete and cement has revolutionized the construction industry by significantly improving mechanical properties. Among these materials, graphene oxide (GO) is a pioneering additive due to its outstanding characteristics, which give concrete mixtures enhanced strength and durability. Research has revealed that the highest addition of GO exhibits a remarkable increase in compressive and flexural strengths. Through the introduction of GO nanosheets in varying weights, research has demonstrated that concrete containing specific dosages, such as 0.02% under standard curing and 0.04% under steam curing, possesses these improved mechanical properties. Such improvements not only strengthen the structure but also make concrete structures sustainable through material reduction, which is fundamental to assisting in the mitigation of environmental concerns in urban development and construction activities (Lambert et al., 2020). Additionally, concrete's mechanical properties can be still improved further by using fly ash when combined with graphene oxide.

This groundbreaking technique not only utilizes the benefit of nanoscale materials but also utilizes waste products to enhance the environment-friendliness of concrete production. Initial findings show that GO-modified fly ash pervious concrete exhibits significant durability against various deteriorating factors, such as freeze/thaw cycles and salt attack, typical in harsh environments. Such enhanced strengths are also exhibited through the improved mechanical properties, thus showing that GO modification impacts the microstructural performance of the concrete. Hence, the use of GO in fly ash mixtures becomes a prospect for developing resilient building materials that can endure the pressures of urbanizing cities without sacrificing structural efficacy (Shi et al., 2018).

The application of nanomaterials like graphene oxide has implications extending far beyond mere mechanical strengthening; it signifies a paradigm shift in construction. With the demand for high-performance concrete growing, the use of these advanced composites can add to reduced environmental effects associated with traditional cement production. The combination of nanomaterials and supplementary cementitious materials points towards a field of principal research where potential is there for achieving stronger and more sustainable and durable concrete solutions. This transition to nanotechnology-based construction not only fills the demands of today's infrastructure but also provides greater goals of sustainability and conservation of resources, emphasizing the necessity of building research in nanomaterials to provide the needs of construction in the future (Lambert et al., 2020; Shi et al., 2018) (Figure 5.8).

Figure 5.8. Nanotechnology in concrete.

Source: https://constrofacilitator.com/wp-content/uploads/2021/02/Nanotechnology-in-Concrete.jpg.

5.3.2. Enhanced Durability and Longevity

The addition of nanomaterials to concrete and cement significantly influences the enhanced durability and prolonged lifespan of structures. The enlarged surface area and reactivity of nanomaterials, coupled with their physical as well as chemical nature, contribute towards increased bonding in the cement matrix. When applied judiciously, the materials are effectively employed to seal pores in the concrete structure, thereby reducing permeability and preventing the intrusion of harmful substances such as chlorides and sulfates. This reduction in permeability not only increases resistance to attack from the environment but also increases the overall structural integrity. Therefore, structures last longer, reducing the requirement for repeated repairs or replacement, ultimately adhering to sustainable construction. Besides increased resistance to chemical attack, nanomaterials are capable of decreasing shrinkage problems, a significant parameter that defines the length of time for concrete. The over-shrinkage is usually attributed to the quest for quick development of strength in today's concrete mix, which has led to increased fineness of cement grains. Utilization of coarse cements with nanoparticles is a novel measure against this problem. The coarse cement is to reduce shrinkage, while nanoparticles account for early strength, avoiding structural performance problems during early curing periods. The concurrent use of these materials presents a significant reduction in shrinkage and subsequent rise in compressive strength, as in the case of coarse cement and nanoparticle mixes that present better performance than conventional mixes (Faltin et al., 2022). Besides this, the use of advanced sensing technology combined with nanomaterial-reinforced concrete introduces a sea change in structural health monitoring. The integration of micro-electromechanical systems (MEMS) into the concrete matrix allows structural parameters to be tracked in real time, and this extends infrastructure lifespan. This preventive approach allows for the early detection of potential failures, followed by timely maintenance and resilience to failure catastrophes. The ongoing assessment of concrete structure condition not only improves safety but also reduces lifecycle costs. This new application underscores the importance of nanomaterials not just for material property improvement but also for life extension through smart, responsive infrastructure maintenance (Buss et al., 2013) (Figure 5.9).

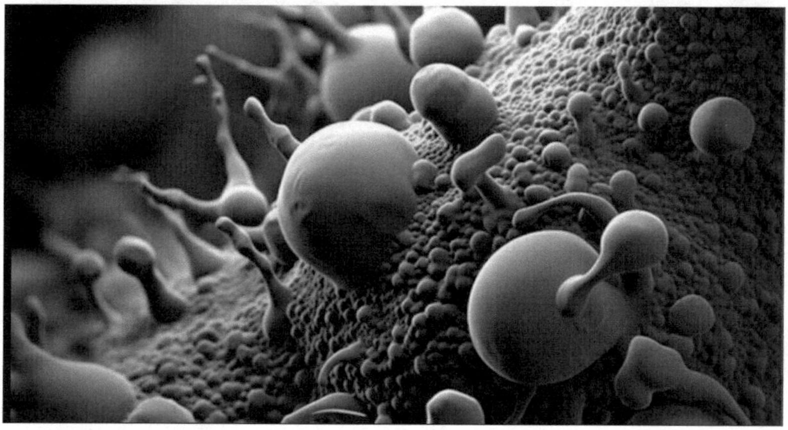

Figure 5.9. Durability and longevity of nanomaterials in cement and concrete.

Source: https://srmprcements.com/wp-content/uploads/molecular-sphere-reveals-futuristic-genetic-research-data_739292-17895.webp.

5.3.3. Increased Resistance to Environmental Factors

Incorporation of nanomaterial such as graphene oxide (GO) in cementitious composites has been shown to be beneficial in enhancing the durability of concrete under various environmental conditions. An example is the development of pervious concrete through modification by GO, using coal fly ash as the predominant binder. Research indicates that not only is this modification beneficial in enhancing the material's durability but also in raising its stormwater infiltration capacity. Notably, the ambient-temperature salt and freeze/thaw resistance of GO-modified pervious concrete has been significantly improved, key parameters of the service life of concrete in aggressive environments. Such enhanced performance in aggressive settings plays a crucial role in sustainable infrastructure with enhanced extended service life and reduced maintenance costs, proof of the critical role played by nanomaterials in modern construction materials (Shi et al., 2018). Furthermore, studies on ultra-high-performance concrete (UHPC) with GO nanosheets emphasize the role of materials in reducing environmental degradation. By introducing different ratios of GO into UHPC mixtures incorporating high volumes of supplementary cementitious materials (SCMs), scientists found that the mechanical performances, including compressive and flexural strengths, were significantly improved. Such enhancement was particularly important at optimal GO contents, which ultimately result in reduction of porosity and enhancement of microstructural stability of the concrete. These factors remain important to render concrete structures weather and chemical-resistant under harsh environmental conditions of urban areas. Therefore, the inclusion of nanomaterials like GO not only aids in attaining improved mechanical performance but also improves the toughness of concrete against various environmental aggressors (Lambert et al., 2020). The impacts of improved resistance against environmental agents by including nanomaterials in concrete extend beyond physical performance to address sustainability goals in construction. With enhanced protection properties of cementitious materials, experts can minimize the environmental impact associated with material degradation and their subsequent repair or replacement. The application of nanotechnology allows for the development of eco-friendly concrete alternatives to aid in circular economy in the construction industry. Through this, the addition of nanomaterials not only addresses near-term structural requirements presented by environmental stress but also sparks wider industry take-up of sustainable technologies. This union of technological innovation and green responsibility positions nanomaterials at the forefront of the creation of long-lasting and sustainable concrete products (Lambert et al., 2020; Shi et al., 2018).

5.4. CHALLENGES AND LIMITATIONS

Application of nanomaterials in cement and concrete is plagued with numerous limitations that must be effectively addressed to facilitate effective implementation. One major limitation is the challenges in manufacturing and incorporating the nanomaterials. The methods for producing nanomaterials are costly and energy-intensive, rendering large-scale application economically unfeasible for most construction endeavors. In addition, uniform dispersion of nanomaterials within the concrete matrix is crucial to their full potential advantages, such as increased mechanical properties and increased durability.

However, uneven mixing could produce inconsistency in performance outcomes, ultimately undermining the integrity of the resultant concrete product. Therefore, the mitigation of such production and integration problems is central to the advancement of nanotechnology in construction. Moreover, the environmental and safety impacts of nanomaterials must not be overlooked. The potential toxicity of nanoparticles introduces severe health risks to the workers in contact with them in the process of production and application. This is accompanied by the absence of scientific understanding regarding their long-term impacts on the environment, especially after they are integrated into infrastructures. As the building sector has to contend with growing concerns for the environment and sustainability requirements, the prospective ecological impact of nanomaterials poses a need to conduct in-depth analysis. An observation is that the building sector has traditionally been resistant to change, as evidenced by low patenting rates and few high-tech startup firms in this sector. This reluctance only contributes to the use and application of developed nanomaterials within the industry (Pacheco-Torgal et al., 2017). Furthermore, the regulatory mechanisms are slow to adapt with technological expansion, so there is more resistance to the application of nanomaterials in cement and concrete. There is no standardized test for measuring the performance as well as safety of concrete enhanced with nanomaterials, hence restricting its widespread application. In addition, inconsistency in regulations among agencies for testing protocols and standards of approval creates doubt among manufacturers and contractors. This regulatory doubt will discourage investment in research and development of the use of nanotechnology in civil engineering and also inhibit innovation. Lastly, addressing these regulatory issues is necessary to make an enabling environment that can facilitate the use of nanomaterials in the building construction sector so that they can transition from laboratory success stories to real marketable solutions.

5.4.1. Cost Implications of Nanomaterials

The application of nanomaterials in cement and concrete is highly poised to enhance performance as well as the lifespan of construction materials. However, the cost of embracing such innovations can be substantial. The nanomaterials production and development require advanced technological processes, and they in most instances translate to higher initial costs for producers and consumers. With the construction sector experiencing widening demand brought about by global population growth—projected to reach more than 11 billion by 2100—it is imperative to examine these cost markers critically. Use of nanomaterials can yield long-term cost savings by way of increased energy efficiency and reduced maintenance, but upfront costs can deter widespread use in a traditionally conservative market (Pacheco-Torgal et al., 2017). Furthermore, the various availability and sourcing of nanomaterials contribute to cost complexities. Different types of nanomaterials, such as silica nanoparticles and TiO_2, have unique properties and costs that influence their feasibility in different applications for concrete (Pintea et al., 2015). Certain nanomaterials enhance durability or self-cleaning but might increase final product cost through supply chain complications or specialized manufacturing methods. Balancing initial costs against long-term benefits of reduced lifecycle cost and lowered environmental impact is a priority for construction companies. This is particularly relevant with the sector facing increased regulation towards sustainability and environmental efficiency, making the economic climate more challenging for the introduction of novel materials into established procedures. Risk management is key to establishing the economic viability of nanomaterials in cement and concrete. Initiatives like the SCAFFOLD project are intended to foster risk management frameworks that lead construction companies through the complexities of applications of nanomaterials. By offering guidelines that incorporate occupational health and safety needs and risk assessment protocols, these frameworks minimize likely costs on safety and compliance concerns (Pintea et al., 2015). With increasing market competition, economic factors, the risks and benefits of innovation, will be important for stakeholders who want to apply nanotechnology in concrete applications. It would not only influence investment decisions but also determine the future of sustainable construction activities (Figure 5.10).

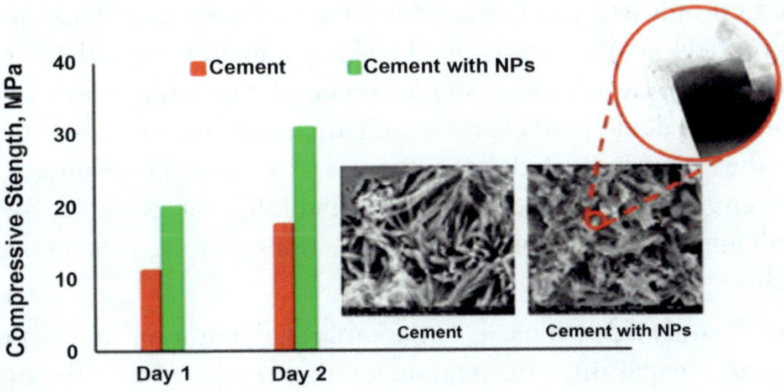

Figure 5.10. Cost implications of nanomaterials in cement and concrete.

Source: https://ars.els-cdn.com/content/image/1-s2.0-S0950061820325010-ga1.jpg.

5.4.2. Potential Health and Environmental Risks

The incorporation of nanomaterials into cement and concrete introduces an enormous range of potential health and environmental risks warranting careful consideration. Nanomaterials such as graphene oxide and various metallic nanoparticles have also been shown to enhance the mechanical strength and durability of concrete. However, their incorporation is of concern regarding toxic leachate into soil and water sources, bringing adverse effects on human health and ecosystems. Research indicates that some nanomaterials, when leached, have the potential to cause increased concentrations of pollutants such as heavy metals and nitrates in stormwater runoff, resulting in serious downstream environmental impacts.

Therefore, to avoid such health and environmental hazards, thorough comprehension of the long-term destiny of such nanomaterials in concrete matrices is important. Besides, nanotechnology use in construction must accompany good risk management methods for workers' safety and environmental protection. The SCAFFOLD project has laid the groundwork for this approach through the development of a Risk Management Toolkit that supports nano-management in the construction sector. This tool combines established standards like OHSAS 18001 and ISO 31000 to develop an exposure scenario assessment framework related to nanomaterials. These standardized risk assessments are significant as they establish laboratory and industrial exposure levels to various nanomaterials used in construction. By the use of advanced analytical techniques and clearly defined exposure ratings, the construction industry can better address uncertainty associated with nano-enhanced products so that envisioned risks might not exceed their benefits (Pintea et al., 2015). Besides that, studies on graphene oxide-modified pervious concrete reflect the dual-sided character of nanomaterials functionality and risk. Whereas studies have indicated that graphene oxide incorporation can improve resistance to environmental agents very substantially, for example, freeze/thaw cycle and salt attack, there are unresolved questions on long-term consequences of such modifications.

The same study identifies a remarkable reduction of toxic pollutants, highlighting the capability of nanomaterials to contribute to stormwater management in a beneficial way (Shi et al., 2018). Nevertheless, skepticism on the environmental effects of long-lasting nanomaterials in water bodies requires monitoring of the environment and the implementation of predictive models capable of simulating the transport and degradation of the substances. As the prospects of nanotechnology in construction look better, a judicious balance between innovation and prudence will be necessary to ensure public health and environmental integrity.

5.4.3. Technical Challenges in Mixing and Application

Adding nanomaterials into cement and concrete has the potential to enhance mechanical performance and durability but also raises tremendous technical challenges for mixing and application. One significant problem is in the uniform dispersion of nanomaterials in the cement matrix. Due to their high surface area and agglomeration tendency, nanoparticles may clump together, producing non-uniform concentrations that can cancel out the achieved benefits of the nanomaterials. Moreover, conventional mixing techniques are ineffective enough in dispersing these materials, and therefore new mixing technologies or additives enhancing dispersion are needed. This challenge is further compounded while dealing with the rheological properties of the final concrete because excessive nanomaterial concentrations have the potential to alter its workability, again leading to pumping and placement issues.

Figure 5.11. Mixing concrete with graphene to build better structures.

Source: https://www.azonano.com/images/Article_Images/ImageForArticle_6059_16479484248828961.jpg.

A second significant technical challenge comes from interaction at the aggregate-cement interface. The latest research indicates that direct incorporation of the nanomaterials into the concrete may not lead to the anticipated development of compressive strength and durability in comparison to other practices. Specifically, the nanomaterial coating of coarse aggregates has been demonstrated to have the potential for optimizing the interface transition zone and thus improving transportation properties without necessarily improving compressive strength (Du et al., 2018). This indicates that a selective approach to the application of nanomaterials is required that considers the special nature of the aggregate-cement interface, which can affect both mechanical performance and durability in the long term. Developing methodologies that optimize the benefit of nanomaterials at this critical interface is vital towards enabling their optimal use in cementitious systems. Furthermore, modification of incorporation of nanomaterials into cement paste can also be controlled by external factors, i.e., addition of magnetic fields, which influence the material's structural properties. Experimental research has shown that the structural development of the cementitious paste can vary significantly with respect to its flow characteristics and mechanical properties under different modes of magnetic fields (Schutter et al., 2019). Such findings have established that the mixing processes need to consider the environmental conditions and external stimuli as well, which may have an effect on the behavior of the nanomaterial-enhanced concrete. With scientists going deeper into these fields, the challenge remains to develop standardized processes for mixing and applying nanomaterials in concrete that can meet both internal and external conditions and deliver consistent performances in actual uses (Figure 5.11).

5.5. CASE STUDIES AND APPLICATIONS

Applications of nanomaterials like graphene oxide (GO) in cement and concrete manufacturing have proven very promising from various case studies. Studies involved the preparation of GO-modified pervious concrete employing coal fly ash as an independent binder. The research sought to examine the stormwater infiltration capacity of the modified concrete alongside its durability using rigorous test processes like freeze/thaw cycling and salt resistance. The results showed that the level of pollution in stormwater was lowered significantly, such as copper and nitrate, thereby showing the environmental benefit. Furthermore, GO usage increased the freeze/thaw strength and durability of concrete higher than before, which was confirmed by using advanced analytical techniques, such as X-ray diffraction. The findings show the contribution of nanomaterials in increasing environmental as well as structural properties of concrete structures (Shi et al., 2018). In another experiment, researchers conducted a test on the effect of nanomaterials on the transition zone interface (ITZ) between the coarse aggregate and cement paste. This process was different from conventional procedures of adding nanomaterials to fresh concrete mixtures, which instead involved coating coarse aggregates with nanosilica or graphene oxide containing mixtures. The case study specifically targeted high-volume fly ash (HVFA) concrete and the compressive strength, splitting strength, and transport properties like water sorptivity and chloride migration. Although the results indicated that there was no significant enhancement in the mechanical strengths of concrete with coated aggregates, the transport properties were significantly enhanced, and this suggested reduced permeability. This is to say that interface alteration can be a strategic route for improving concrete performance, foreshadowing the importance of innovative methods of applying nanomaterials in concrete technology (Du et al., 2018). The implication of these case studies extends beyond technological advancement; it forebodes changes in the construction industry's attitude toward sustainability and durability. The effective use of nanomaterials in cement products represents a significant step towards maximizing concretes sustainability and resilience in various environments. The highlighted case studies provide strong proof that well-directed uses of nanotechnology can yield concrete formulations with enhanced properties, such as long-term durability and reduced environmental impact. As the sector continues embracing such innovations, future research will play a crucial role in establishing standardized protocols for the incorporation of nanomaterials to ensure safety and effectiveness as well as promoting the adoption of such green materials across global construction processes. Such technological advancement attests to the continuous need for research and development in the field of nanomaterials in cement and concrete application.

5.5.1. Use of Nanosilica in High-Performance Concrete

The incorporation of nanosilica into high-performance concrete is a revolutionary step ahead in the evolution of concrete technology. Nanosilica consists of silica nanoparticles and possesses several unique characteristics that significantly enhance the performance characteristics of concrete. Studies have established that when added to the cement matrix, nanosilica increases the mechanical strength and durability of the material, thus addressing most of the deficiencies of typical concrete mixtures. This enhancement is attributed to the ability of nanosilicas to seal the pores in the cement paste, making it a denser microstructure that prevents the penetration of harmful materials. Thus, high-strength concrete containing nanosilica can resist physical and chemical deteriorations more efficiently, thus making it ideal for demanding applications in civil engineering and construction where long-termness is greatly valued. In addition to mechanical strength improvement, the addition of nanosilica to concrete also enhances its resistance to environmental factors. Nanosilica prevents rebar corrosion, which is a critical concern in the lifespan of concrete structures. This reduction in corrosion is significant since it increases the structural integrity and service life of concrete elements exposed to aggressive environments. Besides, the addition of nanosilica enables desirable interactions at the nanoscale to increase the hydration of cement-based composites. Existing literature highlights that the interactions improve not only the physical but also the chemical resistance of concrete, allowing structures to endure exterior stress like freeze-thaw cycles and chemical attack (Du et al., 2019). Therefore, the strategic application of nanosilica enhances the performance of products and also aids in environmentally friendly construction practices by extending the life cycle of concrete structures. The technological innovation brought by nanosilica is part of a larger view of ongoing research on nanomaterials employed in cement and concrete. While the construction industry is facing growing pressure towards high-performance and efficient building materials, the use of nanosilica is progressively integral to its future. Aspects derived from latest research identify the potential of nanosilica and other nanomaterials to introduce multifunctional properties in concrete with the potential of ushering in smart structures featuring embedded health monitoring systems. These developments are a gargantuan leap for the use of nanotechnology in Cementous materials, thereby opening doors to an arsenal of possibilities in future infrastructure. The study of nanosilicas properties accompanies the path of optimizing concrete performance and sustainability, providing researchers and practitioners with new tools to meet construction's evolving challenges (Małgorzata Krystek et al., 2021) (Figure 5.12).

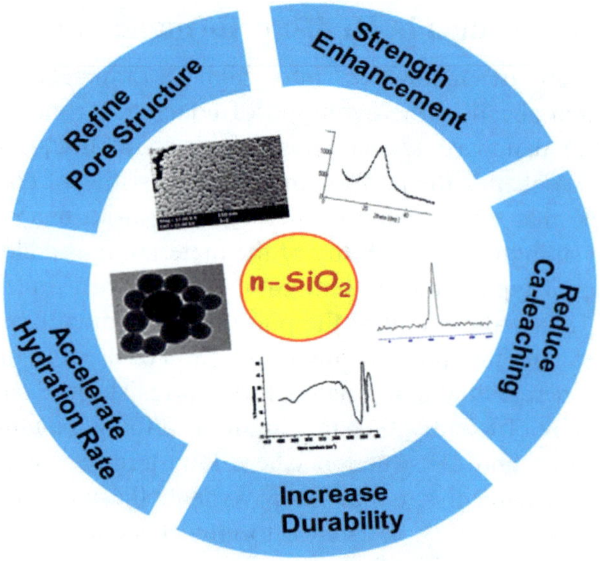

Figure 5.12. Use of nanosilica in high-performance concrete.

Source: https://www.researchgate.net/publication/269984517/figure/fig17/AS:613872494977033@1523370016301/Application-of-silica-nanoparticles-in-cement-and-concrete-128.png.

5.5.2. Carbon Nanotubes (CNTs) in Structural Applications

The use of CNTs in existing cement and concrete represents a paradigm shift in structural application, primarily due to the fact that they possess greater mechanical properties. Carbon nanotubes possess high tensile strength, elasticity, and low weight, thus perfect for use in enhancing the performance of conventional building materials. When added to concrete mixtures, the use of CNTs can significantly improve compressive and tensile strengths, thus contributing to the overall load-carrying capacity of buildings. This improvement is particularly crucial in high-demand settings, where stability of the structure must be maintained for extended durations of time. Secondly, implementation of CNTs would also reduce cracking and fatigue issues, which are common in regular applications of concrete. This enhanced durability makes carbon nanotubes a key component of future sustainable building practices, propelling the world's growing demand for more resilient infrastructure (MORENO et al., 2017). In inclusion studies of nanomaterials in structural applications, multi-walled carbon nanotubes (MWCNTs) have shown promising results when they are used as an addition to cement-based composites. The interaction of MWCNTs with calcium silicate hydrate (C-S-H) particles, a main component in cement mix, has been introduced as one of the richest areas of study. Studies have revealed that the incorporation of MWCNTs has been determined to greatly improve the mechanical properties of concrete, particularly with the presence of C-S-H in favorable proportion. For instance, evidence of lower porosity levels and increased strength values with improvement up to 30% over ordinary concrete has been reported in experiments in some compositions with C-S-H and a Ca/Si ratio of 0.8. These findings demonstrate the capacity of carbon nanotubes not only to serve as reinforcement but to actually increase chemical interaction in the concrete matrix (Duart et al., 2018). Furthermore, the eco-friendly benefits of incorporating carbon nanotubes in concrete mixes cannot be overlooked. With greater durability and strength of their mechanical properties, carbon nanotube-reinforced materials may reduce the overconsumption of resources during construction and hence promote environmental sustainability. The high-strength character of CNT-reinforced concrete could lead to more resistant buildings requiring fewer frequent maintenance and repairs, further limiting the environmental footprint of building operations. In addition, as the construction industry continues to struggle with waste and resource problems, the application of nanomaterials like carbon nanotubes represents a promising path toward more efficient and sustainable building. With ongoing research that continues to uncover the potential applications of carbon nanotubes in structural uses, they are no longer only new-fangled materials but key agents of transformation for the cement and concrete industry (MORENO et al., 2017) (Figure 5.13).

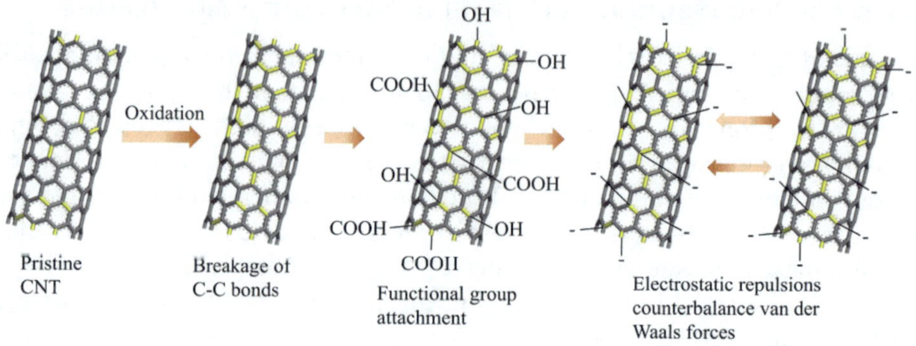

Figure 5.13. Carbon nanotubes in structural applications concrete.

Source: https://www.frontiersin.org/files/Articles/861646/fmats-09-861646-HTML/image_m/fmats-09-861646-g003.jpg.

5.5.3. Nanoclays in Sustainable Construction Practices

Application of nanoclays in sustainable construction strategies is a groundbreaking approach in the area of cement and concrete materials. Being nanoscale in size, nanoclays possess improved mechanical properties, enhanced durability, and improved natural degradation resistance. When incorporated into cement matrices, they not only reinforce the composite structure but also allow the overall cement content required for concrete production to be decreased. This decrease is crucial because the conventional cement production process is notoriously energy-intensive and one of the world's largest carbon-emitting industries.

Thus, the use of nanoclays is a feasible process in enhancing eco-friendliness in the construction sector, in accordance with contemporary sustainability goals aimed at lowering the environmental impact associated with conventional construction processes. Further, the application of nanoclays makes it possible to fabricate high-strength but light concrete, which is essential in sustainable construction. The innovation facilitates the drastic minimization of material use without compromising structural performance. The low weight of such concrete results in lower energy consumption during transportation and simplifies logistical challenges on the construction site. Moreover, the incorporation of nanoclays provides durability to concrete structures, reducing maintenance and enhancing service life, making sustainability much more important. As construction needs become greener towards sustainable alternatives, the role of nanoclays as a green admixture becomes increasingly pertinent. They not only enhance the performance characteristics of concrete but also promote energy efficiency and resource conservation throughout the lifespan of a building. Increased emphasis on sustainability at recent industry conferences reflects how much innovation in nanotechnology, such as nanoclays, matters to construction practice.

For example, the JEC Forum Southeast Asia 2024 emphasized innovation in sustainable materials that are sensitive to the use of natural fibers and recycling technology in composites (Amornsakchai et al., 2025). The discussions shed light on a broader trend in the construction industry with a concern for decreased environmental pressure and greater material efficiency. With the addition of nanoclays, construction practice can utilize these innovations in order to produce material with lower carbon footprints while simultaneously satisfying structural performance needs. This new discovery not only highlights the promise of nanoclays in contemporary construction but also provokes further research and collaborations towards maximizing their effectiveness and applicability in achieving sustainable construction practice in many applications (Figure 5.14).

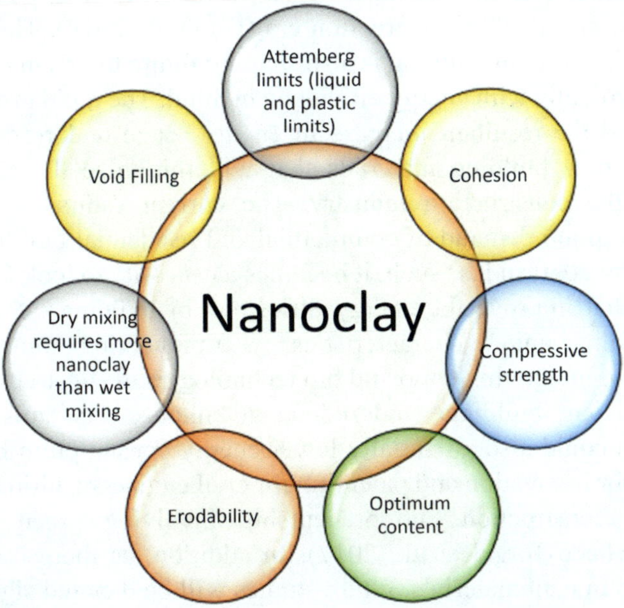

Figure 5.14. Nanoclays in sustainable construction practices.

Source: https://ars.els-cdn.com/content/image/3-s2.0-B9780443133909000084-f08-05-9780443133909.jpg.

5.6. FUTURE TRENDS IN NANOMATERIALS RESEARCH

Future directions in nanomaterials research are continually emerging, and some of these directions are being developed for incorporating the materials into cement and concrete mixtures. An interesting trend is the use of nanomaterials specifically to reinforce the interface between coarse aggregates and cement paste instead of the conventional direct admixing strategy. For instance, recent studies have entailed the application of coarse aggregates with nanomaterial-impregnated cement pastes containing materials such as graphene oxide and nanosilica. Besides aiming to improve the mechanical properties of high-volume fly ash concrete, this approach also targets improving the transportation properties by effectively reducing permeability. The study indicates that while the mechanical properties cannot be significantly enhanced with direct admixture techniques, transport property-related advantages can result in more sustainable, durable, and stronger concrete structures, paving the way towards a new sustainable construction era (Du et al., 2018). The global push towards sustainability in building is also shaping future trends in nanomaterials research, specifically with energy efficiency in mind. The rapid growth of urban population and the resultant increase in energy usage underscore the urgent necessity for new building materials that will minimize the environmental footprint of the construction industry. The current estimates point to the energy consumption demand of commercial and residential buildings as rising significantly by 2050 and, as such, it becomes a necessity to look for innovation in materials that improve the energy efficiency of buildings. Nanomaterials, as a result of their unique characteristics, are a pioneering solution. Research now focuses on integrating nano and bio technologies in construction activities to energy-efficient buildings and reduce greenhouse gas emissions. These advancements could also potentially lower energy consumption but also have the potential for innovation and patenting for civil engineers, ultimately leading to high-tech construction entrepreneurship involving green construction solutions (Pacheco-Torgal et al., 2017). In addition, as there continue to be improvements in nanomaterials, future studies will be dictated significantly by interdisciplinary cooperation. With the fusion of material science, engineering, and environmental study specialists, one can considerably broaden the reach of creating new applications in concrete technology. Scientists will continue exploring how various nanomaterials may synergistically interact to enhance the properties of cement composites, not only mechanical strength but also endurance against environmental aggressions such as freeze-thaw and chemical attack. Systematic understanding of these materials at the nanoscale will be essential to enable new functionality in concrete. This collaborative approach can further enhance the achievement of nanomaterials' potential use in a sustainable way, where the innovations contribute to the fulfillment of global environmental goals and the enhancement of climate change and urbanization resilience of built environments (Du et al., 2018; Pacheco-Torgal et al., 2017).

5.6.1. Innovations in Nanomaterial Synthesis

The manufacturing of nanomaterials has also seen great progress that increases their application in numerous industries, particularly in concrete and cement. Advanced methods such as hydrothermal synthesis and sol-gel allow for the creation of nanoparticles with specific properties such as enhanced strength and durability. This progress allows for the inclusion of nanomaterials in concrete blends to increase the performance characteristics of the end product. For instance, the incorporation of nanosilica can significantly enhance the mechanical strength and chemical resistance to corrosion, addressing major concerns in terms of concrete structure durability and sustainability. Through the optimization of the methods used in nanomaterial synthesis, researchers are able to contribute to more energy-efficient and less destructive building processes, eventually assisting in complementing the overall objective of reducing greenhouse-gas emissions from the construction sector (Pacheco-Torgal et al., 2017). Additionally, the development of green synthesis techniques is increasingly becoming more salient in nanomaterials, particularly as the construction industry is battling the effects of heightened energy consumption. Green synthesis techniques based on bio-based materials, in addition to decreasing the environmental impact, promote the circular economy in construction processes. For instance, utilization of waste byproducts or crop process bio-waste as raw materials in the synthesis of nanomaterials not only provides a green alternative but also reduces dependency on virgin raw materials. This shift towards more environmentally friendly synthesis processes is significant given that the construction industry is set to undergo profoundly increased energy demands by 2050, underscoring the need for energy-efficient means in building construction (Pacheco-Torgal et al., 2017). Through embracing such green methods, the use of nanomaterials in concrete can pave the way for environmentally friendly building solutions that can meet future demands. The application of nanomaterials in cement and concrete is also supported by the ongoing research aimed at emphasizing the use of enhanced properties in these products. Recent studies have focused on managing the size, shape, and surface characteristics of nanoparticles to optimize their effectiveness when they are introduced into concrete matrices. A case in point is the use of graphene oxide, where it increases tensile strength and electrical conductivity while delivering functionalities that surpass the performance of traditional standards. In addition, collaboration between academia and industry in advancing the synthesis of such materials in fostering a high-tech construction startup growth-enabling culture is critical. This synergy is required to initiate innovations that address current environment standards and efficiency requirements, a case in point being nanomaterials' position in revolutionizing sustainable construction in the future.

5.6.2. Development of Smart Concrete Technologies

The technology of smart concrete is a breakthrough technology in the construction sector that has been triggered by the application of nanomaterials. Smart materials are improving the efficiency and longevity of concrete to unprecedented levels of functionality and performance. Technologies such as self-sensing concrete, where nanomaterials incorporated into it allow it to monitor the health of buildings, show the potential of such technologies. Through piezoelectric nanomaterials, smart concrete can detect stress and strain, providing timely information necessary in maintaining structural integrity through time. Aside from possibly reducing maintenance costs, this integration also lengthens the lifespan of concrete structures, which is why nanotechnology is a key driver in the future of construction material development (Malik et al., 2023). Moreover, the fusion of artificial intelligence and smart concrete technologies is transforming construction and maintenance practices. Machine learning algorithms can be utilized to analyze data captured by smart concrete, optimizing resource utilization as well as predictive maintenance activities. The advancement in data capture technologies, including smart sensors and vision systems, facilitates complete monitoring throughout a structure's lifetime from design and construction to operation and ultimate decommissioning. This system-wide consideration not only enhances the durability and resilience of concrete structures but also validates the assumptions of the circular economy through the reduction of waste and energy consumption during the entire life cycle of the concrete (Baduge et al., 2022).

Moreover, the use of nanotechnology in smart concrete ensures environmental sustainability, an ever-growing concern in construction processes today. Nanomaterial-containing new formulations can enhance the durability of concrete against environmental factors such as moisture, thermal fluctuations, and chemical weathering, lowering the number of repairs and resources needed in the long run. Such durability consequently promotes a more sustainable process, minimizing the environmental footprint associated with conventional concrete production and maintenance. As the smart concrete technologies continue to evolve, more work and inter-disciplinary research in areas like material science and environmental engineering will be required to harness the full extent of benefits that such technological developments have in store (Malik et al., 2023) (Figure 5.15).

Figure 5.15. Development of smart concrete technologies.

Source: https://blog.kryton.com/wp-content/uploads/2020/03/Copy-of-IMG_4054-scaled.jpg.

5.6.3. Integration of Nanomaterials with Other Sustainable Practices

The application of nanomaterials in cement and concrete is a groundbreaking way to achieve sustainability in the construction sector. With the addition of the new materials, engineers can enhance the mechanical strength and durability of concrete. The added strength leads to the use of fewer materials, thus minimizing total environmental impact. The integration of nanomaterials allows for the production of high-strength concrete that will be capable of handling harsher conditions, reducing maintenance needs and prolonging infrastructure lifespan. As the world construction sector is being urged more and more to be more innovative and less carbon-emitting as a result of expanding populations and energy needs, these nanomaterial breakthroughs make up a crucial part of sustainable building methods. The synergy of nanotechnology with traditional materials not only builds resource efficiency but also sets the industry on a trajectory toward sustainability and adaptability to the threat of climate change (Pacheco-Torgal et al., 2017). In addition, the use of intelligent technologies combined with nanomaterials enhances the sustainability of cement and concrete applications. Advances in micro-electromechanical systems (MEMS) have introduced new types of sensing capability that can easily be integrated into building materials. This innovation allows for autonomous structural health monitoring (SHM), a critical component for long-term infrastructure integrity maintenance. By combining MEMS technology and nanomaterials, it is now possible to monitor the health of concrete in real-time, facilitating proactive maintenance and prevention of catastrophic failure. Such systems can improve safety, extend service life, and save considerable life cycle costs by optimal utilization of resources. Where stakeholders continue to seek to apply sustainable practices, the advantage of smart structures made possible by nanomaterials is a cost-effective means of addressing critical environmental concerns within the construction industry (Buss et al., 2013). The combination of nanotechnology and traditional sustainability practices further serves to maximize energy efficiency in the built environment. For instance, the incorporation of nanomaterials will lead to advanced thermal insulation properties for concrete to reduce heating and cooling demands within buildings. Since energy consumption by the building sector is going to grow exponentially, it is crucial that nanotechnology be linked with energy-saving measures. The holistic approach of sustainable construction could be complemented by incorporating these advancements with eco-friendly energy resources, i.e., solar panels integrated into concrete buildings. Not only do they make buildings energy-efficient, but they also produce clean energy. With the evolution of innovations that also serve material efficiency and energy consumption, the addition of nanomaterials opens up to a sustainable future for construction activities in response to the needs of a growing population (Pacheco-Torgal et al., 2017).

SUMMARY

The addition of nanomaterials to cement and concrete technology has introduced promising breakthroughs in the field of construction materials. With their unique properties, nanomaterials such as graphene oxide and nanosilica have proved to enhance the cement paste-coarse aggregate interface. This particular use, as compared to direct mixing methods, reflects the innovative techniques that are increasingly employed to optimize material performance. Evidence indicates that nanomaterial coatings not only influence high-volume fly ash (HVFA) concrete's transport properties by reducing permeability but also hold promise to increase durability against environmental stresses. Such advancements represent a shift in how traditional concrete mixes can be adjusted to meet current requirements for resilience and sustainability. Moreover, recent research has confirmed the necessity of fabricating well-dispersed nanocomposites to maximize the performance of carbon-based nanomaterials, including carbon nanotubes (CNTs) and carbon nanofibers (CNFs). *In situ* synthesis of these nanomaterials on matrix particles like silica fume and cement is a robust method for improving mechanical properties and overall material performance. This method not only creates strong composite structures but also embodies the essence of green practice by making optimal use of available materials through advanced nanotechnology. As pointed out in the findings, the process of producing CNTs and CNFs results in chemical reactions that can further be utilized to alter the properties of cement, thereby facilitating the development of new-generation concrete systems. Lastly, the significance of these advancements is two-way, where the applications of nanomaterials in construction address both performance and environmental issues. Whereas some research, like the varied outcomes of compressive strengths of HVFA concrete with coated aggregates by nanomaterials, pave the way for future studies with other combinations of nanomaterials and processes. Greater study into the mechanistic attributes of these materials, as observed by techniques like scanning electron microscopy, will provide additional insight into their advantages and limitations. Therefore, though development within the industry persists, the reinforcement of cement and concrete with nanomaterials is a powerful innovation to enhance durability, sustainability, and efficiency in modern-day construction practices.

The use of nanomaterials, particularly graphene oxide (GO), in cement and concrete has been shown to show notable improvement in their overall performance and durability. Phase I studies to develop graphene oxide-modified pervious concrete with coal fly ash as the sole binder were conducted. Follow-up research aimed at establishing stormwater infiltration capacity, environmental durability in adverse environmental conditions, and the use of advanced analytical instruments to fully explain the modified binders. Results depicted that the addition of GO improved stormwater quality through the reduction of concentrations of numerous pollutants such as copper and nitrates. Besides that, the graphene oxide-modified concrete was very resistant to freeze/thaw and saline environments, and this demonstrates the position that can be played by GO as a necessary additive to enhance cementitious materials in terms of functionality as well as sustainability (Shi et al., 2018). Besides graphene oxide, there are other nanomaterials of similar interest for similar aims. For instance, nanostructured materials have been suggested to be incorporated into asphalt in pavements to improve the mechanical properties of the latter. The sole production of Norwegian asphalt each year indicates the economic importance and necessity for development of this industry. Through analyzing binders, researchers have identified that additives could be employed to enhance tensile strength or flexibility of asphalt mixtures. Having these two capabilities is critical in preventing premature failure of asphalt roads, which crack and get ruts. Interestingly, studies have shown that the addition of graphene in asphalt mixtures can result in enhanced resistance to permanent deformation, and this can have implications in its application not only in cement but also in the performance of road materials (Brcic et al., 2025). The combined body of studies of nanomaterials in cement and asphalt indicates a continued trend towards embracing the use of state-of-the-art materials for the improvement of public infrastructure. As urban areas struggle with increasing pressures from environmental stressors, there is unequivocally a need for new solutions that can extend the life and longevity of building materials. The use of nanomaterials like graphene oxide in pervious concrete and asphalt has major potential to address these goals, leading to improved resistance to climate stresses. Research continues to unravel the undiscovered potential of these nanomaterials, which have the ability to serve as the basis for developing high-performance asphalt and concrete mixtures ideally balancing performance and nature protection (Brcic et al., 2025; Shi et al., 2018).

REFERENCES

1. Ahmad, M., Almeida, M., Angellier, N., Angles, M. N., Araki, J., Ardanuy, M., Arsene, M. A., et al. (2017). Macro and nano dimensional plant fiber reinforcements for cementitious composites. https://core.ac.uk/download/154277082.pdf (accessed on 25th September 2025).
2. Amornsakchai, T., & Goh, K. L. (2025). Outcome and conclusions from the 2024 JEC Forum Southeast Asia, Bangkok. https://core.ac.uk/download/617061919.pdf (accessed on 25th September 2025).
3. Bodnárová, L., Foldyna, J., Foldyna, V., Hlaváček, V., et al. (2017). Effects of continuous and pulsating water jet on CNT/concrete composite. https://core.ac.uk/download/161955767.pdf (accessed on 25th September 2025).
4. Brcic, H. (2025). Investigation of the rheological properties of asphalt binder containing graphene nanoplatelets. https://core.ac.uk/download/pdf/154673111.pdf (accessed on 25th September 2025).
5. Buss, A., Ceylan, H., Ceylan, H., Gopalakrishnan, K., et al. (2013). Highway infrastructure health monitoring using micro-electromechanical sensors and systems (MEMS). https://core.ac.uk/download/38934593.pdf (accessed on 25th September 2025).
6. Carneiro, J. A. O., Fernandes, F., Jesus, C. M. G., Palha, A., et al. (2014). Experimental analysis on the functional properties of rendering mortars with superficial addition of TiO_2 nanoparticles. https://core.ac.uk/download/55630605.pdf (accessed on 25th September 2025).
7. Cwirzen, A., Habermehl-Cwirzen, K., Karppinen, M., Kauppinen, E., et al. (2009). Synthesis of carbon nanotubes and nanofibers on silica and cement matrix materials. https://core.ac.uk/download/80716883.pdf (accessed on 25th September 2025).
8. De Schutter, G., El Cheikh, K., Jiao, D., Lesage, K., et al. (2019). Structural build-up of cementitious paste with nano-Fe_3O_4 under time-varying magnetic fields. https://core.ac.uk/download/227041605.pdf (accessed on 25th September 2025).
9. De Schutter, G., Han, H., Wu, K., Xu, Z., et al. (2019). Supported ITZ modification efficiencies via surface coating nanoparticles on aggregate and its influence on properties. https://core.ac.uk/download/322829250.pdf (accessed on 25th September 2025).
10. Du, S., & Shi, X. (2018). A targeted approach to high-volume fly ash concrete pavement (Phase I). https://core.ac.uk/download/185623461.pdf (accessed on 25th September 2025).

11. Du, S., Wu, J., AlShareedah, O., & Shi, X. (2019). Nanotechnology in cement-based materials: A review of durability, modeling, and advanced characterization. Nanomaterials, 9, 1213–1213. https://doi.org/10.3390/nano9091213 (accessed on 25th September 2025).
12. Duart, M. A. (Marcelo), Mortari, S. R. (Sérgio), & Rodrigues, O. E. (Oscar). (2018). Carbonic nanoparticles and C-S-H insertion into cementitious nanocomposite. https://media.neliti.com/media/publications/237458-carbonic-nanoparticles-and-c-s-h-inserti-0cd4709e.pdf (accessed on 25th September 2025).
13. Faltin, B. (2022). Performance of concrete with different cement fine nesses and nano-activators. https://core.ac.uk/download/533927329.pdf (accessed on 25th September 2025).
14. Hernandez Moreno, S., Solache de la Torre, S. C., et al. (2017). Nano-technological products in architecture and construction. https://core.ac.uk/download/154796653.pdf (accessed on 25th September 2025).
15. Kristombu Baduge, S., Thilakarathna, P. S. M., Perera, J. S., Arashpour, M., Sharafi, P., Teodosio, B., Shringi, A., et al. (2022). Artificial intelligence and smart vision for building and construction 4.0: Machine and deep learning methods and applications. Automation in Construction, 141, 104440–104440. https://doi.org/10.1016/j.autcon.2022.104440 (accessed on 25th September 2025).
16. Krystek, M., Ciesielski, A., & Samorì, P. (2021). Graphene-based cementitious composites: Toward next-generation construction technologies. Advanced Functional Materials, 31. https://doi.org/10.1002/adfm.202101887 (accessed on 25th September 2025).
17. Lambert, P., Liu, C., Wu, Y. Y., Zhang, J., et al. (2020). Effect of graphene oxide nanosheets on physical properties of ultra-high-performance concrete with high volume supplementary cementitious materials. https://core.ac.uk/download/305116797.pdf (accessed on 25th September 2025).
18. Li, Z. (2025). Fly Ash Based Geopolymer for Concrete Infrastructure Development, Characterization, Application, And Life Cycle Assessment. https://core.ac.uk/download/644558338.pdf (accessed on 25th September 2025).
19. Malik, S., Muhammad, K., & Waheed, Y. (2023). Nanotechnology: A revolution in modern industry. Molecules, 28, 661–661. https://doi.org/10.3390/molecules28020661 (accessed on 25th September 2025).
20. N/A. (2017). SciTech News, 71(1). https://core.ac.uk/download/83049835.pdf (accessed on 25th September 2025).

21. Pacheco-Torgal, F. (2017). High tech startup creation for energy efficient built environment. https://core.ac.uk/download/76177771.pdf (accessed on 25th September 2025).
22. Pintea, A., Romero, A., Tabrea, A., Hargreaves, B., Hazebrouck, B., Brouwer, D., BSI, et al. (2015). Strategies, methods and tools for managing nanorisks in construction. https://core.ac.uk/download/80047363.pdf (accessed on 25th September 2025).
23. Robert, M. (2017). Key parameters of pulsed laser deposition for solid electrolyte thin film growth. https://core.ac.uk/download/131211733.pdf (accessed on 25th September 2025).
24. Rodríguez, B., Hewlett, P., Ochoa-Botero, J. C., Quintero, J. H., Mendoza, O., Mendoza-Reales, O. A., Toledo-Filho, R. D., et al. (2017). Influence of MWCNT/surfactant dispersions on the mechanical properties of Portland cement pastes. https://core.ac.uk/download/159381243.pdf (accessed on 25th September 2025).
25. Rodriguez, C. R., Savija, B., Ye, G., Yuan, Y., et al. (2018). Proceedings of the 4th International Conference on Service Life Design for Infrastructure (SLD4). https://core.ac.uk/download/322827780.pdf (accessed on 25th September 2025).
26. Shi, X., & Xu, G. (2018). Environmentally friendly pervious concrete for treating deicer-laden stormwater: Phase II. https://core.ac.uk/download/185623460.pdf (accessed on 25th September 2025).

CHAPTER 6

NANOMATERIALS IN STEEL AND ALLOYS

Contents

INTRODUCTION

There has been an increasing demand for advanced materials in industries, generating significant interest in nanomaterials, particularly where alloy and steel usage is involved. Nanomaterials, characterized by unique features and high surface area-to-volume ratio, have been shown to possess tremendous potential for enhanced performance and longevity of regular steel. With the focus of the industry to enhance mechanical properties while reducing weight, addition of nanostructured components offers transformative opportunities. The unusual properties of materials at the nanolevel provide new technologies such as higher strength, enhanced corrosion resistance, and improved wear characteristics, and nanomaterials are of particular interest for research and development in metallurgy. One of the important research areas is phase transformation in steel systems at the nanolevel. Some of the latest studies have shown that severe plastic deformation can induce a nanocrystalline microstructure in steels, which defies conventional material behavior principles.

Figure 6.1. Nanomaterials in steel.

Source: https://eoxs.com/wp-content/uploads/2022/03/sheets.jpg.

Reverse transformation from body-centered cubic ferrite to face-centered cubic austenite, in specific, has significant manufacturing process and product application implications. Results of high-resolution transmission electron microscopy and atom probe analysis indicate the significant role of nanostructure and shear stresses to these transformation processes, emphasizing the need for more investigations in this field (Fecht et al., 2006). This fundamental paradigmatic shift in the comprehension of nanomaterials positions them as primary agents in advancing steel technology. Nanomaterials' investigation also encompasses the production of composite alloys, wherein nano-reinforcements are added to offer enhanced mechanical performance. Solidification through high shear and equal channel angular extrusion have emerged as promising methods for the production of nanocomposites with homogeneous nanoparticle dispersion. Work on magnesium matrix composites reinforced with hydroxyapatite, for instance, has indicated improved electrochemical and mechanical responses with problems of particle segregation during processing.

The nanocomposites formed possess fine grain microstructures and uniform nanoparticle array distributions, leading to enhanced mechanical properties (Huang et al., 2014). As more research unravels the secret of nanomaterial integration, the future for new applications of steel and alloys is vast, with the potential for next-generation materials in engineering solutions (Figure 6.1).

Definition of Nanomaterials

In nanomaterials, a definition extends beyond the size to encompass the unique properties found at the nanoscale. For the most part, nanomaterials are materials whose dimensions are at least 1–100 nanometers. This size has new physical and chemical properties like more surface area, more reactivity, and novel mechanical properties. For instance, for steel and alloys, the addition of nanomaterials can lead to increased strength and corrosion resistance both of which are essential for many applications such as automotive and aerospace industries. Along with this, creation of nanostructured forms from conventional materials makes it possible to achieve phenomena which are not achievable by conventional processing technologies, hence making nanomaterials important in the advancement of material science. The application of nanomaterials in steel and alloy is brought into perspective by their ability to transform and change the structural characteristics and performance. From studies, it is clear that there can be the inversion transformation of body-centered cubic ferrite into face-centered cubic austenite via severe plastic deformation and the evolution of a nanocrystalline structure. This effect is not seen in regular deformation of coarse-grained irons and steels, which again underscores the significance of nanostructured materials (Fecht et al., 2006). These alterations are significant, as they allow mechanical properties to be enhanced without compromising valuable attributes such as ductility. These developments are typical of the broader picture of nanomaterials, where their unique behavior at the nanoscale applies to increased performance in real-world applications. However, as evident as the future of nanomaterials in steel and alloys is, there needs to be raised concern over potential health and environmental effects as well. Emergence of nanotechnology, characterized by rapid rates of development and numerous applications, requires an urgent address on occupational health and safety control of nanoparticles (Adamec et al., 2019). Workers who work with and produce nanomaterials will be at risk, and this demands tight monitoring and assessment systems. Equilibrium is thus crucial when it comes to incorporating nanotechnology into alloys and steel so that innovation does not adversely affect the environment and human health. This dual focus on progress and responsibility is critical to the sustainable nanomaterial development for applications in modern industry.

Importance of Nanomaterials in Metallurgy

Use of nanomaterials in metallurgy mainly enhances the mechanical character of steel and alloys, leading to increased performance in applications. Nanoscale reinforcement results in better grain structure, which enhances yield strength, ductility, and toughness, characteristics that play an imperative role in the performance of metallic materials. For instance, nanocrystalline techniques create dynamic recrystallization leading to a gradient layer on the surface of iron with grain sizes from micro to nanoscale. This change creates a stability between strength and toughness necessary in materials subjected to high stress and strain conditions.

Figure 6.2. Nanomaterials in metallurgy in steel.

Source: https://external-content.duckduckgo.com/iu/?u=https%3A%2F%2 Fmo-daai.com%2Fwp-content%2Fuploads%2F2022%2F05%2FiStock-939202014-scaled.jpg&f=1&nofb=1&ipt=63768dba4d0be64084f05b40eb2b0dd72782e96411693294e716712cb2451de5.

The requirement of light, high-strength materials in aerospace and automobile industries emphasizes the role of nanomaterials that provide novel solutions to achieve the desired mechanical behavior at lower overall weight. Moreover, the ability of nanomaterials to create solid solutions and solidify heterogeneous phases enhances metallurgical operations such as alloying. Mechanical alloying of iron with zinc has been found through experiments to produce an extremely disordered Fe(Zn) solid solution with nanograins, proving the effectiveness of nanotechnology in maximizing microstructures. The decrease of crystallite size into the nanometer range causes tremendous improvement in properties such as coercivity and magnetization in Fe-Zn alloys. These advances are indispensable to the development of functional materials with specialized applications, such as magnetic materials and catalysts, which require tight control over microstructural properties. By utilizing the unique properties of nanomaterials, metallurgists can design advanced materials to respond to the needs of modern engineering applications. Nanomaterials' contribution to metallurgy goes beyond mechanical innovation; it also enables innovation in material processing technologies and applications. The employment of nanostructured materials reduces raw material requirements and processing temperatures, therefore making metallurgy more eco-friendly. Moreover, the new physical and chemical characteristics of nanomaterials enable ground-breaking applications, for instance, enhanced oxidation resistance and increased corrosion protection. By adjusting microstructure at the nanoscale, researchers are far ahead of creating high-performance steel and alloy replacements. As nanotechnology continues to improve, the connectivity of metallurgy and nanomaterials provides tremendous innovation opportunities that ultimately redefine the material sciences and engineering in the context of evolving industrial requirements (Figure 6.2).

6.1. PROPERTIES OF NANOMATERIALS

The properties of nanomaterials are also very important in the use of nanomaterials in steel and alloy production, particularly by virtue of enhanced mechanical strength and structural integrity. Upon comprehensive plastic deformation of steel, the transformation of body-centered cubic ferrite and face-centered cubic austenite may lead to the evolution of a nanocrystalline structure, not observed in conventional coarse-grained conditions. This phase transformation is triggered by unique shear stresses and inherent properties of nanostructures and plays a critical role in achieving high-performance in alloy systems. The study shows that microstructural property control at the nanoscale can provide a pathway toward triggering new degrees of strength and toughness in steel, and opening the door to innovative use areas demanding more durability and efficiency in materials (Fecht et al., 2006). Furthermore, nanoparticles dispersion in a metallic matrix can greatly enhance composites mechanical properties. As an illustration, hydroxyapatite (HA) particle-reinforced magnesium matrix composites have been found with a dramatic improvement in mechanical and electrochemical characteristics, even eliminating issues like particle segregation. Utilizing advanced fabrication techniques like high shear solidification combined with equal channel angular extrusion (ECAE) has been found to improve grain structures in these composites significantly. This process not only enables the uniform dispersion of nanoparticles but also leads to the creation of real nanoparticles in the magnesium alloy. The created microstructural enhancements directly contribute to better mechanical performance, exhibiting the central role of nanomaterials in driving innovation in metal matrices (Huang et al., 2014). The significance of nanomaterials extends beyond mere mechanical enhancement; they serve to ensure superior corrosion resistance and overall integrity of steel and alloy components. With the increased demand for materials to function in aggressive conditions, inclusion of nanostructured particles has emerged as a feasible approach. Application of nanomaterials enables one to reduce corrosive element diffusion rates, hence the longevity of alloys is considerably enhanced. Besides, the modified surface properties may facilitate enhanced bonding and adhesion in composite materials, thereby leading to enhanced effectiveness in structural applications. Such advancements underscore the groundbreaking potential of nanomaterials in transforming the properties of traditional steel and alloy systems and opening new grounds for material science innovation (Figure 6.3).

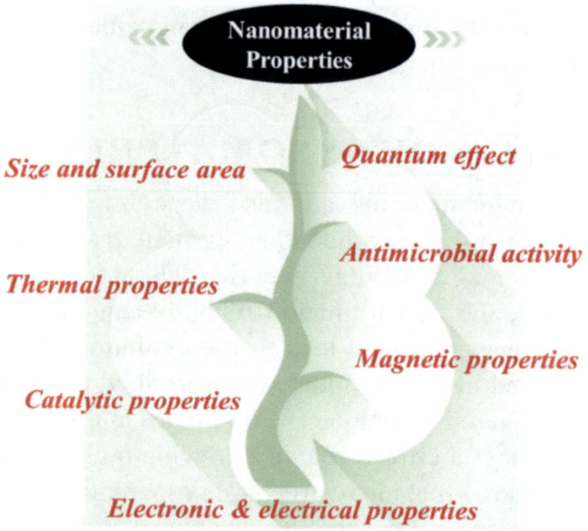

Figure 6.3. Properties of nanomaterials.

Source: https://media.springernature.com/lw685/springer-static/image/chp%3A10.1007%2F978-981-19-1384-6_1/MediaObjects/518919_1_En_1_Fig2_HTML.png.

6.1.1. Size-Dependent Properties

The observation of size-dependent properties in nanomaterials reveals wide-ranging variations in mechanical properties that distinguish them from bulk materials. As sizes of metal structures are reduced to the nanolevel, the influence of surface atoms becomes more consequential, leading to increased strength and altered ductility. For example, nanoporous Copper-Nickel (npCN) consists of a unique bicontinuous network structure, and with its use, materials with feature sizes ranging between 13 and 40 nanometers can be created. Such a variation can be extremely controlled using thermal annealing processes. Notably, continuous compression tests show that npCN also displays ductile deformation behavior but very high strength levels compared to traditional macroporous metal foams, suggesting the possibility of such nanoscale alterations in tailoring the performance of materials in engineering applications (Lührs et al., 2018). In addition to mechanical properties, compositional and thermodynamic properties of alloys like iron-chromium show intricate size-dependent effects. Experiments indicate that the corrosion resistance of the iron-chromium alloy is intrinsically dependent on the concentration of chromium, which must be within a narrow range of atoms between 9 and 13 percent. Surface effects are dominant at the nanoscale, leading to widespread enrichment of chromium on the surface, particularly when the concentration in the bulk is low. Monte Carlo simulation elucidates the complex interdependence between phase separation and local transitions that result in drastic changes in surface composition. This identifies the need to consider attention to nanoscale phenomena in alloy design, especially in pursuing optimal performance under changing environmental conditions (Levesque et al., 2013). In addition to mere performance factors, such size-dependent properties have larger implications. They provide windows to innovative solutions in the construction of advanced nanomaterials that can exhibit tailorable properties for specific applications. For instance, corrosion resistance or mechanical properties can be enhanced without using traditional alloying techniques through control of the morphology at the nanoscale. Such advancements hold promise in the majority of industrial applications, e.g., aerospace and automotive, where material durability and functionality are paramount. As researchers continue to uncover the secrets of size-dependent properties, the introduction of nanostructured materials into steel and alloys offers an evolutionary path for redefining engineering materials and their applications in the contemporary world (Levesque et al., 2013; Lührs et al., 2018).

6.1.2. Enhanced Mechanical Strength

Application of nanomaterials in steel and alloys has been extensively studied for the ability to enhance mechanical strength, producing materials with better performance compared to their conventional analogs. Among the greatest strengths of nanomaterials is that they can enhance microstructures, thereby leading to improved mechanical properties, such as increased yield strength and improved ductility. For example, in magnesium matrix composites, high-shear solidification processes can achieve a uniform and fine grain structure with a homogeneous nanoparticle dispersion of hydroxyapatite. The fine-grained microstructure not only improves strength but also minimizes the unwanted effects of particle segregation, a common problem in composite materials (Huang et al., 2014). This refinement process ultimately results in a potent matrix that displays enhanced mechanical strength, needed for demanding structural applications across many industries. In attempting to identify the mechanisms for the enhanced strength of nanostructured materials, several recent studies have provided insights into the role of severe plastic deformation in reorganizing material microstructures. Especially, it has been exhibited by studies that the reverse transformation of body-centered cubic ferrite to austenite face-centered cubic under massive plastic deformation can yield a nanocrystalline, finer structure. This is not usually seen in conventional deformation of coarse-grained steel. Instead, they require special conditions—namely, the application of large shear stresses and the existence of a nanostructured form of iron, which are critical factors to facilitate such transformations (Fecht et al., 2006). Such an understanding of nanomaterial behavior has far-reaching consequences for engineering applications, where mechanical strength and ductility are of the utmost concern. Moreover, the enhancement of mechanical properties through nanomaterials is not confined to a single process or technique but rather encompasses a variety of new methods, including the development of new alloy compositions. Through manipulation of alloying elements and employing nanoscale reinforcement techniques, scientists have been able to design mechanical properties of steel and alloys to fulfill required performance specifications. For instance, incorporation of nanostructured particles has the ability to create an increased interfacial bond density in the matrix, thereby enhancing load transfer efficiencies and overall material toughness. Additionally, synergistic effect of blending multiple nanoscale reinforcements can create new mechanical properties that are not possible with conventional materials alone. Overall, nanomaterials research in steel and alloys is a pioneering effort toward the realization of enhanced mechanical strength in various applications in engineering, leading toward the material science of the future.

6.1.3. Improved Corrosion Resistance

Addition of nanomaterials to steel and alloys has significantly enhanced corrosion resistance, a key consideration for increasing the working life and dependability of the materials. Nanostructured materials such as hydroxyapatite (HA) and carbon nanotubes not only improve mechanical properties but also provide good corrosion barriers against corrosive species. Recent findings indicate that nanocomposite materials are capable of creating more homogeneous microstructures which do not permit the propagation of corrosion channels. For instance, Mg matrix nanocomposites with HA particles have good corrosion resistance because they have finer microstructures from advanced processing techniques like equal channel angular extrusion (Huang et al., 2014). Increased particle distribution homogeneity means fewer weak areas that can be potential sites for corrosion initiation, making such nanocomposite materials applicable for applications in harsh environments. Moreover, corrosion resistance in metal matrix composites is also increased by synergistic effects imparted by some nanomaterials. Carbon nanotube (CNT) reinforcement is found to decrease the wear rate and friction coefficient of aluminum composites as compared to unreinforced pure aluminum, particularly under varying load conditions. These features reveal that CNTs not only enhance mechanical properties but also serve as a corrosion protective agent when utilized in aluminum matrices (Basu et al., 2016). The ability of these nanomaterials to distribute stress leads to corrosion susceptibility reduction in localized areas.

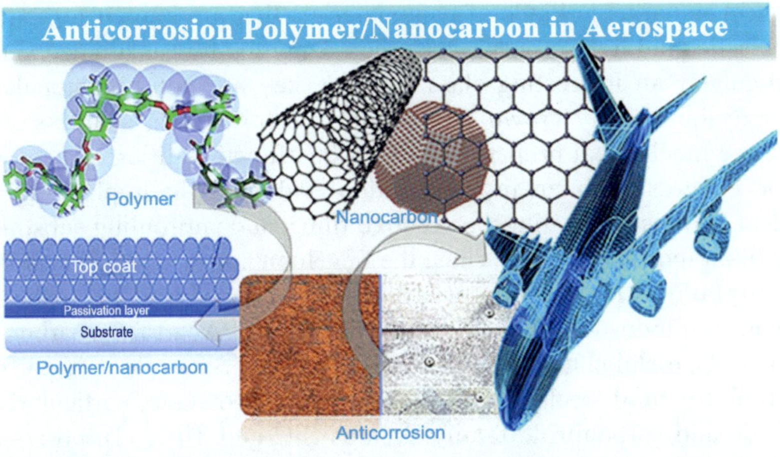

Figure 6.4. Improved corrosion resistance of nanomaterials in steel and alloys.

Source: https://pub.mdpi-res.com/applnano/applnano-04-00008/article_deploy/html/images/applnano-04-00008-ag.png?1684224811.

By minimizing environmental wear and degradation, CNT use lies within the greater goal of preserving durability in applications ranging from automobile to aerospace where life and performance are paramount. Testing of different nanomaterial reinforcements continues to demonstrate the potential benefits of using nanotechnology to combat corrosion. Current research indicates that the high shear solidification with nanoparticle reinforcements results in improved corrosion resistance materials via a denser and homogenous microstructure. Addition of HA nanoparticles into Mg alloys has been reported by researchers to not only enhance the mechanical properties but also substantial corrosion resistance improvement due to a tightly packed structure that blocks the penetration of corrosive media (Huang et al., 2014). This evidence underscores the revolutionary impact of nanomaterials and suggests that optimum selection of reinforcement can revolutionize the metallurgical industry. With further development, embedment of nanomaterials can redefine corrosion limits in steel and alloys, hence addressing some of the problems encountered in various industrial environments (Figure 6.4).

6.2. TYPES OF NANOMATERIALS USED IN STEEL AND ALLOYS

Utilization of nanomaterials in steel and alloys has made significant contributions to their mechanical and functional properties. Carbon nanotubes (CNTs) are one of the most remarkable kinds of nanomaterials that are known for extremely strong, elastic, and light properties. Carbon nanotube forests aligned vertically are particularly an interesting class because they are multifunctional. These forests may be directly grown on metallic substrates, i.e., stainless steel, to enhance the mechanical properties and surface functionalities of steel. CNTs' behavior towards high energy absorption and super-hydrophobicity make them ideal options for coatings, protective films, and antifouling substrates for devices like pipes and blades. Thus, the development of CNT forests that are specifically tailored for steel improves durability along with adding to the overall performance of industrial-alloys (Daraio et al., 2018). Apart from carbon-based nanomaterials, metal nanopowders also play an important role in imparting the features of structural steels and alloys. These nanopowders, particularly from the carbide and carbonitride family such as SiC and Ti(C, N), possess great potential for transforming low-alloyed steels. Investigations have shown that introduction of these nanodispersed mixtures via plasma-chemical synthesis induces an excellent refinement of grain structures, which can enhance strength and toughness. Specifically, the shifts have resulted in 2–3.5 times the reduction of grain size, promoting a stable alloy structure. Through improvement in crystallization characteristics and the creation of macro and microstructures, these nanopowders facilitate the production of high-quality steel components that satisfy strict mechanical specifications (Dzhur et al., 2017). The efficiency and flexibility of nanomaterials in steel and alloys are not merely restricted to reinforcing structural strength; they also contribute to the augmentation of functional properties. Nanostructured materials, when strategically used, can lead to applications such as super-hydrophobic coatings that reduce friction or corrosion, eventually increasing the lifespan of steel components in various environments. The intricate relationship between the nanomaterials and the substrate characteristics highlights the importance of conducting extensive research into how these materials can be tailored for a particular purpose. As these linkages are increasingly well understood, the potential for using advanced nanomaterials in pilot projects toward high-stakes markets, such as aerospace and automotive, will further expand, pushing the next generation of steel and alloy production toward greater efficiency and sustainability (Daraio et al., 2018; Dzhur et al., 2017) (Figure 6.5).

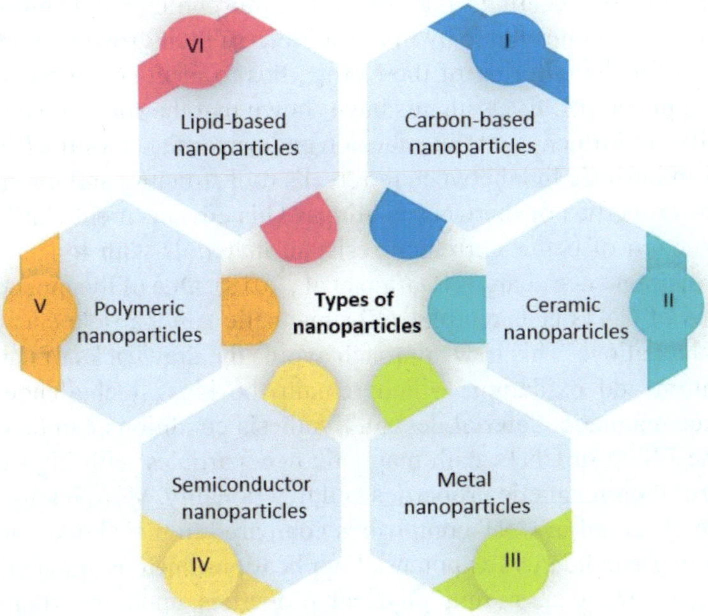

Figure 6.5. Types of nanomaterials used in steel and alloys.

Source: https://www.researchgate.net/publication/348793693/figure/fig3/AS:1023134325014532@1620945640625/Commonly-used-types-of-nanoparticles.jpg.

6.2.1. Carbon Nanotubes (CNTs)

Over the past few years, CNTs have emerged as key contributors to reinforcing steel and alloy materials, demonstrating enormous potential with their outstanding structural and functional properties. Their outstanding strength-to-weight ratio, good thermal conductivity, and outstanding electrical properties render them valuable additions in metallic matrices. CNTs have the potential to increase the mechanical properties of steel as reinforcement materials, which produces composites with greater tensile strength and durability. These are crucial in applications such as aerospace and automotive, where material efficiency is synonymous with performance gains and fuel economy. Furthermore, the multicompability of CNTs makes them of great interest as they can be synthesized using various methods, making it easy for them to be inserted into metal matrices. CNT forests growth on metallic substrates, e.g., stainless steel, is a recent strategy to enhance surface features. These vertical arrays of CNTs are multifunctional in nature, being characterized by high energy absorption, compressive capacity, and super-hydrophobicity, and are thus ideal choices as protective coatings and layers. The morphology and composition of the CNTs are both controllable by the conditions of their growth, and this offers the potential for the tailoring of their properties to meet specific demands for particular applications. Experiments have shown that the substrate morphology has a significant influence on the adhesion and performance of the CNT forests and hence an intrinsic link between the steel's root structure and the subsequent mechanical properties of the CNT coatings. This advancement can lead to the mass production of better corrosion-resistant materials with longer lifecycles and lower maintenance charges (Daraio et al., 2018). One of the most promising applications of CNTs is as templates for magnetic nanoparticle encapsulation, e.g., Fe1-xNix alloys. This new approach avoids the drawbacks of nanoparticle agglomeration and oxidation, which remain the biggest challenge in high-performance magnetic material design. Synthesis conditions can be controlled to fine-tune filling of CNTs with magnetic nanoparticles with high selectivity and optimize their magnetic properties and functionality. More intense coercive fields were observed in CNT composites compared to their bulk counterparts, thus becoming candidates for potential applications such as magnetic storage devices. Apart from upgrading physical properties upon incorporation into magnetic nanoparticles, CNTs impart new functionalities that are in demand in future technologies, thus propelling nanomaterials to have increased prospects in metal and alloy modifications (Büchner et al., 2018).

6.2.2. Nanoparticles of Metals

The application of metal nanoparticles in steel and alloys is a significant breakthrough in material science that enhances the mechanical properties, including strength, ductility, and corrosion resistance of the materials. Nanoparticles due to their large surface-to-volume ratio hold good promise to influence the mechanical properties of bulk materials in a positive manner when properly incorporated. The incorporation of nanoparticles into a metal matrix, for instance, can lead to improved grain refinement, which is crucial to achieve enhanced mechanical performance. Experiments have demonstrated that nanoparticle dispersion has an impact on the microstructural properties, contributing to the hardness and toughness of the resulting alloy. This can be linked to enhanced wear resistance because nanoparticles may inhibit dislocation mobility, hence enhancing the material's overall resilience in harsh environments. Thus, valorization of nanoparticles in metallurgical applications is a future prospect for the development of advanced structural materials. Aside from this, methods for achieving the most dispersion of metal nanoparticles in steel matrices have also garnered interest for their capacity to reveal new material capabilities.

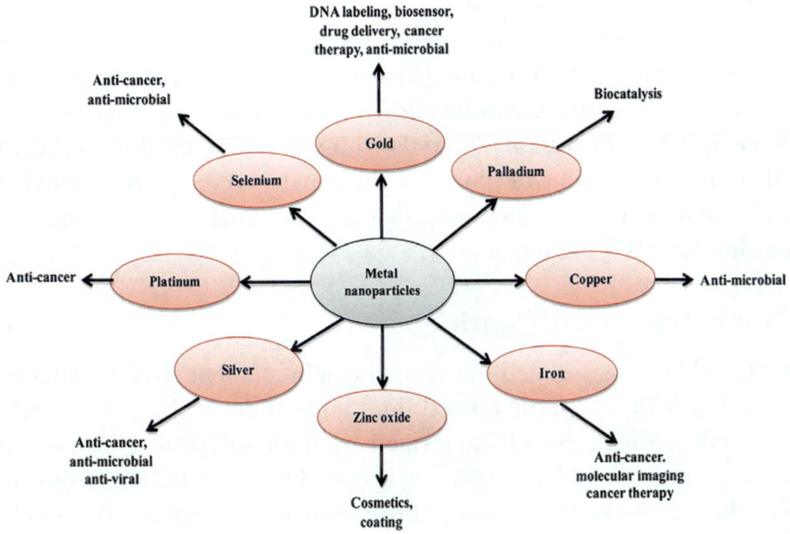

Figure 6.6. Nanoparticles of metals.

Source: https://www.researchgate.net/publication/306944495/figure/fig6/AS:399363303067648@1472227039051/Different-types-of-metal-nanoparticles-and-their-applications-13.png.

Techniques such as high shear solidification and equal channel angular extrusion (ECAE) have been applied in attempts to achieve a homogeneous nanoparticle distribution, critical to ensure their beneficial effects are maximized. For example, studies of magnesium alloys processed using hydroxyapatite nanoparticles have been found to exhibit significant improvements in microstructural homogeneity and mechanical properties following effective processing techniques (Huang et al., 2014). These advancements indicate the importance of processing techniques in attaining metal nanoparticle effectiveness due to their influence on metal interactivity between nanoparticles and metal matrix. Hence, the advancement in strategic fabrication processes remains significant for successful integration of nanoparticles in steel and alloy matrices. In addition to the enhancement of mechanical properties, metal nanoparticles are also crucial for improving the electrochemical performance of alloys, particularly for applications in energy storage. The outstanding characteristics of nanoparticles can lead to outstanding charge transfer rates, enhancing the efficiency of many battery systems. Experiments conducted on alloy materials such as single crystalline structure alloy material covered with carbon nanoparticles have shown significant improvement in reversible capacity and cycling stability, a decisive parameter for lithium-ion batteries (Seo et al., 2011). Composition optimization of alloys to allow such performance improvement can be focal in exploring advances in energy storage devices. With growing materials science, exploring the addition of various metal nanoparticles into alloys offers promising directions for the development of next-generation materials with tailored properties for a variety of industrial applications. The work in this field is on the verge of creating visionary technologies that address the mounting demands for efficiency and endurance in modern technology (Figure 6.6).

6.2.3. Nanostructured Coatings

Nanostructured coatings constitute a major achievement in the enhancement of steel and alloys properties used in many industrial applications. The nanostructured coatings are characterized by their microstructure, usually in the form of grain sizes in the nanometer range. The nanostructure significantly influences the mechanical, thermal, and chemical properties of the substrate material. For instance, the increased hardness and wear resistance through nanostructured coatings makes them useful in high-performance applications where material reliability and durability are extremely crucial. The versatility of nanostructured coatings is also illustrated through their application in biomedical implants, where they can both increase osseointegration and antibacterial activity and thus avoid complications with implant rejection and infection. Production of nanostructured coatings is typically accomplished through advanced deposition techniques such as magnetron sputtering or chemical vapor deposition (CVD). An example employing the magnetron sputtering approach was recently shown to be effective for large-area implant coating, producing coatings with properties mirroring the laboratory findings and the scalability required by industrial applications. The new method showed comparable osteoblast proliferation and antibacterial properties, demonstrating the high potential of nanostructured coatings to satisfy stringent biomedical requirements. The innovation demonstrates not just the high potential for enhanced functionality but also the ability to pave the way for the large-scale production of superior-quality nanostructured materials, ultimately bridging the gap between laboratory exploration and practical application in various sectors, including healthcare and production (Arcos et al., 2019). Further, the inherent nanostructure of coatings can be optimized by designing processing conditions. The studies have indicated that annealing treatments significantly impact nanostructure through grain size and impurity segregation at interfaces, which have a crucial role to play in defining the mechanical properties of coatings. For example, coatings subjected to thermal annealing processes showed improved hardness of 18-20%, attributed to reorganization of nanograins upon annealing and thus enhanced mechanical strength. Techniques such as X-ray diffraction and scanning electron microscopy are utilized to investigate such structural changes, capturing information about the performance properties of the coatings. Therefore, continuous research and development in deposition conditions and post-treatment is required to advance the progress and implementation of nanostructured coatings in steel and alloy materials, ultimately resulting in material science breakthroughs (Demianenko et al., 2013) (Figure 6.7).

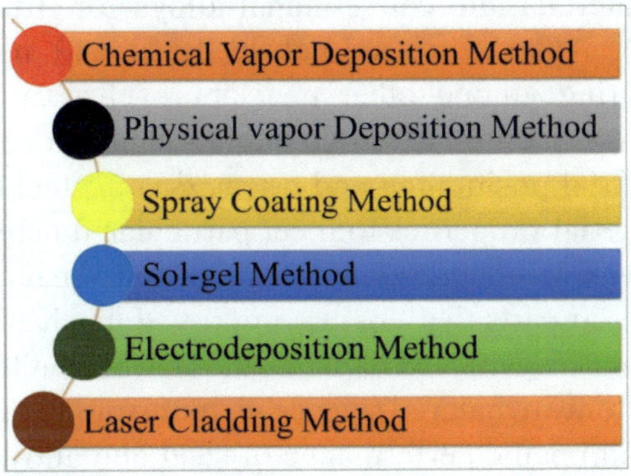

Figure 6.7. Nanostructured coatings.

Source: https://www.mdpi.com/nanomaterials/nanomaterials-12-01323/article_deploy/html/images/nanomaterials-12-01323-g001.png.

6.3. METHODS OF INCORPORATING NANOMATERIALS

Insertion of nanomaterials into steel and alloys is performed through a variety of methods that significantly enhance the material's properties. One technique is applying the sol-gel process that employs precursors in the preparation of nanostructured films. The process has the ability to produce hybrid films containing inorganic and organic phases, e.g., tetraethoxysilane (TEOS) with laminar nanoclays as a coating material. Such coatings have been reported to improve the mechanical properties, corrosion resistance, and barrier performance through the new nanoscale interactions. The enhancement brought by this process demonstrates the promise of sol-gel processes to improve the functionality of traditional materials in demanding applications, paving the way for steel and alloy formulation innovations (Ceré et al., 2016). Another effective method of incorporating nanomaterials is mechanical mixing, in which nanoscale material is combined with the bulk matrix of steel or alloy. The process makes it possible to disperse nanoparticles into the metal matrix evenly, leading to improved strength and toughness. Interference in the motion of dislocations by the interaction of the nanoparticles with the surrounding metal matrix improves the mechanical properties of the alloy. For instance, the incorporation of metallic nanoparticles or carbon nanotubes into steel has a dramatic rise in yield strength and ductility. These are critical in high-performance material applications where such a need exists, a testament to the necessity of effective methods of infusing nanomaterials into conventional alloys for structural applications (Seo et al., 2011). Apart from sol-gel methods and mechanical mixing, additive manufacturing among other new approaches is gaining more and more popularity in the addition of nanomaterials. This process allows sensitive control of nanomaterial positioning and orientation in steel or an alloy matrix, thus modifying material properties to meet particular functional requirements. Through 3D printing techniques, researchers can embed nanocomposites within the matrix, resulting in parts that possess improved mechanical and functional properties that cannot be achieved using traditional manufacturing techniques. This is a great milestone in material science, demonstrating the power of frontier technologies to optimize the performance of steel and alloys through strategic incorporation of nanomaterials, ultimately creating new applications for the materials in various industries (Ceré et al., 2016).

6.3.1. Mechanical Alloying

Mechanical alloying is an all-around solid-state processing technique that plays a pivotal role in nanomaterial synthesis for steel and alloys. Through the application of a series of high-energy ball milling operations on an elemental powder mixture, the process causes intimate mixing of the components at the nanometer scale. Ongoing fracturing and welding of particles create an increasingly finer microstructure that strengthens the mechanical properties of the final product. In processing, the materials shift during processing from the conventional coarse-grained microstructure to ultrafine-grained or nanocrystalline structures that yield improved strength and hardness, crucial in various engineering applications. The transformation is beneficial towards optimal improved mechanical properties, making mechanical alloying a fascinating process for the preparation of advanced materials for high-demanding conditions in industries ranging from the automobile industry to aerospace. Furthermore, mechanical alloying has been observed to aid the synthesis of composite materials with enhanced properties, harmonizing with increasing demands for superior performing materials in numerous technological applications.

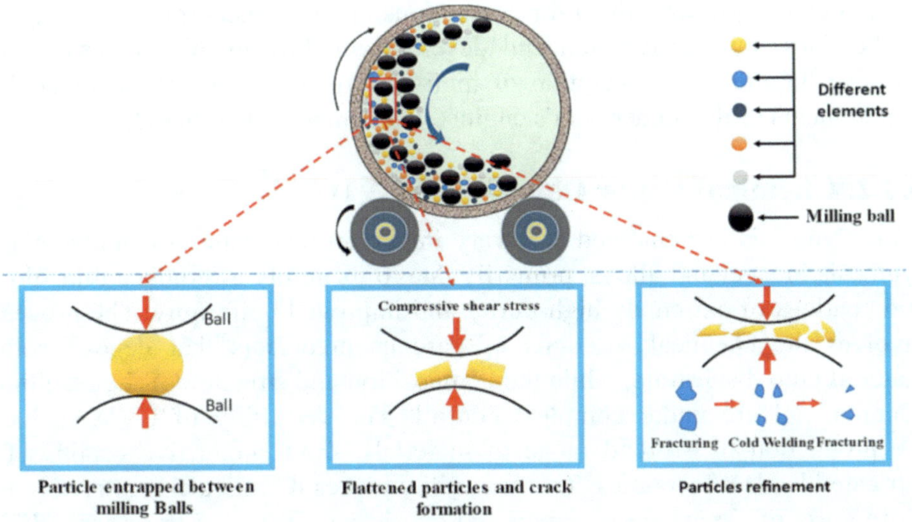

Figure 6.8. Mechanical alloying.

Source: https://ars.els-cdn.com/content/image/1-s2.0-S1526612523011441-gr7.jpg.

The process makes it possible to introduce several alloying elements through a single powder blend, which can bring forth desired characteristics depending on the specific application. For instance, mechanical alloying has been used to synthesize Al-Mn-Ce nanocomposites, and this demonstrates the method's capacity for synthesizing materials that take advantage of the unique mechanical properties in the nanoscale domain. In accordance with observations in recent research, nanomaterials exhibit super strength, increased diffusivity, and enhanced thermal as well as electrical conductivity compared to their traditional counterparts (Rai et al., 2008). These benefits identify the critical importance of mechanical alloying in the development of conventional materials into the area of advanced nanostructured composites. Lastly, the impact of mechanical alloying extends beyond basic material synthesis due to its offering new prospects for functional application in advanced technologies. The ability to tailor the material properties by controlling milling environment and time leads to significant innovations in fields such as electronics and energy storage. For example, the development of composite electrodes with nanostructured materials demonstrates improved electrochemical performance parameters in lithium-ion batteries, where the innovative architecture of the material enhances capacity and cycling stability (Seo et al., 2011). This reveals how mechanical alloying not only improves the physical and mechanical properties of steel and alloys but also prepares them for innovations in next-generation technological applications. Further research and development through this process holds promise for the next generation of high-performance materials to meet the requirements of the modern era's engineering demands (Figure 6.8).

6.3.2. Chemical Vapor Deposition (CVD)

CVD Synthesis has emerged as a very important technique in nanomaterials synthesis in steel and alloys, primarily due to its ability to produce thin films and coatings of extremely high purity and improved uniformity. The process involves the chemical reactions of gaseous precursors that deposit solid material onto a substrate, while the composition and structure of the resulting nanomaterial are under complete control. The versatility of CVD enables the production of a broad range of materials, from conductive coatings to intricate blends of ceramics. This versatility makes it particularly attractive to those areas of use where accuracy and consistency are most important. With advancing technologies, the scalability of CVD processes becomes enhanced continuously, which paves ways for wide applicability in enhanced steel and alloy systems towards meeting the increasing performance and efficiency needs in the majority of engineering applications. Furthermore, the application of CVD is crucial for enhancing the mechanical features of nanostructured steel and alloys. For instance, the incorporation of nanoscale coatings holds promise for improved wear and corrosion resistance, which is essential for increasing component lifespan under severe operating conditions. Additionally, as nanomaterials synthesized through CVD generally exhibit better mechanical strength and toughness, their capacity for strengthening conventional steel and alloys stands as a newfangled method. Experiments have indicated that the microstructural properties of these kinds of nanomaterials, including grain refinement and phase transformation, are entirely controllable through the CVD method, leading to unprecedented levels of performance in structural applications. Incorporation of CVD-synthesized nanomaterials within steel and alloy mixtures thus not only enhances their in-built properties but also aligns with overall sustainable strategies to maximize material utilization and reduce resource consumption. The significance of CVD in the area of nanomaterials is demonstrated through its ability to facilitate innovations at the nanoscale architecture with immediate implications for the performance of steel and alloy systems. Current research in this area is seen to center on the identification of interactions between deposited films and the substrate, ultimately leading to the synthesis of composite materials with synergetic characteristics. For example, the combination of CVD coatings with nanowires has been found to exhibit better electrochemical performance, making them suitable for applications in batteries and other energy-saving devices. The innovation fits with the role of nanomaterials in meeting contemporary challenges facing emerging industries. As technology advances, integrating CVD with new casting and manufacturing techniques will further enhance the potential and applications of nanomaterials in steel and alloys towards a new materials science that promises to transform many sectors of the economy, as seen in available literature (Rai et al., 2008; Seo et al., 2011).

6.3.3. Sol-Gel Processing

Sol-gel processing has now evolved into a fundamental process for the synthesis of nanomaterials, particularly in steel and alloys. This process is the conversion of a liquid sol (colloidal solution) system into solid gel state, enabling mixing of several constituents at the molecular level. By making use of this unique ability to customize the microstructure and chemistry of materials, researchers can produce well-acclimated nanocomposites with enhanced mechanical and physical characteristics compared to their bulk counterpart. The nanomaterials synthesized through sol-gel synthesis were found to be more mechanically strong and thermally stable, and thus found to be of great application in the steel and alloys sector, where performance is a matter of utmost concern. Sol-gel process capability to be tailored allows the incorporation of multiple metal oxides, further increasing the applications of nanomaterials in advanced manufacturing. The sol-gel process is also beneficial in the aspect that it incorporates low processing temperatures, which are vital to maintaining the metal matrix integrity throughout the fabrication of nanocomposites. Traditional methods involve high-temperature processes that lead to undesired phase transformation or grain growth. However, sol-gel processing is used to prepare nanomaterials retaining useful properties but getting incorporated into steel and alloy systems. Hybrid nanocomposites prepared by sol-gel processes were found to have outstanding wear resistance along with enhanced strength, for instance. Recent research has emphasized the contribution of sol-gel synthesized nanomaterials in wear-resistant coatings and corrosion protection layers and identified their importance in industrial applications. This new method not only enhances the quality of materials but also achieves sustainability goals by lowering energy used in production. In addition to enhancing mechanical properties, sol-gel processing is central to the development of functional materials that improve steel and alloys' performance in extreme conditions. The use of sol-gel synthesized oxides, such as alumina and zirconia, is applied for incorporating beneficial properties, such as thermal stability and corrosion resistance, into metal matrices. For instance, incorporating zirconia into aluminum oxide through sol-gel methods has been shown to cause a significant reduction in sintering temperatures to create stable nanocomposites. The process enables the production of high-temperature and severe chemical condition-resistant advanced materials, thereby increasing their lifespan. Increasing demand for high-performing materials, sol-gel provides a simple and effective means of achieving desirable features for nanomaterials used in steel and alloys and, in turn, driving innovation in the industry (Ganin et al., 2018; Rai et al., 2008) (Figure 6.9).

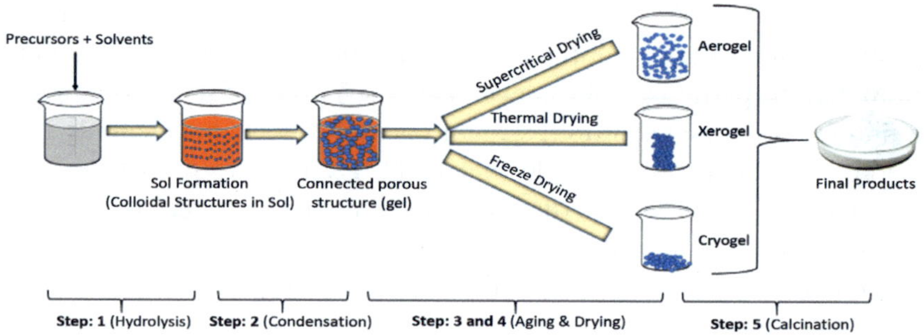

Figure 6.9. Sol-gel processing of steel and alloy.

Source: https://media.springernature.com/lw685/springer-static/image/art%3A10.1007%2Fs10854-020-02994-8/MediaObjects/10854_2020_2994_Fig2_HTML.png.

6.4. APPLICATIONS IN INDUSTRY

Applying nanomaterials to boost steel and alloy production has revolutionized industrial applications, particularly healthcare and building. One such example is the development of a magnetron sputtering technique to deposit titanium nanocolumns on large implants, showing the application of nanotechnologies to enhance biomedical material functionality. This new method, attempted in a semi-industrial reactor, ensures coatings formed on bone plates not only exhibit similar osteoblast proliferation efficiency but also possess antibacterial activity similar to that produced in the lab. It implies that the proposed approach can be effectively scaled up for industrial use, opening up viable alternatives that enhance the implant's durability and performance in surgical procedures (Arcos et al., 2019). The flexibility of nanomaterials also extends into the area of protective coatings for metal infrastructure, specifically by the use of vertically aligned carbon nanotube (CNT) forests. The nanomaterials possess multifunctional properties that greatly contribute to the increase in durability and functionality of metallic devices, such as pipes and turbine blades. Its new mechanical properties, including high compressive strength and energy absorption, coupled with super-hydrophobicity, render CNT forests ideal prospects as protective films. Direct growth of CNTs on metals, with industries benefiting from the tunable functionalities of such nanostructures, renders them an attractive option for single applications. Systematic studies of the structural characteristics and mechanical behavior of CNTs have established the fundamental interplay between substrate characteristics and the resulting efficiency of the aforementioned nanomaterials, and thus important implications for industrial applications (Daraio et al., 2018). In addition, the emerging field of nanotechnology could make a critical impact on not just steel-making efficiency but also sustainability. As companies transition to decreasing environmental impacts, nanomaterials present the possibility of enhancing recycling and resource efficiency. For example, engineered nanomaterials can be used to enhance steel alloy microstructure for better mechanical properties and energy efficiency in production processes. It not only increases the service life of steel products but also reduces waste, which aligns with international sustainability goals. With the advancement of knowledge and application of nanomaterials in steel and alloys, industries can develop new solutions not only to satisfy market demands but also to contribute positively to environmental responsibility.

6.4.1. Automotive Industry

The inclusion of nanomaterials in the motor vehicle industry has stimulated colossal innovation in automobile design and production processes. Use of nanostructured materials in steel and alloys makes them stronger, more durable, and lighter, which becomes critical in fuel efficiency maximization and emission minimization. For instance, the inclusion of nanomaterials makes it possible for carmakers to design lighter vehicle bodies without compromising safety. Programs such as the EVolution program demonstrate emphasis on composite and aluminum alloys, with novel body structure ideas that utilize the newest nanomaterial applications. As vehicle manufacturers pursue greener ways, the benefits of nanotechnology in reducing the weight of vehicles and increasing efficiency become more apparent, cementing its role in driving industry direction in the years to come (Alcalde et al., 2017). Not only does the application of nanotechnology enhance the mechanical properties of car materials but also increases their functional capabilities. For example, more recent nano-improved coatings provide additional weathering, wear, and corrosion resistance, which are of paramount importance when applied to exterior vehicle parts. Aside from aesthetics, these developments are driving the industry towards more durable and environmentally friendly solutions. Moreover, the increasing use of nanomaterials results in the formation of multifunctional components, serving different roles throughout the vehicle, from structural support to energy efficiency.

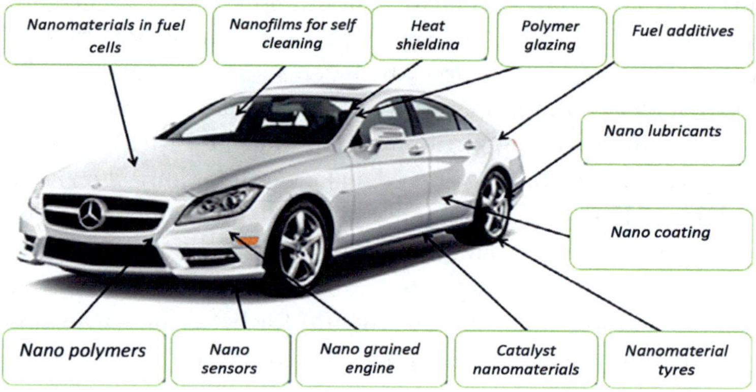

Figure 6.10. Nanoparticles in automotive industry.

Source: https://i.ytimg.com/vi/q6gtEoHIoNA/hqdefault.jpg.

Manufacturers are finding increasingly that adopting such technologies will achieve great long-term cost savings, as well as minimizing environmental effects and lifecycle expenses. With further development of these technologies, the automobile sector is about to undergo revolutionary change based on the incorporation of nanomaterials within steel and alloys (MORENO et al., 2017). The application of nanomaterials by the automobile sector is also in accordance with the overall technological progress in different sectors. As mentioned by research, the majority of nanoproducts not only enhance the performance of material but also conserve energy in applications from production to end use.

This is proof that there is a convergence of interests whereby the car industry can learn from what other industries have achieved, such as in construction and electronics. By adopting the use of nanomaterials that increase strength and functionality, automakers can make cars that meet the demands of consumers today for performance and eco-friendliness. With the sector in the midst of challenges related to engine efficiency and emission control, strategic application of nanotechnology in material science can ensure competitiveness. In total, ongoing application of nanomaterials in the automobile industry is a significant milestone toward more efficient, innovative, and green manufacturing practices (MORENO et al., 2017) (Figure 6.10).

6.4.2. Aerospace Applications

The aerospace industry is always on the lookout for innovations that will make it function better and lighter, and stronger. Nanomaterials, particularly nanomaterials derived from steel and alloys, are spearheading such innovations. Their unique properties—greater tensile strength and corrosion resistance—make them highly useful in aeronautical engineering. When aircraft manufacturers work hard to meet stringent fuel efficiency and safety requirements, employing nanostructured materials can lead to significant weight savings without compromising structural integrity. This is highlighted in current research reports and conferences where the discussion with respect to nanomaterials has gained immense popularity with a demonstration of their potentiality to bring about revolutionary transformation in aerospace technology (Langdon et al., 2014). The ability to create such materials on the nanoscale facilitates technological advancements beyond present technologies. The addition of carbon nanotubes into aluminum composites is an example in the capability to enhance material properties in the aerospace sector. Such composites, created by sophisticated techniques such as powder metallurgy, exhibit noteworthy enhancement of tribological properties under various operational conditions. Reports indicate that carbon nanotube addition results in increased hardness and wear resistance, key aspects when the parts undergo high-stress conditions while in flight (Basu et al., 2016). The effectiveness of these nanomaterials to resist wear during moderate operating conditions makes them suitable replacements for typical materials in application. With the high standards of aerospace use, the versatility of carbon nanotubes in alloys is a new concept for optimizing aircraft performance and longevity. Beyond even performance increases, the thinking of nanomaterials even promises the potential for enhancing manufacturing processes in the aerospace industry. The drive towards integrating nanotechnology in alloys and steel results in far-reaching changes in the way components are manufactured, potentially leading to cost-effective manufacturing and reduced material wastage. With the industry embracing techniques such as severe plastic deformation, highlighted in recent conferences, it results in enhanced understanding of nanomaterial behavior and advances aerospace engineering further (Langdon et al., 2014). These relationships between nanomaterials and the effectiveness of manufacturing not only facilitate the production of improved components but also support sustainability. For this reason, the aerospace industry is about to experience a materials revolution as a result of the extensive application of nanostructured steel and alloys.

6.4.3. Construction Materials

Application of nanomaterials to construction material is now an essential trend in modern architecture and construction, offering new possibilities for enhancing durability and sustainability. With the problems presented to the sector by environmental degradation and demands for resource-conserving solutions, breakthroughs in nanotechnology have introduced a series of nano-products that significantly increase the performance of standard materials. These advancements are not limited to alloys and steel but extend up to concrete and polymers as well. The introduction of nanomaterials can lead to the development of more resilient structures and coatings, as reported by a number of studies in terms of the capacity of nano-engineered mixtures to mitigate issues of wear and degradation. Against this backdrop, stakeholders in the construction industry need to appreciate the myriad uses of such innovative materials (MORENO et al., 2017). In the field of steel and alloys, material manipulation at the nanoscale has resulted in dramatic enhancements in strength and performance. This includes strategic design of nano-arrays and the meticulous management of defects and compositions at an atomic scale. For instance, recent research has indicated that by the use of core-shell nanoparticle morphologies in iron alloys, their mechanical properties can be radically enhanced. That is, their strength can be enhanced from 500 MPa–1 GPa upon prolonged heat treatment, demonstrating vividly the radical impact nanostructuring can have on conventional materials (Gault et al., 2019). This enhancement is not only critical for the stability of structures but also for extending the lifespan of constructions and, therefore, reducing maintenance and replacement costs. Furthermore, the application of nanomaterials in construction not only addresses the issue of performance but also aligns with the goals of sustainability in the construction industry. Proper application of nanotechnology can extend to more sustainable construction practices in the form of reduced waste and improved performance of individual materials. For example, nano-enhanced sealants and coatings not only enhance weather resistance but also make buildings energy efficient through heat reflection and minimizing the overall building energy consumption. With architects, builders, and material scientists working together to research and adopt these cutting-edge solutions, the potential for changing construction processes keeps increasing. Hence, a deeper understanding and application of nanomaterials can become a significant role in more sustainable and more resistant architectural designs (MORENO et al., 2017) (Figure 6.11).

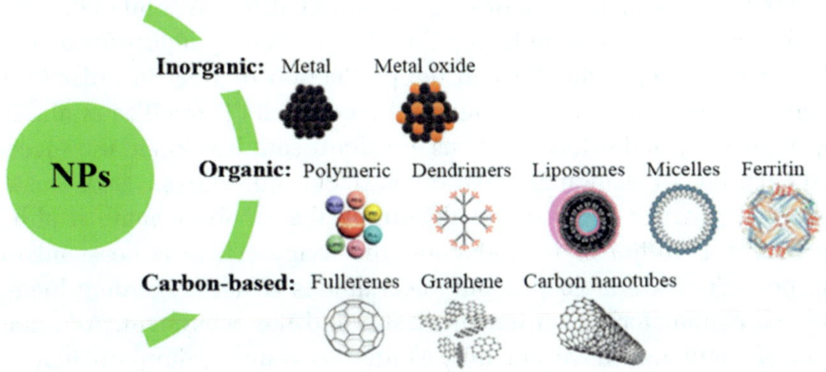

Figure 6.11. Nanomaterials into construction materials.

Source: https://www.mdpi.com/jcs/jcs-07-00067/article_deploy/html/images/jcs-07-00067-g001.png.

6.5. CHALLENGES IN IMPLEMENTATION

Application of nanomaterials in steel and alloys encompasses numerous issues that hinder their application on a large scale in industrial processes. One of the biggest challenges is uniform distribution and stability of the nanoscale particles in a bulk metallic matrix. Nanomaterial properties are very sensitive to distribution, and reproducibility of uniform microstructure is essential for the achievement of maximum mechanical performance. Besides, changes in processing conditions such as temperature and pressure have a tendency to produce agglomeration or phase separation, which is counterproductive to the use of nanomaterials. Given this, resolving these processing issues becomes paramount to make real practical applications of the performance boosts nanomaterials have been said to offer. This complexity demands stringent experimental techniques in order to facilitate the reproducible synthesis of steel and alloys with nanostructures, triggering further investigation in this field. Apart from technical challenges, there are economic concerns that have a major impact on the application of nanomaterials in steel and alloys. The manufacturing processes involved to include nanomaterials are also generally more expensive compared to traditional methods, potentially limiting economic feasibility for use in competitive markets. Furthermore, access requirements for specialized equipment and raw materials create disincentives for manufacturers to embrace these advanced materials. Furthermore, the hefty cost of research, development, and quality assurance would also have to be incorporated into any strategy for commercialization of nanostructured alloys. A solid understanding of cost-benefit analyses should, therefore, be pursued by industries that would want to employ nanotechnology in the production of steel in order to avoid sabotaging efforts in case of no financial incentives (Torrecillas et al., 2009). Lastly, regulatory and safety concerns are significant barriers to the successful implementation of nanomaterials onto steel and alloys. New materials require stringent test protocols to determine the environmental, health, and safety concerns within their production and usage. There is no standardized testing procedure available yet, and uncertainties remain regarding long-term impacts of nanomaterials on human health and the ecosystem. Additionally, regimes of regulation have not fully shifted to framing the particular issues of nanotechnology, and manufacturers are therefore unclear on compliance. Regulation convergence and thorough safety assessment are needed to trigger public confidence and enable industries to navigate the complexity of nanomaterials. Consequently, framing the regulatory challenges is necessary to propel the use of nanotechnology into the overall framework of steel and alloy applications.

6.5.1. Cost of Production

Cost of production in nanomaterial synthesis, particularly in steel and alloys, has a significant impact on their commercialization and application potential. Its intrinsic complexity in the fabrication processes, such as precise control over nanostructure development and requirement of tailored equipment, is to blame for high production costs. For instance, preparation of nanoporous Copper-Nickel (npCN) with feature sizes that can be precisely controlled in the 13-40 nm range involves advanced thermal annealing techniques. These technological needs impose economic expenses in terms of energy consumption and personnel skilled in technology. Therefore, optimizing the production steps not only makes the economic viability better but also can yield materials with improved mechanical properties, being one of the areas that must be kept on the R&D map in the case of nanomaterials in steel and alloys. Raw material choice is the other major factor in determining the cost of production. Nanomaterials often require the application of high-purity metals and specialized additives to produce maximum properties, which are costly and difficult to Source: An example is carbon materials being applied as effective additives maximizing the metal hydrides' hydrogen storage capacity, as studied in recent studies. These materials possess the drawback of pushing costs unaffordable when acquired in the forms necessary for best integration with nanostructured metals. The balance between manufacturing high-performing materials and aligning costs during production is a continuing chief challenge. Scientists continue to pursue innovative methods of establishing cost-effective replacements or synthetic routes that can ensure reduced material costs without sacrificing functionality and structural integrity in applications such as energy storage and automobile components. Because of the economic realities of producing nanomaterials, industries are increasingly finding it more imperative to adopt modular and scalable approaches to manufacturing. This change aims to reduce initial capital outlay as well as operational costs associated with producing advanced nanostructured alloys. By employing new methods such as powder metallurgy and additive manufacturing, businesses can achieve simplifying their production processes and become more effective. Importantly, the integration of novel technologies with conventional alloying techniques provides opportunities for cheaper paths to accessing high-performance materials. New research indicates that although having higher initial costs, lifecycle cost saving through nanomaterials' enhanced strength and toughness is yielding high cost advantage, particularly for aggressive applications. Therefore, surmounting the manufacturing cost is important in advancing nanomaterials in steel and alloys and enhancing business competitiveness and sustainability.

6.5.2. Scalability Issues

The study of nanomaterials in alloys and steel has a number of scalability challenges that have extensive impacts on their practical application as well as usage. The fabrication of nanomaterials typically requires processes that are complex and can be challenging to scale while maintaining the materials' distinctive properties. For instance, usage of techniques like Equal Channel Angular Pressing (ECAP) has a tendency to form smaller-scale structures, and scaling up from laboratory-scale synthesis to industry-scale manufacturing raises issues regarding homogeneity and consistency, which are critical to the delivery of reliability in performance. The case of AA-5083 aluminum alloy reveals that different processing routes produce different crystallite sizes, thus pointing to scalability issues arising due to differences in process optimization (González Crespo et al., 2013). With increasing scale, it is necessary to maintain the favorable mechanical properties of nanomaterials, necessitating ongoing research into scalable methods that are effective. Commercialization of nanomaterials in steel and alloy production must also address scalability regulatory and environmental concerns. Regulatory bodies are often slow to adapt to technological advancement, presenting barriers for industry players to commercialize new nanostructured materials. The intricate character of nanomaterials needs thorough testing and qualification requirements to establish safety and efficacy, further complicating scalability. Second, the potential environmental price of large-scale processes—e.g., emissions and waste management—adds complexity to the overall scalability strategy. For instance, production of shape memory alloys (SMA) illustrates the challenges in producing materials that are both efficient and green. While MA coils can significantly enhance measurement of performance—offering high force at smaller sizes—scaling the processes of manufacturing to be applied on a large scale requires a delicate balance of efficiency and environmental stewardship (Holschuh et al., 2014). Lastly, the economic impacts of scalability issues of nanomaterials on steel and alloys cannot be overlooked. The expense of the upfront investment into advanced manufacturing technology is generally a barrier for businesses wishing to enter the nanostructured material business. With increasing production of nanomaterials, economies of scale must be established in terms of saving costs and enhancing competitiveness. However, the cost-effective means of production must be balanced with the preservation of nanoscale properties. Suffering the quality for the sake of cost will lead to inferior material performance, negating the very purpose of the use of nanomaterials. Lastly, the solution to the scalability problem in nanomaterials needs interdisciplinary collaboration between materials science, engineering, and economic foresight so that the exciting attributes of nanostructured steel and alloys can be harnessed without sacrificing performance or sustainability (Figure 6.12).

Figure 6.12. Scalability issues of nanomaterials in steel and alloys.

Source: https://www.azonano.com/images/Article_Images/ImageForArticle_6126_16528996350644068.jpg.

6.5.3. Environmental and Health Concerns

Incorporating nanomaterials into steel and alloys brings with it a full array of potential health and environmental implications that need to be carefully examined. While the technology for nanotechnology improves, these substances have singular properties that enhance performance capabilities; however, simultaneously, their very small size and reactivity create severe safety issues. Evidence is presented that nanoparticles can act in unforeseen ways with biological systems and cause unforeseen injury to health, including respiratory issues and cellular toxicity. In addition, their life cycle from production to disposal is environmentally toxic. Emission of nanomaterials into the environment can lead to bioaccumulation within organisms and have unforeseen impacts on food webs and biodiversity. These problems require regulatory frameworks to properly assess and address the dangers of nanomaterials in business applications, especially regarding public health and environmental protection. Scientific societies and regulatory bodies are increasingly acknowledging the need for an overarching definition and management strategy to nanomaterials. Current legislation, like that of the European Union's REACH and directives on nanomaterials, aims to facilitate good communication of potential danger to both customers and employees.

The lack of clear definition makes it harder as various groups perceive the properties of nanomaterials differently. There is a need for an integrated framework explaining what constitutes a nanomaterial in order for guidelines to be developed on risk-free handling and environmental protection. The need for ongoing dialogue and modification of regulations is paramount to complete any widening knowledge gaps, as new evidence continues to rebut previous assumptions regarding the safety and effectiveness of nanomaterials (Peter et al., 2010). Aside from immediate occupational exposures, also of concern are the health effects of nanomaterial exposures. Chronic respiratory disease and other systemic effects due to long-term airborne nanoparticle exposure are growing issues among scientists, with studies indicating potential effects. The environmental impacts of release of nanoparticles into water and air streams are also worrisome; these substances may be altered, altering their bioavailability and impacts on ecosystems. Compared to traditional pollutants, environmental behavior by nanomaterials is unpredictable due to their unique properties. Such uncertainty strengthens the necessity for a precautionary approach, suggesting cautious risk assessment before widespread commercialization of nanomaterials in alloys and steel. Lastly, addressing these health and environmental challenges is paramount in realizing the full potential of nanotechnology while safeguarding public health and ecological balance.

6.6. FUTURE TRENDS IN NANOMATERIALS RESEARCH

The future of nanomaterials research will revolutionize the field of steel and alloys, particularly enhancing their mechanical properties and their resistance to wear. One of the likely evolving trends is the development of new composites, for instance, aluminum oxide-zirconium dioxide (Al_2O_3–ZrO_2) nanocomposites, which are capable of reducing sintering temperatures by quite a large amount. A treatment using microwave hydrothermal processing has been shown to generate such materials with higher structural coherence by suppressing deleterious grain growth and thus stabilizing important phases at high temperatures. This innovation heralds the way towards more energy-efficient processing methods that maintain the target material properties while minimizing energy consumption. Sintering temperature reduction makes high-performance alloys amenable to processing, eventually leading to greener methods in material engineering as well as new application horizons in aerospace and automotive sectors (Ganin et al., 2018). Another emerging trend is metallic surface modification by employing advanced deposition techniques, e.g., chemical vapor deposition (CVD). CVD allows the deposition of alloys and steel, introducing considerable resistance against wear, oxidation, and corrosion. The versatility of CVD techniques enables them to be adapted to different types of powders and substrates, emphasizing the importance of innovative reactor designs. This also entails the development of fluidized and vibro-fluidized beds to ensure maximum contact between powder particles and reactive gases so that coating efficiencies are improved. The possibility of making tailored coatings using CVD is a huge materials science achievement that can potentially improve the functional characteristics of steel and alloy components in such applications like microelectronics and photovoltaics (Angelopoulos et al., 2006). Moreover, the future studies on nanomaterials will likely involve recyclability and sustainability as fundamental principles of research. Researchers are increasingly keen on developing materials that not only perform extremely well but also economically and environmentally. The move towards green nanotechnology in steel and alloys involves the use of bio-based materials and biodegradable additives that minimize harm to the environment. With the growing demand for sustainable manufacturing processes, enhancing collaboration between academia and industry will be essential in getting theoretical innovations to practical applications. By rethinking conventional approaches through the lens of nanotechnology, researchers can create a next generation of materials not just with added complexity and functionality but also that will enhance a more sustainable future for the metallurgy sector.

6.6.1. Innovations in Synthesis Techniques

Progress in synthesis techniques has impacted nanomaterials development in steel and alloys applications considerably. The developments allow more control over the size, shape, and composition of nanoparticles and therefore their mechanical properties. Some of these include techniques such as sol-gel processing and hydrothermal synthesis to synthesize nanoparticles with preferred morphologies to enhance steel composite strength and hardness. These developments are significant because they help to improve the performance of structural materials under a variety of conditions. Additionally, these processes enable the incorporation of functional additives that offer corrosion protection and wear lifetime. As the construction and engineering industries demand more sustainability and durability, developments in synthesis techniques pave the way for creating next-generation nanomaterials with such high expectations. Moreover, the application of green synthesis protocols in the manufacture of nanomaterials has emerged as a vital innovation within the field. Through the use of sustainable solvents and the reduction of harmful reagents, these procedures not only reduce environmental footprints but enhance the safety profile of building and manufacturing materials as well. For example, researchers have begun studying biological processes that utilize plant extracts and microbes to synthesize metal nanoparticles with the environmental integrity preserved. The green methods allow the creation of nanomaterials that can effectively enhance the strength of steel and alloys with environmental sustainability. In addition, this shift to sustainable synthesis fits into broader trends in the building industry, whereby the demand for durable materials with emphasis on energy efficiency and longevity stands above everything else, emphasizing more the importance of environmentally friendly practices in nanotechnology innovation. Lastly, the synergy of additive manufacturing with nanomaterial synthesis is yet another frontier in innovation. Techniques such as selective laser melting and 3D printing can employ nanoscale materials to create structures that are of geometrical complexity beyond the capabilities of traditional fabrication methods. This capability is highly desirable for the fabrication of steel components with engineered properties for specific applications. For instance, the capability of creating gradient structures with high strength and light weight can lead to innovation in aerospace and automotive industries. Moreover, such additive processes also offer the ability to monitor properties of the material in real time during production, ensuring reliability and consistency. Through strategically synthesized nanomaterials via additive production, scientists are well-positioned to revolutionize the utilization of steel and alloys, answering the demands of evolving engineering and construction projects today (Moreno et al., 2017).

6.6.2. Potential for Smart Materials

The promise of smart materials in nanomaterials in steel and alloys has been widely recognized for its revolutionary potential for engineering applications. Smart materials can be described through their ability to dynamically respond to environmental stimuli, such as temperature, stress, and electric or magnetic fields. The addition of nanomaterials to traditional steel and alloy compositions can result in advanced material properties, such as enhanced strength, durability, and responsiveness. As explained in current research, innovation in new applications using the materials has the ability to revolutionize the structural reliability and functional performance of engineering parts and eventually result in design breaks and efficiency breakthroughs. The emphasis on applying nanotechnology to exploit such smart capabilities is towards a future where materials not only serve static roles but also adaptively modify themselves with respect to evolving conditions, enabling unparalleled versatility in various industrial applications (Langdon et al., 2014). Additionally, the synthesis of nanomaterials allows for precision tailoring of the microstructure that enhances the functional properties of steel and alloys. For instance, advances in powder metallurgy techniques can enable the uniform dispersion of carbon nanotubes (CNTs) within aluminum composites, thus significantly improving their mechanical properties. Studies have demonstrated that by using optimal loading of CNTs, such composites exhibit improved hardness and reduced wear rates, thus qualifying them as probable choices for smart materials in high-performance applications (Basu et al., 2016). Such performance improvement is just one example of the intricate interplay between macroscopic performance and nanoscale materials, and it indicates that smart material systems can be designed to satisfy certain performance requirements. Through improvements in methodologies and large-scale production, incorporation of nanomaterials can transform the way steel and alloys are used across all industries, from the automotive to the aerospace sector. Essentially, the potential of nanomaterial-based smart materials in steel and alloys is a significant step toward the development of more responsive and functionally richer materials. In the future, as technology progresses, there will be greater efforts to optimize the synthesis, processing, and deployment of such materials so that they remain beneficial under varying operating conditions. Continuing innovation in smart material technology promises to provide innovative products that not only enhance efficiency but also support sustainability goals through waste minimization and energy conservation. With the education and industrial setting embracing these innovations, the engineering practice of the future will be inevitable with the promise of smart materials to provide stronger and more responsive infrastructures (Langdon et al., 2014) (Figure 6.13).

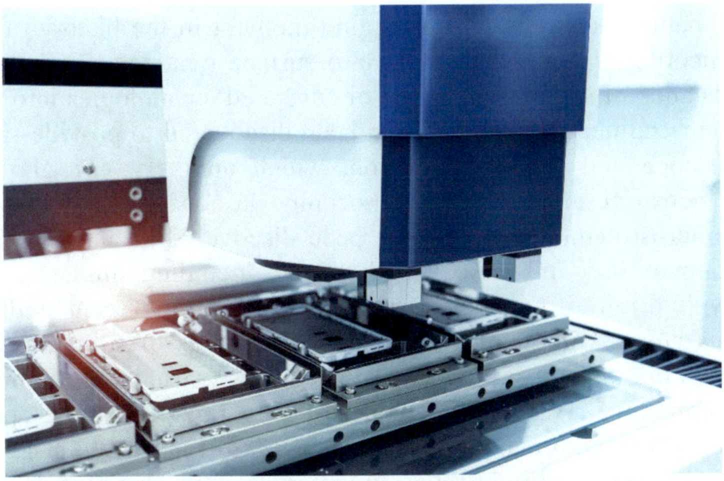

Figure 6.13. Potential for smart materials of nanomaterials in steel and alloys.

Source: https://www.azonano.com/images/Article_Images/ImageForArticle_5402_ 15790880987774725.png.

6.6.3. Integration with Other Advanced Technologies

The integration of nanomaterials with advanced technologies significantly enhances the performance characteristics of steel and alloys, particularly in construction and aeronautical engineering. Nanomaterials, owing to their unique mechanical, thermal, and electrical properties, can be utilized to produce high-strength steel as well as corrosion-resistant alloys. For instance, carbon nanotube addition to steel matrices has been established to provide enhancements in tensile strength and ductility, in that it is now essential for use in products that require both strength and elasticity. Innovations such as these not only provide increased durability in structural materials but reduce maintenance cost, further adding legitimacy to the adoption of nanotechnology in standard metallurgical processes. Moreover, the synergistic interaction between nanomaterials and manufacturing technologies fosters innovation, namely in applying additive manufacturing techniques. With the possibility of designing complex geometries with high accuracy comes the capability of producing light alloy components that are stronger. The techniques involved, such as selective laser melting, are enhanced by the unique features of nanomaterials and are made capable of tailoring material properties to precise application requirements. By optimizing such additive processes, companies can produce components with bespoke microstructures, ultimately leading to improved performance and reliability under hostile operating conditions. This convergence of nanotechnology and advanced manufacturing is a leapfrogging moment in the history of material science. Incorporating nanomaterials into surface treatments and coatings is another instance of such convergence of advanced technologies into steel and alloys. Nano-engineered coatings have been discovered to provide significant wear resistance and friction reduction, which are very crucial properties in most industrial applications. Nanocomposite coatings, synthesized by different microstructural design, have been discovered to enhance durability and biocompatibility in medical applications, including implants, to prove the genotypic origin of nanomaterials. As highlighted by recent studies, these innovations not just improve the mechanical features of structural elements and implants but also create the path towards more sustainable solutions that meet contemporary sustainability goals, thereby broadening the application of steel and alloys with nanomaterials (Torrecillas et al., 2009). The potential of this blend foretells a bright future in materials engineering, underlining the significance of interdisciplinary approaches to technological development.

SUMMARY

The integration of nanomaterials into steel and alloys is a revolutionary advancement in materials science that elevates both their mechanical and physical properties to new dimensions. Through reconfiguring materials at atomic levels, researchers have demonstrated to exhibit impressive improvements in toughness, longevity, and wear resistance. Most significantly, the creation of extremely orderly nano-arrays in Fe-based alloys, as discussed, had a sharp increase in tensile strength, illustrating how targeted changes can effectively enhance an materials' properties. This advancement creates the potential for development in other sectors aside from structural application, such as possible breakthroughs in medical implants in biocompatible materials and many more, where the overall relevance of nanostructuring technologies is seen in day-to-day applications (Gault et al., 2019). Aside from nanocomposite development, with differences in microstructural design, it emphasizes the function in developing tough and longer-duration materials. Nano-nano and macro-micro-nano composites produced are the testimonial to the sophistication level of material engineering that can endure environmental stressors while maintaining its integrity. The synergy of multiscale mixtures of materials enables greater mechanical strength and biocompatibility, which are essential to the industry sector ranging from construction to medicine. Hence, as technology and science increasingly converge, the future of material development will flourish on the unique advantages offered by nanomaterials, capable of revolutionizing existing paradigms in material usage and application (Torrecillas et al., 2009). Lastly, the trajectory of nanomaterials in steel and alloys indicates a sustained impetus for revolution in materials engineering. As the causes of nanoscale enhancement become more and more uncovered through research, opportunity for even more advanced materials can be anticipated. Future research aimed at exploring new formed nanostructures can bring record amounts of strength and function in alloys, opening up potential for application under the most extreme conditions. These developments not only have the potential to drive standards of material performance higher but also make sustainable contributions towards creating engineering solutions that are capable of coping with the dynamics of an ever-changing technological environment. Thus, the importance of nanomaterials stems not just from their immediate payoff but also from their ability to define the future of material science (Gault et al., 2019).

The study of nanomaterials in alloys and steel shows significant advantage in mechanical performance, especially in strength and fatigue life. Research indicates that materials with ultrafine grain structure, such as those produced using Equal Channel Angular Pressing (ECAP), show excellent improvements in fatigue limits compared to coarse-grained materials. For instance, one study indicated that the fatigue limit of ARMCO iron was enhanced greater than 250% following eight ECAP passes, increasing it from 274 MPa–717 MPa. That steep increase testifies to the potential that nanostructured materials have to transform applications requiring enhanced strength and durability in industries like aerospace and automotive, where mechanical performance is a major concern. The findings reveal that an exceptional microstructure can be achieved through advanced processing techniques and, as a result, enhanced mechanical properties, thereby establishing a critical link between nanostructure and performance in steel alloys (Marrero et al., 2017). In addition to heightened mechanical properties, the unique properties of nanomaterials provide increased electrochemical performance in application areas like energy storage, for instance, in the use of nanocomposite electrodes. Recent findings show that single crystalline germanium nanowires, when surface-functionalized with amorphous carbon, show remarkable electrochemical characteristics beneficial for lithium-ion battery applications. In experiments, these composite electrodes showed reversible capacity of 963 mAh/g at coulombic efficiency of 90% and showed capacity of 700 mAh/g even at high discharge rates. All of this performance can be attributed to the structural advantages of nanowires, which enable greater stability and charge transfer on cycling. This is intended to highlight the variety of applications for nanomaterials not only in traditional metallurgy but also for new applications such as clean energy, where their unique properties can lead to staggering performance improvement (Seo et al., 2011). Furthermore, the synthesis and integration of nanomaterials into steel and alloy systems provide prospects for innovative design capabilities that influence prospective material selection and fabrication methods. Materials scientists and engineers can design properties for specific functions by optimizing grain structures and microstructural attributes, maximizing the functional operation of various systems. With ongoing research into the mechanical properties of such materials, coupled with increasing knowledge about their electrochemical processes, there is vast potential for creating high-performance alloys. Such innovations could lead to more effective, lighter, and safer engineering structures. Generally speaking, with ongoing advances in the nanomaterials sector, its application in steel and alloy design and utilization will most likely redefine practice and industry norms (Marrero et al., 2017; Seo et al., 2011).

REFERENCES

1. Adamec, V., Köbölová, K., Pavlíková, S., Procházková, D., et al. (2019). Current approaches of occupational and safety health management in work environments containing nanoparticles. https://core.ac.uk/download/287648860.pdf (accessed on 25th September 2025).
2. Adelhelm, P., & De Jongh, P. E. (2011). The impact of carbon materials on the hydrogen storage properties of light metal hydrides. https://core.ac.uk/download/56345104.pdf (accessed on 25th September 2025).
3. Alcalde, E., Cabulis, P., Cischino, E., DeClaville Christiansen, J., et al. (2017). Evolution FP7 funded project: Body structure design strategies using new composite and aluminum materials and enabled technologies. https://core.ac.uk/download/157765183.pdf (accessed on 25th September 2025).
4. Alleg, S., Bensalem, R., Bououdina, M., Guergueb, A., et al. (2013). Structure and microstructure properties of ball milled Fe-Zn. https://core.ac.uk/download/20360474.pdf (accessed on 25th September 2025).
5. Angelopoulos, G. N., Caussat, B., Serp, P., Vahlas, C., et al. (2006). Principles and applications of CVD powder technology. https://core.ac.uk/download/12039263.pdf (accessed on 25th September 2025).
6. Arcos, D., Fernández-Martínez, I., García-Martín, J. M., García-Valenzuela, A., et al. (2019). Antibacterial nanostructured Ti coatings by magnetron sputtering: From laboratory scales to industrial reactors. https://core.ac.uk/download/288002820.pdf (accessed on 25th September 2025).
7. Basu, S., Elayaperumal, A., Le, H., Manikandan, M., et al. (2016). Micro/nanostructure and tribological characteristics of pressure less sintered carbon nanotubes reinforced aluminum matrix composites. https://core.ac.uk/download/46171324.pdf (accessed on 25th September 2025).
8. Büchner, B., Damm, C., Ghunaim, R., Hampel, S., et al. (2018). $\mathrm{Fe}_{1-x}\mathrm{Ni}_x$ alloy nanoparticles encapsulated inside carbon nanotubes: Controlled synthesis, structure and magnetic properties. http://arxiv.org/abs/1809.10500 (accessed on 25th September 2025).
9. Byakova, A. V., Milman, Y. V., & Yurkova, A. I. (2011). Structure and mechanical properties of iron after surface severe plastic deformation under friction with simultaneous nitrogen saturation: I. Structure formation. https://core.ac.uk/download/47236887.pdf (accessed on 25th September 2025).

10. Cabrera Marrero, J. M., Higuera Cobos, O. F., & Muñoz Bolaños, J. A. (2017). High cycle fatigue of ARMCO iron severely deformed by ECAP. https://core.ac.uk/download/81580717.pdf (accessed on 25th September 2025).

11. Ceré, S., Pellice, S. A., Pepe, A., Santana, A., et al. (2016). Hybrid sol-gel coatings containing clay nanoparticles for corrosion protection of mild steel. https://core.ac.uk/download/159296239.pdf (accessed on 25th September 2025).

12. Daraio, C., Diamantopoulou, M., Johanns, P., Parcianello, D., et al. (2018). Characterization of vertically aligned carbon nanotube forests grown on stainless steel surfaces. https://core.ac.uk/download/211389522.pdf (accessed on 25th September 2025).

13. Demianenko, A. A., & Rogoz, V. M. (2013). Influence of thermal treatment on the structure, physical and mechanical properties of nanostructured coatings (Ti, Hf, Nb, Si)N deposited by C-PVD method. https://core.ac.uk/download/20360566.pdf (accessed on 25th September 2025).

14. Dzhur, Y. (Yevgen), Grekova, M. (Maryna), Guchenkov, M. (Michael), Kalinin, A., et al. (2017). Investigation of the influence of nano dispersed compositions obtained by plasm chemical synthesis on the crystallization processes of structural alloys. https://core.ac.uk/download/301580505.pdf (accessed on 25th September 2025).

15. Fecht, H. J., Ivanisenko, Y., MacLaren, I., Sauvage, X., et al. (2006). Shear-induced $\alpha \rightarrow \gamma$ transformation in nanoscale Fe-C composite. https://core.ac.uk/download/1395045.pdf (accessed on 25th September 2025).

16. Ganin, A., Kimmel, G., Koltsov, I., Małysa, A., et al. (2018). Mechanism of reduced sintering temperature of Al_2O_3–ZrO_2 nanocomposites obtained by microwave hydrothermal synthesis. https://core.ac.uk/download/157812874.pdf (accessed on 25th September 2025).

17. Gault, B., Han, J. C., Kim, J. E., Na, Y., et al. (2019). Core-shell nanoparticle arrays double the strength of steel. https://core.ac.uk/download/160231358.pdf (accessed on 25th September 2025).

18. Gibson, P., Loëvestam, G., Putaud, J. P., Rauscher, H., Roebben, G., Sokull-Klüttgen, B., & Stamm, H. (2010). Considerations on a definition of nanomaterial for regulatory purposes. https://core.ac.uk/download/38620969.pdf (accessed on 25th September 2025).

19. González Crespo, P. A., Hughes, D. J., Pérez, C. L., Turrillas, X., et al. (2013). Strain mapping and nanocrystalline size determination by neutron diffraction in an aluminium alloy (AA5083) severely plastically deformed through equal channel angular pressing. https://core.ac.uk/download/19444165.pdf (accessed on 25th September 2025).

20. Hernandez Moreno, S., Solache de la Torre, S. C., et al. (2017). Nanotechnological products in architecture and construction. https://core.ac.uk/download/154796653.pdf (accessed on 25th September 2025).
21. Holschuh, B. T., & Newman, D. (2014). Low spring index, large displacement shape memory alloy (SMA) coil actuators for use in macro- and micro-systems. https://core.ac.uk/download/78062230.pdf (accessed on 25th September 2025).
22. Huang, Y., & Li, J. (2014). Microstructure and mechanical properties of an Mg-3Zn-0.5Zr-5HA nanocomposite processed by ECAE. https://core.ac.uk/download/29139708.pdf (accessed on 25th September 2025).
23. Langdon, T. G., & Valiev, R. Z. (2014). Report of International NanoSPD Steering Committee and statistics on recent NanoSPD activities.
24. Levesque, M. (2013). Anomalous surface segregation profiles in ferritic FeCr stainless steel. http://arxiv.org/abs/1302.2838 (accessed on 25th September 2025).
25. Lührs, L., & Weissmüller, J. (2018). Nanoporous copper-nickel–Macroscopic bodies of a strong and deformable nanoporous base metal by dealloying. http://arxiv.org/abs/1806.08610 (accessed on 25th September 2025).
26. Rai, S., & Roy, D. (2008). Sintering characteristics and characterisation of Al-Mn-Ce nanostructured material prepared by mechanical alloying. https://core.ac.uk/download/53186928.pdf (accessed on 25th September 2025).
27. Seo, M. H. (2011). Ge nanowires anode sheathed with amorphous carbon for rechargeable lithium batteries. https://core.ac.uk/download/79702833.pdf (accessed on 25th September 2025).
28. Torrecillas, R., Moya, J. S., Díaz, L. A., Gómez, B., et al. (2009). Nanotechnology in joint replacement. https://core.ac.uk/download/pdf/36093046.pdf (accessed on 25th September 2025).

CHAPTER 7

NANOMATERIALS IN POLYMERS AND COMPOSITES

Contents

INTRODUCTION

The emergence of nanomaterials has brought tremendous revolutions in material science and engineering, particularly for polymer and composite development. These materials harness the unique properties of nanoscale building blocks to enhance the property and functionality of traditional polymers. With sizes ranging from 1–100 nanometers, nanomaterials exhibit remarkable properties such as superior strength, lower weight, better thermal stability, and superior electrical conductivity. The combination of these materials with polymers could lead to new applications in various industries, including electronics, automobile, and healthcare. With industries continually demanding materials beyond the conventional state of the art, the work on nanomaterials in polymer composites is not only timely but also essential for advancing technology and sustainability. Unraveling the intricate interaction between nanomaterials and polymers is crucial in order to realize their full potential benefits. For instance, incorporation of carbon nanotubes into polymer matrices enhances the mechanical and electrical properties of ensuing composites significantly.

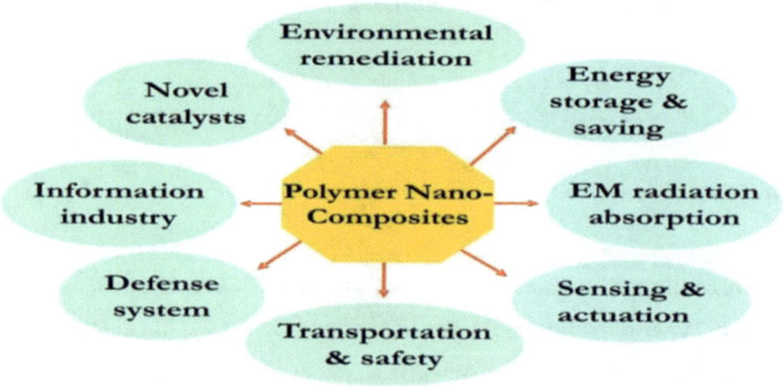

Figure 7.1. Nanocomposite materials.

Source: https://external-content.duckduckgo.com/iu/?u=https%3A%2F%2 Fcd-nintech.com%2Fmedia%2Fchapter%2F72636%2F1512345123%2Fmedia%2FF4.png&f=1&nofb=1&ipt=abd755f30e6966c137fdf015f76598d64a899df42d10337beecae606a177dbb5.

This is evident in research that confirms the filler concentration dependence of electrical conductivity in multi-wall carbon nanotube-polydimethylsiloxane (MWNT-PDMS) nanocomposites. With a varying percentage of the filler, electrical conductivity of the composites can be easily controlled with very high accuracy, thus emphasizing the ability of nanomaterials to design flexible pressure sensors and stretchable interconnects that are quite essential for advanced electronic applications (Dahiya et al., 2015). There are new possibilities for applications that could not be met before using standard materials. The rapid-emerging nanomaterials branch of science also includes some related fields like synthesis, characterization, and study of properties required for their inclusion within polymers and composites. Recent research has explored the stabilization and synthesis of noble metal nanoparticles, gold and silver specifically, which widens their application horizon in nanomedicine and sensor technology.

These chemically stabilized metal nanoparticles have unusual optical and physical features that can significantly improve the properties of polymer-based composites (Fratoddi et al., 2017). The investigations that are ongoing in this area reveal not only the amenability of nanomaterials but also the significant role that these are playing in shattering boundaries around traditional usage, postulating a revolutionary impact on upcoming technologies that are based on creative material characteristics (Figure 7.1).

Definition of Nanomaterials

Nanomaterials, with their unique properties and size typically between 1 and 100 nanometers, represent a principal frontier for material science. The size matters since materials possess size-dependent properties, i.e., at the nanoscale, physical, chemical, and biological properties may significantly differ from their bulk counterparts. The capability of nanomaterials to react to light, heat, and other stimuli is likely to result in increased strength, lower weight, increased chemical reactivity, and improved electrical conductivity. The high surface area-to-volume ratio of the nanomaterials also decides their properties in various applications and thus can be incorporated in numerous fields including electronics, medicine, and environmental science. The emergence of nanomaterials has particularly revolutionized the processing and properties of polymers and composites. It becomes feasible to enhance mechanical characteristics, thermal stability, and barrier properties by doping these matrices with nanomaterials, opening the window to new applications in packaging, automobile, and aerospace industries. Some nanomaterials such as carbon nanotubes and metal nanoparticles make a material contribution to composite performance through mechanisms such as stress transfer that improve the overall strength and durability of the composites. As industries seek materials that can withstand ever-increasing demands for enhanced performance, the role of nanomaterials cannot be overemphasized, ushering in a revolutionary period in material technology. Besides performance enhancement, nanomaterial application also extends to new solutions to complex issues in medicine and environmental sensing. For instance, recent advances show how nanomaterials can be engineered for site-specific interactions with biological systems with the assistance of breakthrough nanotechnology applications in drug delivery systems and biosensing. Incorporation of nanomaterials into medical devices, such as the biochips here described, facilitates real-time monitoring of human health indicators, meeting critical demands for real-time diagnosis and therapy. Incorporating such technologies, however, demands close attention to biocompatibility, reliability, and confidentiality, all of which are significant to the applied feasibility of using these advanced systems. This juxtaposition of benefits and limitations is reflective of the evolving definition and scope of nanomaterials in today's science and engineering (Baj-Rossi et al., 2012).

Importance of Polymers and Composites

The importance of polymers and composites in today's materials science cannot be overstated, particularly as we step further into the realm of nanomaterials. Polymers due to their high flexibility and green features are the foundation of most applications in the automobile, aerospace, and biomedicine industries. Due to their light weight and capacity to be engineered to desired properties, they make ideal candidates for composite materials. When nanomaterial such as metal nanoparticles is incorporated into polymers as reinforcement materials, they can be provided with increased mechanical strength, thermal stability, and better electrical conductivity. The introduction of these nanofillers not only improves the end-product performance but also facilitates more efficient, green process manufacturing, thereby addressing timely sustainability issues. The addition of nanomaterials into polymers has also led to the door of revolutionary technologies in electronic devices and biosensors.

In medicine, for instance, the development of continuous monitoring biochips is a byproduct of employing polymer-based systems with nanomaterial supplementation. These innovations can potentially revolutionize patient care through real-time analysis of biomarkers, critical to early disease diagnosis and treatment. However, fabrication of such sophisticated devices calls for answers to questions regarding long-term biocompatibility as well as data transmission security. The efficient use of nanomaterials in composite structures can lead to devices that not only function at optimal levels but also according to very stringent health and safety regulations, offering possibilities for future healthcare applications (Baj-Rossi et al., 2012). Beyond their functional uses, polymers and composites also help towards significant environmental issues. With greater emphasis on sustainability, the development of biodegradable polymers and nanocomposites ensures that high-tech material is not responsible for long-term environmental contamination. Recent research has examined the synthesis and characterization of biodegradable polymer matrices with noble metal nanoparticles, which have promising results in terms of performance without compromising ecological accountability (Fratoddi et al., 2017). These advances are crucial since industries are facing growing pressure to reduce carbon footprints. Continuing research into polymers and composites, enabled by nanotechnology, thus has the capacity to revolutionize material performance while setting the pace in environmental sustainability in manufacturing and product design (Figure 7.2).

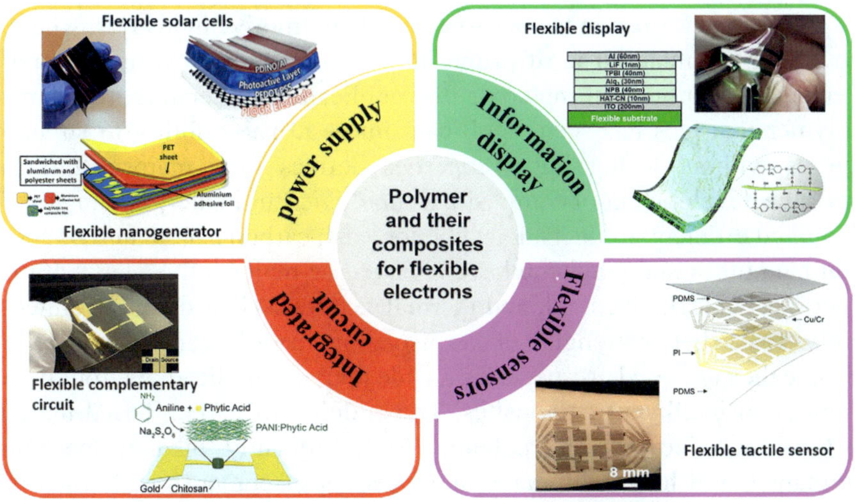

Figure 7.2. Polymers and composites.

Source: https://external-content.duckduckgo.com/iu/?u=https%3A%2F%2Ftse3.mm.bing.net%2Fth%2Fid%2FOIP.WdlvnDR3DxihHefxCrUhFAHaEW%3Fr-%3D0%26pid%3DApi&f=1&ipt=f6646d9a80ab2ae20d67e5da1625c9f442b589ca80ee85debf6c6dc92ff74c7e.

7.1. TYPES OF NANOMATERIALS

Nanomaterials are typically categorized according to their structure and composition, which in turn define their specific characteristics and applications. Metallic nanoparticles, such as gold and silver, are among the most deeply researched nanomaterials. The nanoparticles exhibit certain optical, electronic, and catalytic properties, which make them suitable for various applications, such as drug delivery, imaging, and sensors. The size and shape of such nanoparticles can be controlled with high precision through various synthesis methods in order to facilitate the formation of materials having predetermined functionalities. Such work in recent years has aimed at stabilization and size control of such metal nanoparticles, highlighting their promising potential in emerging fields such as nanomedicine and environmental monitoring to determine their role as key building blocks in composite materials (Fratoddi et al., 2017). Apart from metallic nanoparticles, ceramic-based nanomaterials are also another important class that is recognized by hardness and thermal stability. Titanium dioxide and silica nanoparticles are common examples that find applications to enhance mechanical and thermal properties of polymers and composites. Ceramics are prone to have fine resistance to heat and chemical corrosion and, therefore, find applications for use in the aerospace and automotive industries, where the longevity of the material is of prime importance. With growing demand for lighter but more powerful materials, the application of ceramic nanoparticles in polymer matrices has seen significant interest. Their ability to strengthen structures without losing light properties defines their major significance towards the development of composite technologies, ultimately leading to more resilient and efficient products. Last but not least, carbon nanomaterials such as carbon nanotubes and graphene have emerged as major influencers with their distinct mechanical, electrical, and thermal features. Their unique architecture not only creates better strength but can also allow record electrical conductivity, and hence they are good prospects for a wide range of applications from flexible electronics to intelligent composites. The addition of carbon-based materials to polymer matrices has the potential to significantly enhance mechanical performance, and hence the material properties of composites. Furthermore, collaborative work like that of the 2BFUNTEX project aims to accelerate industrial uptake of such advanced materials by bridging academia-industry gaps, prompting the development and implementation of new textile structures and materials that are attuned to current performance needs (Kiekens et al., 2012).

7.1.1. Carbon-Based Nanomaterials

The advancement in the area of carbon-based nanomaterials has been followed by enormous research on their deployment in composite and polymer matrices mainly due to their ultra-mechanical and electrical properties. Graphene and carbon nanotubes (CNTs) are the best cases in point, demonstrating super-high tensile strength and conductivity, respectively, which make them highly suitable for enhancing the performance of other polymers. To take an example, the incorporation of multi-wall carbon nanotubes into PDMS has been researched quite extensively. Studies have demonstrated that the electrical conductivity of such composites can be regulated through control of filler concentration, presenting a linear correlation between structural design and functional behavior. Thus, the composites promise revolutionary applicability in flexible electronics and wearable electronics with the help of the universality of carbon nanomaterials to universalize the functional properties of traditional polymer matrices (Dahiya et al., 2015). Aside from their electrical applications, carbon-based nanomaterials also dictate the mechanical properties of polymer composites. The application of exfoliated graphite, for instance, in polyamide 6 (PA_6) speaks well of the effectiveness of these nanomaterials in tailoring the behavior of polymers. By initiating a simple process of dispersion, researchers have significantly improved the mechanical properties of the resultant composite. Compatibility and quality of dispersion between the polymer precursor and the graphite surface may be analyzed using atomic force microscopy, which provides informative details regarding their compatibility and quality. The major analysis by scanning and transmission electron microscopy supports our understanding of how graphite structure interpenetrates the polymer matrix, showing how these modifications lead to composites that are stronger and durable (Bocchini et al., 2012). The incorporation of carbon nanomaterials into polymers is not merely a matter of optimization but also drives innovation in multicomponent composite design. The unique characteristics of carbon materials provide novel possibilities for application in thermal management, energy storage, and pollution remediation. By harvesting the high thermal conductivity of CNTs and graphene, researchers are looking for ways to produce composites that optimally regulate heat transport and thermal stability. In addition, their surface properties enable functionalization according to desired applications, e.g., to improve absorbent character in environmental application. With the development of technology, the ability to use and manipulate the behavior of carbon-based nanomaterials will undoubtedly lead revolutionary advances in the context of polymers and composites to enhance their applications in different industries (Figure 7.3).

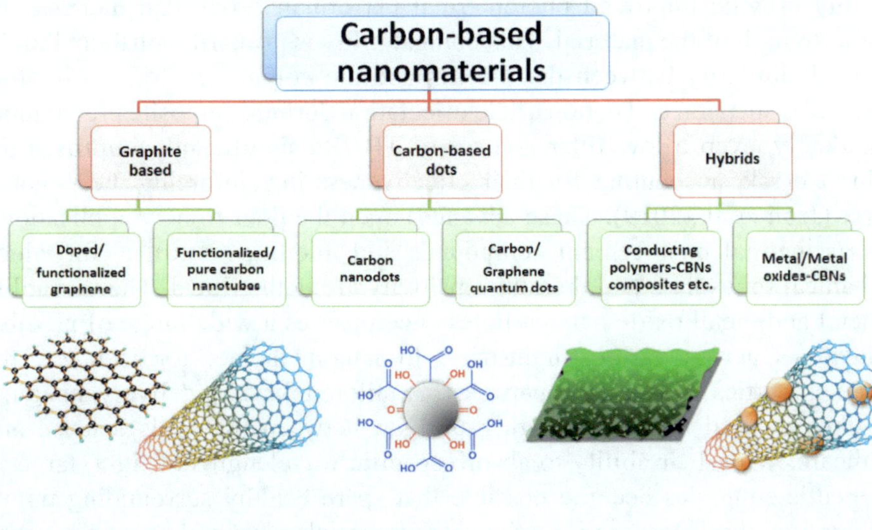

Figure 7.3. Carbon-based nanomaterials.

Source: https://external-content.duckduckgo.com/iu/?u=https%3A%2F%2Fwww.mdpi.com%2Fsensors%2Fsensors-17-02587%2Farticle_deploy%2Fhtml%2Fimages%2Fsensors-17-02587-g003.png&f=1&nofb=1&ipt=ca85d632d12fd0cb3aac299431f3551141f66da3d9a7a811e3054908ccc26b13.

7.1.2. Metal and Metal Oxide Nanoparticles

Metal and metal oxide nanoparticle use in polymer matrices tremendously enhances the physical and chemical properties of composite materials. These particles, with a high surface area-to-volume ratio, exhibit new optical, electrical, and mechanical properties different from their counterpart bulk material. This differential property is particularly evident in noble metal particles like gold and silver, which can be stabilized and synthesized using a variety of ligands to achieve desired properties in composites. Current research has indicated how effective synthesis of such nanoparticles could be, with demonstrations of how the control of parameters such as size and morphology can be utilized to engineer their properties for use in applications in fields of nanomedicine and sensing technologies. An example is that stability and uniform size of such nanoparticles could lead to enhanced functionality in optical devices, with new applications reliant on plasmonic properties (Fratoddi et al., 2017). Additionally, polymer nanocomposite production via metal oxide nanoparticles has been extremely promising regarding boosting mechanical and thermal properties. Tungsten bronze nanorods (TBNRs) are a good case in point. TBNRs, when utilized with ethylene propylene diene monomers (EPDM), not only provide improved photothermal performance but also increase the tensile strength of the material. This compatibility is primarily attributed to the chemical similarity between the TBNRs and the polymer matrix, facilitating improved dispersion and nanoparticle orientation during composite preparation. Remarkably, even a low filler content of TBNRs significantly improves the strain at break, accounting for their effectiveness in reinforcing the polymer matrix (Jeon et al., 2019). These advances open the door to new applications, such as thermal management compounds and intelligent textiles, for which mechanical robustness and thermal sensitivity are both critical. The versatility of metal and metal oxide nanoparticles encompasses a wide range of possible applications across various industries. In nanomedicine, for instance, the unique properties of these nanoparticles are tailored for drug delivery, imaging, and photothermal therapy, bridging the gap between materials science and healthcare. With their ability to absorb specific wavelengths of light, targeted therapeutic strategies become possible that spare healthy surrounding tissue. Furthermore, the incorporation of such nanoparticles into polymers can result in bio-compatible composites that play a vital role in medical implants and devices. Continued research into optimizing the synthesis of nanoparticles and dispersion in polymer matrices has the potential to deliver material that is both functional and designed for specific biomedical applications, underlining the significance of metal and metal oxide nanoparticles in fueling the advancement of nanomaterials (Fratoddi et al., 2017) (Figure 7.4).

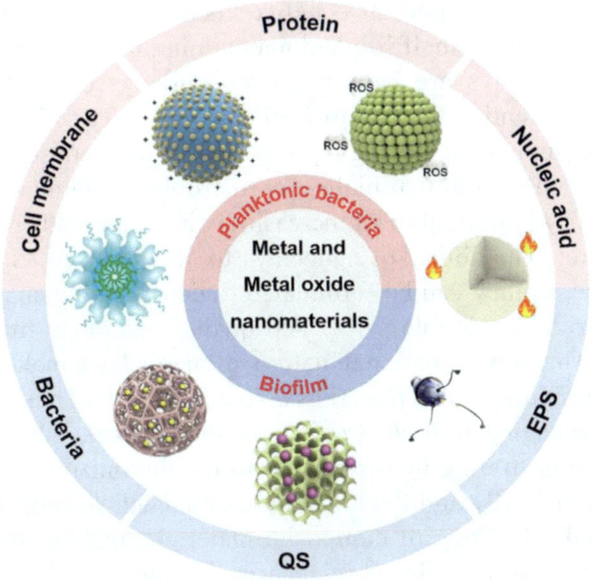

Figure 7.4. Metal and metal oxide nanoparticles.

Source: https://external-content.duckduckgo.com/iu/?u=https%3A%2F%2Fencyclopedia.pub%2Fmedia%2Fitem_content%2F202302%2F63db2accd942fijms-23-11348-g001.png&f=1&nofb=1&ipt=221a09c1f467b3656317464fa5f4c4d99278a11dd6ccf9dda4ff9de1c7b6d7ba.

7.1.3. Nanoclays and Their Applications

Nanoclays application in polymers and composites is a young but rapidly expanding technology with significant potential for improvement of material property and function. Halloysite, sepiolite, and laponite are only some of the nanoclays with good nanofiller quality, which can markedly improve mechanical and thermal behavior. A crucial example is seen in the formulation of bioplastics like Mater-Bi, which is now applied more extensively for food packaging. Research on Mater-Bi/nanoclay composites has revealed that the incorporation of halloysite significantly enhances the mechanical properties of such films without adversely affecting their thermal behaviors, suggesting that such composites can match or even excel conventional petroleum-based plastics in some performance metrics. Such a discovery, therefore, is rich with potential for green material creation, especially in applications that demand environmentally friendly approaches without compromising much on quality or functionality (Cavallaro et al., 2018). Aside from their mechanical properties, nanoclays also possess important benefits when employed in food packages, particularly in extending shelf life and maintaining the quality of foods. Used in packaging, nanoclays are barriers to moisture, gases, and light that are responsible for preventing the loss of freshness and safety of foods. Nanoclays are utilized to complement packaging materials at the nanoscale to contribute functionalities such as antimicrobiality for microbial contamination protection. This is particularly important in an increasingly hygiene- and consumer health-conscious market. The ability of nanoclays to also be used as food additives widens their use, as they can be formulated to detect pathogens or report food quality, thus providing double safety and quality assurance functions (Samal et al., 2017). The multifunctional nature of nanoclays makes them a key component in designing novel materials across industries. Their applications go beyond food packaging into electronics, pharmaceuticals, and environmental remediation, demonstrating their potential to revolutionize more than a single industry. The ecofriendly and flexible nature of nanoclays enhances the overall performance and efficiency of composite materials, making them invaluable in modern manufacturing. Since science is continuously looking for newer approaches to work with nanoclays, the synergy of nanotechnology and material science is able to pioneer the way to new products that balance performance and environmental accountability. Not only does this synergy of technologies duplicate the industry's trend towards greener solutions but also responds to the increasingly stringent consumer interest in multibenefit and health-promoting material (Cavallaro et al., 2018; Samal et al., 2017).

7.2. MECHANISMS OF INCORPORATION

Incorporation of nanomaterials into polymer matrices is a sophisticated interplay of physical and chemical mechanisms that has a marked impact on the properties of the final composites. Knowledge of these mechanisms is essential to achieve the maximum performance of nanocomposites. One key area of incorporation is the dispersion of nanoparticles in the polymer matrix that is determined by particle size, shape, surface chemistry, and viscosity of the polymer. The homogenous distribution of nanomaterials increases properties such as mechanical strength and thermal stability, which are crucial for applications across different sectors. Furthermore, inadequate dispersion may cause nanoparticle agglomeration and adversely affect intended performance improvements of materials. Therefore, investigating successful techniques for the addition of nanomaterials is central to improving the field of composites (Fratoddi et al., 2017). Besides dispersion, the interface between nanomaterials and the polymer matrix is crucial in defining the composites properties. Such interactions give rise to the creation of novel interfacial properties that have a strong effect on mechanical and thermal behavior. For example, in the case of titanium dioxide nanoparticles in an epoxy matrix, there are novel synergistic effects formed that are different from those found with traditional fillers, e.g., greater charge mitigation. The nano-sized particles allow for more effective charge distribution within the matrix relative to larger particles, which are prone to interfacial polarization effects. This fact shows that when material is examined in the nanoscale, the filler properties change, and this enhances performance characteristics in nanocomposites. The study of these interactions is crucial for adapting material properties to fulfill particular application needs (Alison et al., 2004). In addition, dynamics of incorporation mechanisms can also be related to the particular loading concentrations of nanoparticles in the polymer matrix. Studies have demonstrated that there is an optimal loading range of nanoparticles that optimizes the advantage with minimal disadvantage through reduced elasticity or enhanced brittleness. For instance, studies suggest that the addition of around 10% by weight improves the electrical strength of the composite enormously. If more nanoparticles are added above this critical concentration, the performance levels out or even drops due to overcrowding of the matrix. Thus, careful control of the concentration of nanomaterials is critical to realize enhancements in mechanical, thermal, and electrical properties. Examination of these loading dynamics offers significant insight into developing next-generation composites with specific functionalities (Fratoddi et al., 2017).

7.2.1. Methods of Synthesizing Nanomaterials

Synthesis of nanomaterials involves a range of methodologies that take advantage of distinctive physical and chemical principles to manipulate material at the nanoscale. Among these processes, top-down and bottom-up are the most well-known. Top-down synthesis is a mechanical reduction of bulk materials into nanoscopic particles, depending on methods such as milling, lithography, and etching. Top-down synthesis can control size and shape but can add structural defects. Conversely, bottom-up synthesis is concerned with the assembly of materials atom by atom or molecule by molecule using processes like chemical vapor deposition, sol-gel synthesis, and hydrothermal procedures. This method tends to produce purer and more homogeneous nanoparticles, an essential requirement for uses in advanced composites and polymers where the homogeneity of nanomaterials can have a direct impact on material properties. A good example of synthetic approaches in noble metal nanoparticles involves numerous stabilization methods that facilitate their use in various areas. For example, gold and silver nanoparticles synthesized via citrate reduction or seed-mediated growth show great stability and polymer compatibility. These syntheses enable tunable particle size, which can be critical for use in nanomedicine and optical sensing, where nanoparticle optical properties depend linearly on size and shape. In addition, syntheses of these nanoparticles are frequently coupled with surface modification techniques, including ligand attachment, to enhance dispersion in polymer matrices. By integrating such altered nanoparticles into polymer composites, the mechanical, thermal, and electrical properties of the resulting materials can be improved, thus increasing the functionality and performance of the resultant material (Fratoddi et al., 2017).

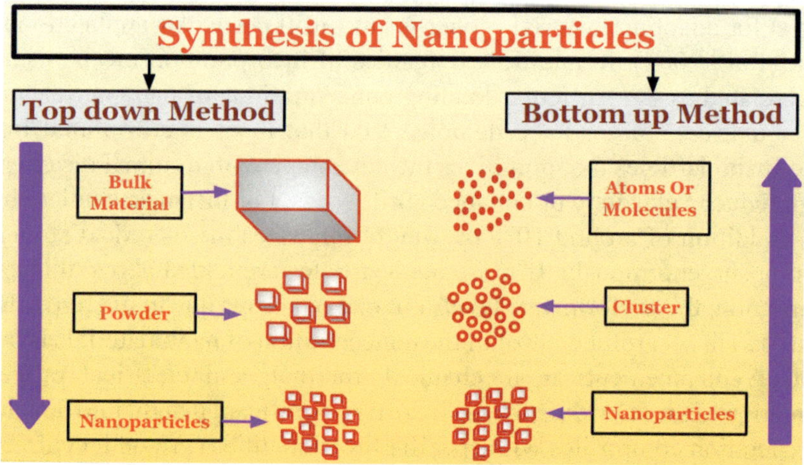

Figure 7.5. Methods of synthesizing nanomaterials.

Source: https://external-content.duckduckgo.com/iu/?u=https%3 A%2F%2 Fblogger. googleusercontent.com%2Fimg%2Fb%2FR29vZ2xl%2FAVvXsEjCPIsRBs ZfiToJGExxIYAf6ZFHxsEIdcQe5qeh34PYUdGL9V40pk1k57YclV9o8C4PANn 6K-uVNuLsI2jl-A6ow_7u5TSFHmSl3Bk0RFByye0UCySikvL1L1jyQwu7Z-0kEzsuDuog-cspdTUqlM5fXwGnGKAHZgl_tk2n7bl5hU9fQHTgfOm3mptQ-sg MY%2Fs1267%2FSynthesis%2520of%2520Nanoparticles.jpg&f=1&nofb=1&ipt =16db3baebacee9ffe3b499556a51627db165399abf3ffc39a9fac8f81aed1965.

Moreover, cooperative frameworks among multiple stakeholders—ranging from research bodies to industrial allies—form an essential aspect in moving nanomaterial synthesis forward. Examples such as 2BFUNTEX illustrate the role played by interdisciplinary collaboration in bridging technological gaps in functional materials. Through collaboration, such projects promote rapid development and industrial adoption of novel nanomaterials customized for precise functionalities. The multidisciplinary nature of such projects guarantees that not only are the synthesis procedures scientifically valid, but they are also economically feasible. By having the project specify technology requirements and ensure knowledge sharing, it facilitates the advancement of new fabrication processes that can be exploited in the manufacture of advanced composites and textiles. Finally, such collaboration creates synergy that places research and development at the forefront of the nanomaterials field (Kiekens et al., 2012) (Figure 7.5).

7.2.2. Techniques for Integrating Nanomaterials Into Polymers

The integration of nanomaterials into polymers offers several techniques that improve the mechanical, thermal, and electrical features of the resultant composites. One common approach is the solution mixing method, where the nanomaterials are mixed with a solvent and the polymer to create a uniform blend before precipitating. This approach provides improved control over nanomaterial dispersion within the polymer matrix, resulting in a dramatic enhancement of properties like tensile strength and elasticity. Moreover, scientists have noted that the electrical conductivity of MWNT-PDMS nanocomposites is very sensitive to the concentration of fillers, making it plausible that precise tuning of these parameters will result in materials appropriate for uses such as flexible pressure sensors and electronic skin elements (Dahiya et al., 2015). The approach demonstrates the criticality of having precise formulation when creating high-performance nanocomposite materials. Another significant method is the utilization of melt blending, where nanomaterials are mixed with molten polymers. This is preferred due to its effectiveness and ease, especially in industrial utilization where there is a need for large-scale fabrication. Through the application of high shear forces to the mixture, melt blending facilitates sufficient dispersion of nanomaterials in the polymer matrix, as well as the interaction between the two phases. The nanomaterial's compatibility with the polymer is important; surface chemistry modification of nanomaterials improves their dispersion and compatibility. As implied in project research such as 2BFUNTEX, building alliance between academia and industry can result in the recognition of technological gaps, allowing the creation of materials with unique functionalities and performance improvements (Kiekens et al., 2012a). Such a multidisciplinary study is crucial in advancing the study of nanocomposites. Finally, *in situ* polymerization is an advanced method that is also employed in including nanomaterials in polymers. In this technique, the polymerization process itself is performed in the presence of suspended nanomaterials so that there is a closer interaction between the nanofillers and the polymer chains. Advantages of this method include greater efficiency of load transfer and customized properties by the control of the conditions of polymerization, e.g., temperature and pressure. Studies show that thermal stability and mechanical strength are better in nanocomposites fabricated by *in situ* methods compared to nanocomposites prepared by other methods. As the need for multifunctional materials increases, cross-disciplinary investigation of integration methods becomes increasingly important for maximizing the properties and usefulness of nanomaterials in polymers and composites, thereby widening their application scope across a wide range of industries (Figure 7.6).

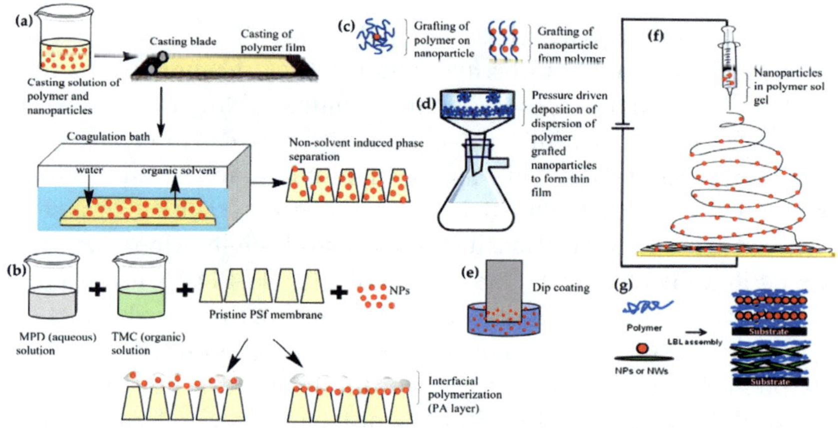

Figure 7.6. Techniques for integrating nanomaterials into polymers.

Source: https://www.researchgate.net/publication/367346921/figure/fig1/AS:11431281409623044@1745834848326/Various-methods-of-integrating-nanoparticles-with-polymer-to-form-polymer-nanocomposite.tif.

7.2.3. Challenges in the Incorporation Process

The addition of nanomaterials to polymers and composites has many-sided challenges that need to be carefully addressed in order to obtain desired material behavior. One of the major issues is the uniform dispersion of nanoparticles within the polymer matrix. Uniformity is essential; however, the presence of agglomeration and incompatibility between the polymer and nanoparticles can act as obstacles to this objective. For example, when LDHs are added to other polymers, the polymer type influences the performance and dispersion of the LDHs. As shown, ZnAl performs better in polyethylene whereas MgAl performs best in poly(methyl methacrylate), which shows that knowing intrinsic interactions is critical for maximizing performance. Unless the compatibility problems are solved, even the most hopeful nanomaterials can't improve the properties of the composite structure, posing great obstacles to practical applications. Another great difficulty is the thermal stability and longevity of the nanocomposites. The addition of nanomaterials may change the polymer thermal degradation processes, as demonstrated in research with flame properties associated with the kind of LDH incorporated. Whereas the use of oleate-loaded LDHs can enhance thermal stability and minimize peak heat release rates in polyethylene, this is not the case with every polymer. In particular, poly(ethylene-co-butyl acrylate) fails to provide adequate interactions with specific LDHs, resulting in poor improvements in material properties. This demonstrates that the choice of nanofillers must not only account for compatibility but also for the probable influence on thermal and mechanical properties. In-depth knowledge of these interactions is key towards further developing the application of nanocomposites, and thus intensive empirical studies are required. Finally, stabilization and synthesis of nanomaterials themselves pose further challenges in the integration process. Forming defined nanoparticles demands advanced methods that can yield uniform sizes, morphologies, and stabilizing ligands to preserve their functionality in mixed environments. Consider metal nanoparticles, for example, that difficulties in terms of uniformity and stability can finally determine their performance in applications from nanomedicine to sensors. Although important progress has been made in the synthesis of gold and silver nanoparticles, more study and development in terms of stabilization tactics continue to be a top priority. Herein, the literature offers a wealth of examples describing improvements, but determining the best conditions for various nanomaterials is still a daunting task that highlights the voluminous amount of research needed to overcome the challenges of integrating them into polymer matrices.

7.3. PROPERTIES ENHANCED BY NANOMATERIALS

The incorporation of nanomaterials into polymers and composites greatly improves their inherent properties, thus becoming more desirable for many industrial uses. One of the most significant advances relates to thermal stability. For example, application of layered double hydroxides (LDHs) of zinc and magnesium has been demonstrated to shield polymers against thermal oxidation. In research, oleate-functionalized LDHs were blended into polyethylene and poly(methyl methacrylate) to create nanocomposites that showed significantly enhanced thermal and fire performance. The distinctive packing of the oleate anions in the LDH galleries promotes better dispersion of the nanoparticles in the polymer matrix, ultimately resulting in better thermal performance. Additionally, the varying morphologies of ZnAl and MgAl LDHs enabled enhanced properties based on the type of polymer, giving an indication of the effectiveness and potential of nanomaterial modifications in improving thermal behavior and ignition resistance of polymer composites (Hossenlopp et al.,

2009). The mechanical properties of polymers are also significantly improved with the addition of nanomaterials. Gold and silver nanoparticles have been highly researched for their tensile strength-enhancing and flexibility-improving capabilities in composite materials. Using diverse techniques of synthesis, scientists have been successful in achieving well-defined size and morphology of nanoparticles, which are necessary for allowing optimum interfacial bonding with the polymer matrix.

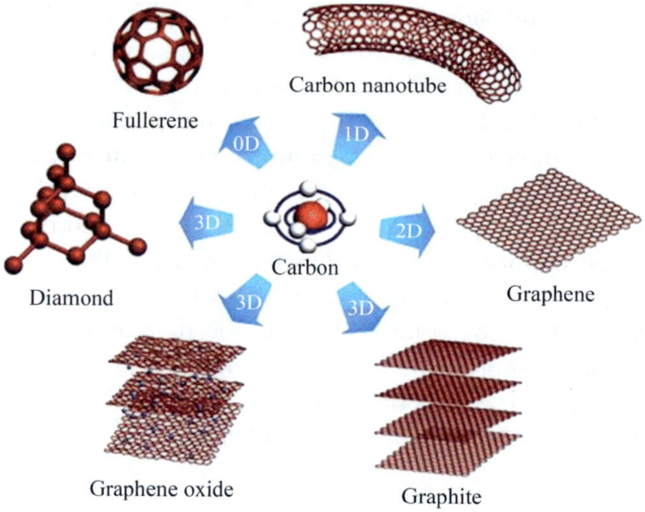

Figure 7.7. Nano-enhanced polymer composite materials.

Source: https://www.mdpi.com/polymers/polymers-17-00893/article_deploy/html/images/polymers-17-00893-g017-550.jpg.

Their distribution has a direct impact on the mechanical stability of the composite; an even distribution results in enhanced load-carrying ability and toughness. In addition to that, the novel properties of metallic nanoparticles, due to their surface plasmon resonance, also contribute to changes in mechanical responses, thus promoting a range of applications in areas like nanomedicine and sensing, as illustrated by recent publications (Fratoddi et al., 2017). Lastly, adding nanomaterials significantly improves the fire retardancy of polymer composites, an area of significant research and application. The involvement of nanoparticles in flame suppressive mechanisms is of critical significance; for instance, layered double hydroxides have a strong ability to suppress peak heat release rate in polymers such as polyethylene. This decreased flammability is due to char protection provided through the nanomaterials, a barrier against heat transport and fire propagation. Moreover, the compositional variations in nanomaterials, like the specific interactions of the LDHs with the polymers, provide different fire safety properties in different composites. These observations highlight the customized design for nanocomposite development, where the choice of the right nanofiller can greatly improve fire resistance without sacrificing other critical properties, thus broadening the range of applications of such advanced materials in safety-critical situations (Hossenlopp et al., 2009) (Figure 7.7).

7.3.1. Mechanical Properties Improvement

The addition of nanomaterials to polymer matrices greatly improves their mechanical characteristics, which is essential to grow their application in other industries. Recent studies have proved that layered double hydroxides (LDHs), e.g., zinc aluminum (ZnAl) and magnesium aluminum (MgAl), can greatly influence the performance of polymers such as polyethylene and poly(methyl methacrylate). These LDHs not only enhance the mechanical strength but also offer thermal protection, as noted by the improvement in thermal and fire properties from their use in polymer nanocomposites. The unique morphological interactions between the LDHs and various polymers create selective mechanical durability enhancements, indicating that attention must be paid to the selection of nanomaterial when used for targeted applications so as to maximize performance features and desired effects in polymer blends (Hossenlopp et al., 2009). Application of exfoliated graphite in polyamide 6 (PA_6) illustrates the immense improvement in mechanical features possible using nanomaterial addition. The process of preparation includes dispersion of graphite in the monomer, ε-caprolactam, and subsequent polymerization, leading to a well-dispersed composite system. Techniques such as atomic force microscopy and scanning electron microscopy have found that the involvement of specific interactions between the polymer and graphite is responsible for enhanced mechanical strength and toughness of the composite material. This improvement in mechanical characteristics is critical for engineering and manufacturing applications, wherein increased durability and resistance to deformation are frequently most important (Bocchini et al., 2012). In addition, not only does the incorporation of nanomaterials aim for the enhancement of mechanical properties, but it also addresses issues concerning material flammability and thermal degradation. The differences in the properties of different nanomaterials, as observed with the difference in dispersion effectiveness of ZnAl in polyethylene and MgAl in poly(methyl methacrylate), reflect the necessity of formulating composites according to needs. For example, the strong suppression of peak heat release rate in zinc-based LDH-filled polyethylene composites is consistent with improved fire resistance, while the magnesium-based LDH ensures enhanced fire properties in polar polymers. This focused strategy in the application of nanomaterials offers a guide in designing superior polymer composites that are in equilibrium between mechanical strength, safety, and thermal stability (Hossenlopp et al., 2009).

7.3.2. Thermal Stability and Conductivity

Thermal stability and conductivity of nanomaterials are key elements in improving the performance of polymers and composites. The use of metal nanoparticles, for example, gold and silver, has been widely investigated for their potential to enhance thermal performance. These metal nanoparticles possess features that enable enhanced thermal dissipation and thermal stability enhancement of the host polymer matrix. Their synthesis and stabilization have been extensively proven in various studies to achieve favorable thermal characteristics, leading to composites that function best in various applications. Through the control of nanoparticle-polymer matrix interactions, researchers can design thermal properties similar to desired specifications, thus broadening the scope of possible applications in nanomedicine and electronics (Fratoddi et al., 2017). Besides thermal stability, nanomaterial thermal conductivity has significant impacts on the advancement of sophisticated composites. Nanofillers like carbon nanotubes or graphene have demonstrated excellent thermal conductivity, greatly improving the heat management properties of polymer-based materials. Through strategic addition of these nanofillers, one is able to produce composites with superior thermal performance as well as mechanical integrity. Such improvement is vital in applications involving efficient heat dispersal, for example, aerospace and automotive sectors. The capacity to engineer and manufacture composites with customized thermal behavior is, as such, a powerful development enabled by the employment of nanomaterials (Fangueiro et al., 2015). Such developments highlight the revolutionary capability of marrying advanced materials science with application in engineering. Additionally, sustained research is imperative to maximize the potential of nanomaterials in polymer composites. The area of study goes on changing with new methods being engineered to optimize thermal characteristics while overcoming issues like dispersion of nanoparticles within polymer matrices. Future research needs to concentrate not just on maximizing thermal stability and conductivity but also the impact of environmental conditions and long-term strength of such materials. Understanding the mechanisms of interaction between nanomaterials and their polymers will help optimize their functioning under different conditions. As such, the study of nanomaterials promises the development of multifunctional composites, which can have significant effects on different industries, such as construction, medical devices, and energy storage devices (Fangueiro et al., 2015; Fratoddi et al., 2017) (Figure 7.8).

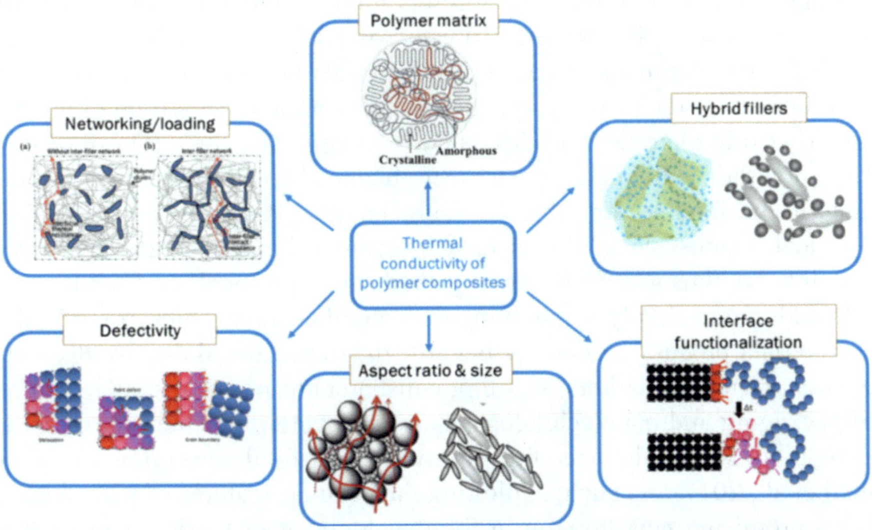

Figure 7.8. Thermal stability and conductivity by nanomaterials into polymers.

Source: https://www.researchgate.net/publication/352302770/figure/fig2/AS:1143 1281394012525@1745427062377/Main-factors-influencing-the-thermal-conductivity-in-polymer-nanocomposites.tif.

7.3.3. Barrier Properties and Permeability

Permeability and barrier properties of materials play significant roles in assessing their effectiveness, especially for polymers and composites that involve the use of nanomaterials. Polymers by nature have different levels of permeability depending on their molecular structure, which may affect their usefulness in applications demanding airtightness or moisture resistance. The incorporation of nanomaterials, including nanoparticles and nanofibers, reinforces these barrier properties with a more tortuous pathway for permeating molecules. This structural enhancement not only enhances the physical performance of polymers but also significantly contributes to improving shelf life and safety of packaged products, hence meeting consumer needs for longevity without quality compromise. The use of nanotechnology in polymers is not just for the enhancement of mere structures; it includes functional enhancement, most especially in food packaging where barrier properties are critical. For example, adding nanomaterials can result in the creation of active packaging systems, which actively engage with the packaged food products. Such systems have the ability to counteract oxidative degradation and regulate moisture transfer, providing optimal conditions for preservation of environmentally sensitive substances, such as beverages. Efficacy of such systems is demonstrated through the development of packaging materials that protect against nutrient loss and preserve flavor integrity, such as innovations focused on eliminating aromatic profile alterations in drinks such as wine and juice (Garrigós et al., 2015). In addition, the consequences of enhanced barrier properties are most notable in the food sector, where consumer protection and product durability are critical. Nanoscale additives have proven to be able to repel microbial contamination, thus greatly improving food safety protocols and product shelf life. In addition to simply maintaining the natural characteristics of foods, these nanomaterials possess a two-way benefit: they have the ability to also act as freshness or quality markers, warning consumers of possible spoilage. Where food packaging and nanotechnology therefore overlap lies the critical area of investigation, setting the pace for improved, safer food preservation measures (Samal et al., 2017). As such, exploiting the peculiar features of nanomaterials offers a promising new horizon in the development of barrier properties and permeability in polymers and composites.

7.4. APPLICATIONS IN VARIOUS INDUSTRIES

The use of nanomaterials within polymer and composite matrices is transforming applications across various industries, most notably food technology. Nanotechnology enhances food processing through the addition of nanoparticles, which perform a double role as additives and as packaging materials. Nanoparticles greatly enhance the shelf life of the product by preventing microbial contamination and generally improving hygiene. As stated, use of Nano particles as food ingredients makes the food to stay out of microbial contamination thereby extending the shelf life (Samal et al., 2017). Such capability not only takes into account safety factors but also enhances consumer confidence in food quality. Additionally, nanotechnology development enables the production of functional foods that have the potential to deliver enhanced nutritional profiles, hence addressing today's demands for health-oriented products. The ongoing development in this field eventually gives a competitive edge to the manufacturers in a progressively health-conscious market. Going in search of sustainable resources, the development of cellulose nanomaterials from renewable resources is an example of the innovative applications of nanotechnology in various industries. The biodegradable materials hold the potential for numerous environmental and economic benefits in that they provide substitutes to conventional plastics and composites. Through the utilization of trees' cellulose, industries are able to manufacture light, but strong and adaptable, products in alignment with the growing need for sustainable solutions. The report by the U.S. Forest Services outlines the capacity for creating hundreds of thousands of jobs while framing the restoring of forests as urgent through this new process (Brun et al., 2014). While stakeholders from academia, industry, and government come together to discuss research priorities and commercialization roadblocks, it is evident that cellulose nanomaterials are a breakthrough toward not only fulfilling market requirements but also improving green practices in numerous industries. Additionally, the versatility of nanomaterials transcends their applications into industries like construction and automotive, where property enhancements of materials are becoming more critical. Application of nanoparticles can improve mechanical characteristics, thermal stability, and even fire resistance of composites. This translates to products that can withstand more extreme conditions and reduce the weight of buildings and cars overall, thus enhancing fuel efficiency and performance. Increased lifespan offered by nanocomposites also reduces the need for frequent high-frequency replacements and hence makes lifecycle costs favorable. These advances not only answer the industry's requirements for high-performance materials but also assist in fulfilling the broad agenda of sustainability through a global optimization of product lifetimes. The promise of nanomaterials thus promotes innovation in the majority of industries and positions nanotechnology at the forefront of shaping the destiny of material science.

7.4.1. Automotive Industry Advancements

The automotive industry is experiencing revolutionary advances led by the application of nanomaterials in polymers and composites. Such materials possess improved mechanical and thermal characteristics that are ideal for applications to enhance vehicle performance and sustainability. Nanocomposites used during production processes offer an impressive weight savings without compromising structural strength. The weight savings are significant because they assist in improved fuel efficiency and lower emissions, which are consistent with global standards for cleaner transportation technologies. As car makers become increasingly concerned with becoming environmentally compliant, the application of nanomaterials becomes increasingly important to the future of automobile production. Moreover, the light weight of these materials also makes it possible to produce electric and hybrid vehicles, which are extremely reliant on efficient energy use and advanced engineering design in an effort to remain competitive within an increasingly dynamic market space. Furthermore, the application of cellulose nanomaterials in industry provides a new path to sustainability for the automobile sector. Derived from renewable resources, the biobased nanomaterials are used as reinforcing additives in polymers to construct properties and improve biodegradability. Recent workshops organized by the U.S. Department of Agriculture and the National Nanotechnology Initiative highlighted commercialization of inexpensive, renewable cellulose nanomaterials, potentially triggering substantial job creation on a massive scale as well as beneficial environmental impacts. Research into these materials reinforces the automotive industries growing focus on sustainability while maintaining competitive positions.

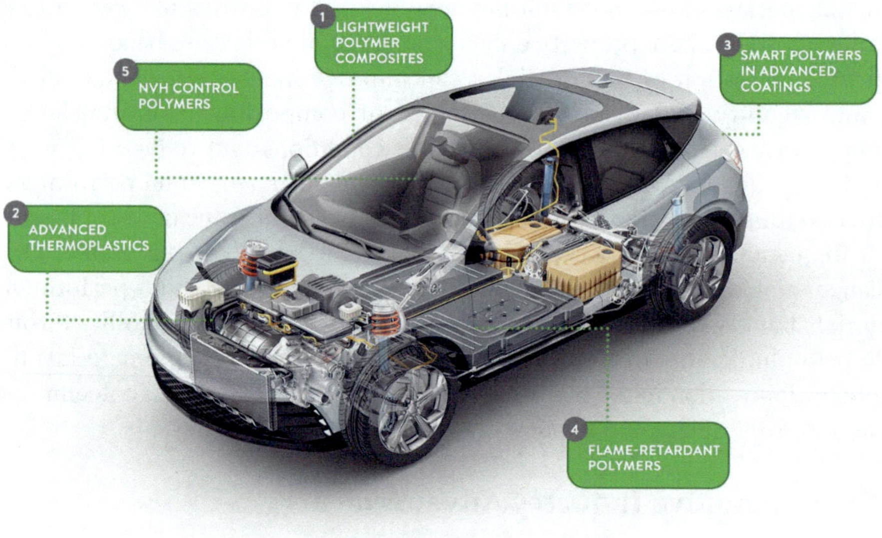

Figure 7.9. Automotive industry advancements of nanomaterials into polymers.

Source: https://www.plasticsengineering.org/wp-content/uploads/2024/09/Auto-Polymers_Infographic.jpg.

By advancing and enhancing research and development of cellulose nanomaterials, the industry is offered a unique opportunity to move traditional manufacturing practices towards sustainable means towards the final establishment of a circular economy in the automotive sector (Brun et al., 2014). Bio-based and nanotechnology-reinforced materials have extremely critical mechanical properties for utilization in the automotive sector. These materials possess high strengths in terms of strength-to-weight ratios and enhanced durability under environmental stressors. Specifically, research revealed that incorporation of nanofillers can enhance mechanical properties like tensile and flexural strength, which are of utmost importance in parts of motor vehicles.

The challenge lies in overcoming manufacturing technique barriers as well as ensuring batch-to-batch uniformity in quality in different production runs. With focus on developing these qualities using advancements in nanocomposite technology, the automotive industry can further integrate green aspects without losing performance. As environmentally friendly vehicles become increasingly in demand, creative solutions to the application of nanomaterials in polymers and composites will play a significant role in shaping the future automobile industry (Barbhuiya et al., 2025) (Figure 7.9).

7.4.2. Aerospace and Defense Applications

The defense and aerospace sectors have in recent years increasingly turned to nanomaterials in polymers and composites in order to enhance performance and reduce weight. The introduction of nanomaterials, such as graphene and carbon nanotubes, to the typical polymer matrices creates composites that have superior mechanical properties such as enhanced tensile strength and improved impact resistance. This imparts strength to military and aviation vehicles, which are subjected to severe stresses in use. Moreover, the light weight of these new materials enhances fuel efficiency, resulting in lower operational costs and lower environmental impacts. Continued research in the field demonstrates a strong shift towards sustainable aerospace operations, highlighting the importance of material science innovations for use in the future in the aerospace industry. In addition to mechanical enhancements, nanomaterials also exhibit inherent properties beneficial for space use. For instance, certain nanocomposites exhibit superior thermal and electrical conductivity, which is suitable for use in advanced electronic systems, such as avionics and communication. With these material components, one can engineer devices that are not only lighter but also efficient and robust even under harsh conditions. Apart from this, the potential to incorporate multifunctional properties—such as self-healing behavior and damage sensing—adds another level of security and life to aerospace parts. Such properties are of greatest importance to defense systems, in which mission success would generally rely on equipment availability and survivability in adverse environments (Brun et al., 2014). Furthermore, economic benefits of nanomaterial application in aerospace and defense extend beyond the improvement in performance to the potential for cost savings. Economical production of cutting-edge composites can reduce maintenance cycles and improve aircraft and military vehicle lifecycle costs. Focusing on utilizing renewable resources, such as cellulose nanomaterials, has also become a focus area, aligning with overall sustainability goals. The U.S. Agriculture Department and the U.S. National Nanotechnology Initiative have assisted in debating commercialization pathways for these materials, indicating growing recognition of their significance in aerospace industry. Through cross-sectoral partnerships, such efforts aim to promote innovation while responding to market needs, thus contributing towards a greener aerospace and defense technology future (Brun et al., 2014) (Figure 7.10).

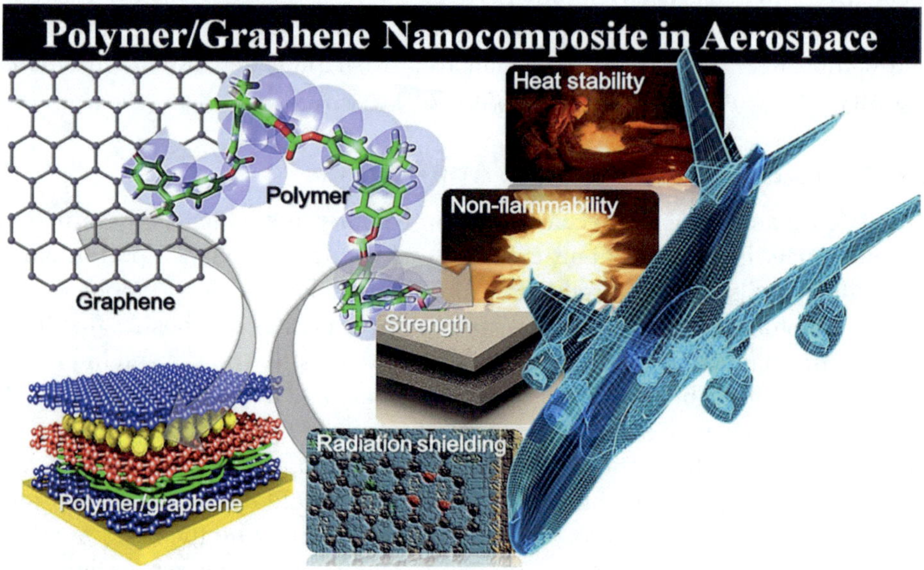

Figure 7.10. Aerospace and defense applications of nanomaterials into polymers.

Source: https://pub.mdpi-res.com/carbon/carbon-09-00029/article_deploy/html/images/carbon-09-00029-ag.png?1677845023.

7.4.3. Biomedical Uses of Nanocomposites

Nanocomposites are revolutionizing the biomedical sector due to enhanced properties at the nanolevel, which offer far-reaching advantages over traditional materials. The inclusion of nanomaterials, e.g., noble metal nanoparticles of gold and silver, into polymer matrices yields composites with enhanced biocompatibility, mechanical strength, and drug-release properties. Recent studies reveal that these nanocomposites can be systematically designed using certain synthetic pathways to obtain desired characteristics, such as optimal size, shape, and stability for particular biomedical purposes. The accuracy in such design is of importance since it offers the possibility of developing new therapy, diagnostic agents, and ultimately improves patient recoveries during medical interventions. The promise of such superior materials has incited comprehensive overviews of their multifaceted medical uses, most particularly in the area of nanomedicine, where precision and efficacy are critical in therapeutic and diagnostic strategies (Fratoddi et al., 2017). Moreover, the unique properties of nanocomposites have created possibilities for their application to smart drug delivery systems. By placing nanoparticles in polymeric carriers, researchers can design systems that are sensitive to external stimuli—pH, temperature, or light—and release therapeutic agents in a controlled manner. This not only enhances the effectiveness of the drugs but also reduces side effects by releasing medication primarily at the site of interest. Additionally, the blend of biocompatible nanocomposites with imaging agents can find applications in improved diagnostic capability. This integrated functionality, wherein therapeutics and diagnostics are combined together, is a critical innovation in personalized medicine that further enhances real-time monitoring of treatment efficacy and aligns with dosages (Kiekens et al., 2012b).

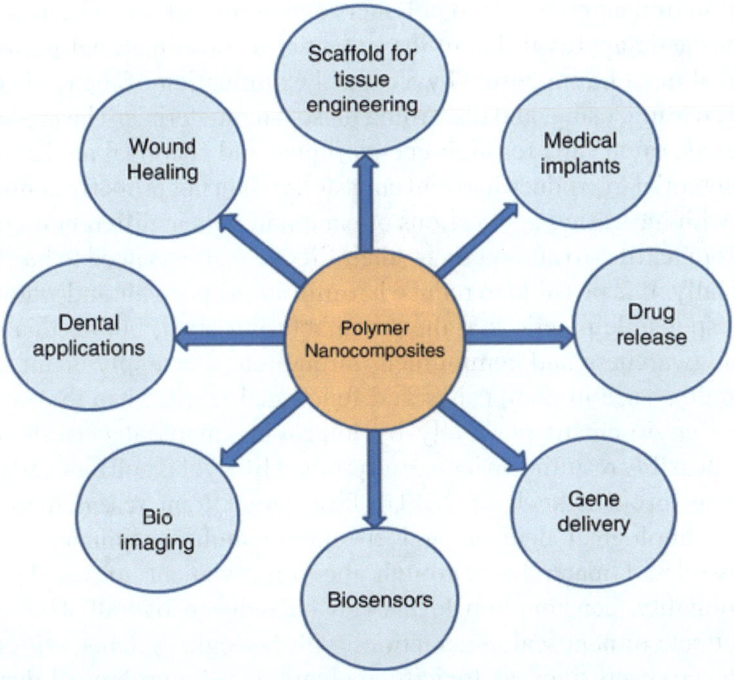

Figure 7.11. Biomedical uses of nanocomposites.

Source: https://media.springernature.com/lw685/springer-static/image/chp%3A10.1007%2F978-3-662-53517-2_4/MediaObjects/301952_1_En_4_Sch1_HTML.gif.

The eventual applications of nanocomposites exceed current uses and move into the design of new materials cutting across different fields of science. Collaborative efforts of research institutions and industrial stakeholders are vital to realizing the complete potential of nanocomposites in biomedical applications. By facilitating a multidisciplinary set-up, scientists can identify areas of technology gaps and hasten the journey from laboratory research to commercial products. These collaborative efforts would enable realization of innovations towards the production of the next generation of composite materials with further enhanced functionality and performance benefits to medical implants and devices. With continuing research, the synthesis and characterization of such nanocomposites are likely to change, shifting focus towards new applications with potential to revolutionize healthcare provision (Fratoddi et al., 2017; Kiekens et al., 2012b) (Figure 7.11).

7.5. ENVIRONMENTAL IMPACT

The addition of nanomaterials to polymers and composites has been welcomed with enthusiastic approval due to the potential to raise material performance. But this zeal must be supported by a careful examination of the environmental impact of creating, using, and discarding these nanomaterials. The processing of nanomaterials often calls for high-energy inputs and chemical reactants, which have the potential to produce harmful emissions when not properly controlled. In addition, with increasing applications of nanomaterials in different industries—ranging from textiles to automotive—their lifecycle effects need to be evaluated systematically. It is possible to reduce harmful impacts on air and water quality through responsible practices at the manufacturing level, but all this needs to come with awareness and commitment throughout the supply chain. Besides, nanotechnology use in composites and functional textiles has the capacity to impact the environment positively by improving material performance and longevity, possibly resulting in increased product lifecycles and reduced wastage. For instance, projects such as 2BFUNTEX target joint research to identify areas of technological demand and encourage multidisciplinary innovation in textiles-related materials. Through the improvement of textile lifespan and functionality, consumption levels can be reduced overall. However, the biologic effects of nanoscale interactions with biologic systems, with possible biopersistence, need rigorous toxicity evaluations. The prolonged duration of exposure to nanoparticles, when emitted during use or after disposal, is a matter of serious concern about their ecological impact cumulatively. Last, the shift to mass production of 3D multifunctional composites presents the prospect for a paradigm shift in the production of ecologically benign material. The 3D-LightTrans project thereby demonstrates this change by seeking to develop efficient, adaptable technologies to produce high-performance, lightweight composites. The innovation aims not only to reduce the material loss during manufacturing but also to maximize energy efficiency for application in different areas in industries ranging from transportation to healthcare. But as the technologies become increasingly popular, continuous monitoring of their impact on the environment will be critical so that advancement is not achieved at the expense of environmental sustainability. Maintaining a balance between the benefits of nanomaterials and the environmental cost will be critical to determining the future of polymer and composite applications.

7.5.1. Lifecycle Assessment (LCA) of Nanomaterials

LCA of nanomaterials is essential to understand their environmental impact, especially as their usage grows in numerous products, including polymers and composites. LCA provides an organized assessment of the environment and possible impacts associated with a product during its entire life cycle—from raw material harvesting, production, and use, to end-of-life processing. As nanomaterials promise to significantly enhance the properties of polymers and composites, possessing reliable LCA methodologies facilitates decision-making on their sustainability. Such special properties of nanoscale materials, however, make these decisions challenging as standard LCA methods might not be able to accurately portray their interactions along with nanolevel hazards, thereby necessitating the development of new evaluation schemes that can effectively handle such intricacies. Addition of nanomaterials into products generally proposes enhancement of durability, efficiency, and functionality, especially in the field of polymers and composites. While such improvements may contribute to decreased material use and energy consumption during the lifetime of the product, possible environmental hazards and health effects need to be evaluated holistically. Raw material extraction and processing of nanomaterials can cause environmental disturbances, whereas end-of-life impacts like recycling or disposal remain priority-focused issues. Additionally, concerns related to the unascertained health impacts of engineered nanomaterial exposures—as with the case of green technology—require meticulous investigation in LCA models. This holistic approach is vital in the development of safe and sustainable applications of nanomaterials in the larger economic sectors and regulatory guidelines for ensuring protection of human health and the environment (Hodson et al., 2014). It should also involve setting the lifecycle impact of nanomaterials with stakeholders like workers who are working within the nanotechnology sectors. The manufacturing, management, and disposal steps involved in the production of nanomaterials have been known to harbor a variety of risks whose management needs strategic management measures. Having well-detailed health surveillance schemes as well as training protocols is imperative to defend workers against occupational hazards as well as deriving the benefits of nanotechnology at the same time. This knowledge facilitates intelligent safer nanomaterials design as well as public acceptance of emerging applications. Lastly, the integration of LCA with rigorous safety protocols not only enhances the sustainability of nanomaterials in polymers and composites but also identifies with growing trends towards a green economy to harmonize technological advancement and environmental conservation (de Jong et al., 2014) (Figure 7.12).

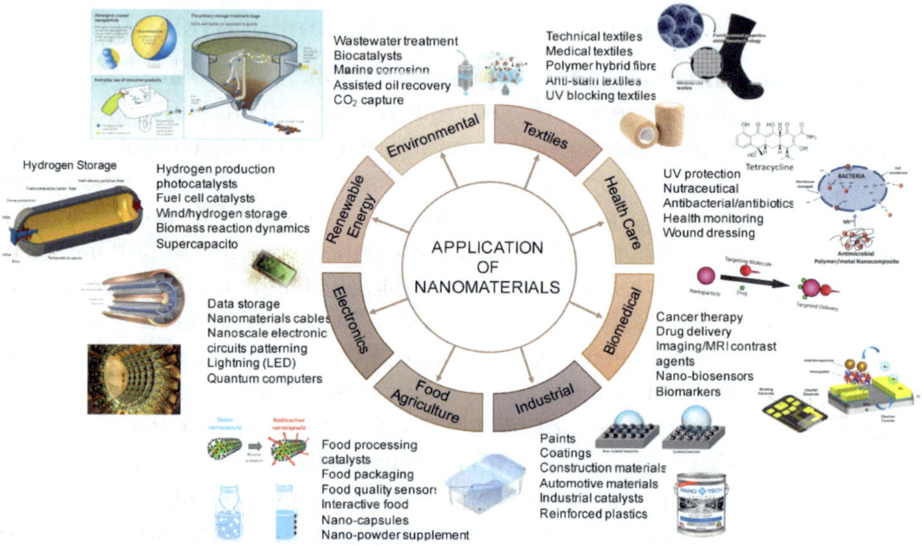

Figure 7.12. Lifecycle assessment of nanomaterials.

Source: https://www.mdpi.com/nanomaterials/nanomaterials-11-03324/article_deploy/html/images/nanomaterials-11-03324-g001.png.

7.5.2. Potential Toxicity and Safety Concerns

Nanomaterials impregnation into composites and polymers has revolutionized various industries with tremendous material property enhancements such as strength, durability, and lightness. While this rapid progress is flourishing, enormous safety issues are invoked on the likely nanoscale particle toxicity. Researches have indicated that the unique characteristics of nanomaterials due to size and surface traits can lead to unanticipated interactions in biological systems. Because these materials interact with human tissues and the surrounding environment, cytotoxicity is a likely possibility, which may lead to adverse health effects. Nanomaterials' intricate behavior must be carefully studied and characterized to determine their ability to do harm, which makes it obligatory to implement stringent safety protocols in their manufacturing and application processes. This preventive measure aims to guard both workers and consumers alike from possible health risks associated with using nanomaterials in products. In addition, the effects of nanomaterials extend to environmental health and safety concerns as well, particularly their life cycle from production to disposal. Nanoparticles can enter the environment through various routes, potentially building up in soils and water bodies and thus disturbing ecological balance as well as posing risks to wildlife. Studies have generated controversy regarding such materials' persistence in the environment as well as their bioaccumulation within living beings. Policy inclusion towards safe disposal and handling of nanomaterials is essential towards mitigating their environmental impact. In addition, green nanotechnology is envisioned as a viable alternative, emphasizing the production of eco-friendly applications with sustainability as an objective (Samal et al., 2017). However, finding a balance between technology and nature motivates researchers and policymakers to adopt responsible means with balancing technological advancement and environmental conservation. Finally, the complexities in the determination of health risks nanomaterials present emphasize the importance of clear-cut regulatory mechanisms and policies. The lack of standardized testing procedures for determining nanomaterial toxicity compounds risk assessment and management. Collaborative input from researchers, industry stakeholders, and government agencies is necessary in formulating valid safety practices that take into account the unique properties of nanomaterials. The collaborative action must include employee training in safe handling techniques, use of personal protective gear, and establishment of health monitoring programs for maintaining occupational exposure under control. Enhanced communication of risks and safety protocols is necessary in order to create an informed workforce capable of addressing the challenges presented by nanotechnology in polymers and composites. While the technology develops further, continued research and debate are essential to maintain development without threatening public health and safety (Hodson et al., 2014; Samal et al., 2017).

7.5.3. Sustainable Practices in Nanomaterial Production

Manufacturing nanomaterials is a significant chance to drive green manufacturing in the overall backdrop of composite and polymer production. By exploiting the unique properties of nanomaterials, manufacturers are able to produce products that are not only lighter and resilient but also more energy-efficient. This new approach avoids the consumption of traditional materials, which tend to absorb enormous resources and have the possibility of polluting the environment. The introduction of green practices in this sector is aimed at increasing resource productivity and minimizing waste during the manufacturing process. Also, by emphasizing the utilization of renewable resources and green practices, the sector is part of the broader push towards sustainability. Thus, the creation of nanomaterials integrated into polymers and composites provides a way to create material that is environmentally sustainable but retains performance levels at a significantly reduced negative environmental impact. Apart from the potential benefits, incorporating nanomaterials into sustainable processes is also raising critical issues regarding the environmental and occupational health risks. Addressing such concerns entails creating frameworks that ensure secure production and deployment of nanomaterials. By putting in place safety protocols and standards that give priority to the quality control of nanomaterials, producers can minimize potential risks arising from exposure of workers and environmental contamination. It should be ensured that there is ongoing investigation into the hazardous nature of such materials and establish robust communication strategies to educate workers and stakeholders about potential hazards. Investing in rigorous health surveillance programs is also a priority, as it can help ensure that appropriate protection strategies are being followed to protect both workers and the environment. These initiatives support the green economy ideology that aims to incorporate sustainable practices throughout the production ecosystem (Hodson et al., 2014). Additionally, the sector should cooperate with the regulatory bodies to establish standards that direct the sustainable process within the production of nanomaterials. Creating an open framework that will call for the evaluation of sustainability will guarantee that subsequent processes and materials are designed with respect to the environment. Cooperation between academia, industry, and government bodies can enhance the exchange of best practices while promoting innovation in green nanotechnology. Knowledge- and resource-sharing initiatives can create environmentally friendly production processes that are economical as well as sustainable. These methods not only enhance the overall sustainability of the business of nanomaterial production but also sustain the long-term sustainability goals of the polymer and composites sector. Therefore, sustainable nanomaterials production can be a basis for manufacturing materials that are in balance with both ecological and economic systems.

7.6. FUTURE TRENDS AND INNOVATIONS

The use of nanomaterials in composites and polymers is a transformative era with new uses across industries. One of the trends includes the application of nanotechnology to food, where nanoparticles are increasingly applied as additives for ensuring food safety and quality. Nanoscale ingredients can extend the shelf life of products by inhibiting microbial contamination, thereby offering health advantages to the consumers. With the increasing demand for fresh, convenient, and healthy food products, the food processing sector is hoping to look up to nanotechnology in order to retain their leadership status.

All these advancements illustrate the multidimensional role of nanomaterials in improving existing products along with pioneering innovative uses that are aligned with evolving consumer needs and health requirements (Samal et al., 2017). Besides uses in food, use of nanomaterials in the production of high-performance composites mirrors leading trends in the material science future. Polymer-based nanocomposites, produced by introducing nanoparticles, are strengthening mechanical properties, thermal stability, and barrier performance. All these developments place such composites at the forefront in a wide range of industries, from automotive to aerospace, electronics, and more. As makers seek light materials that are strong enough to meet severe performance specifications, advances in nanocomposites present a positive direction for research and business expansion. Industries are investing more dollars in researching these substances due to the potential for reduced manufacturing costs and added functionality, which would further transform product design and use in everyday applications (Debackere et al., 2025). Also, the future of polymers and composites with nanomaterials is going to be determined with the use of smart technology. Smart polymers, which respond to stimuli, have been invented and this is bringing with it the possibility of making adaptive materials that heal themselves or can change their shape or color. All these innovations have promising implications for efficiency and safety in most applications, but most importantly in the health industry where nanocomposites can be rendered biocompatible and therefore enable development in drug delivery systems and tissue engineering. The more functions revealed by ongoing research, the more likely it is that nanomaterials will assist in endeavors toward sustainability through the development of green polymers. In total, the integration of nanotechnology and lifecycle sustainability is a promising path to future advances in material science (Samal et al., 2017).

7.6.1. Emerging Nanomaterials and Technologies

Research on emerging nanomaterials and technologies is a primary sector for the development of the properties and functionality of polymers and composites. As industries increasingly demand materials with enhanced properties, nanomaterials offer an unprecedented ability to tailor functionalities at the molecular level. These advancements are driven by improvements in synthesis techniques, which allow the control over nanoparticle, nanotube, and nanosheet production, leading to particles with predetermined properties. Tailoring in this case extends beyond the miniaturization of size; it also involves modifications of mechanical, thermal, and electrical properties, opening up potential applications across industries like aerospace, automobile, and biomedical engineering. The insertion of nanomaterials into polymer matrices has thus become a central research focus area, facilitating the fabrication of materials which are light in weight but with unparalleled strength and durability, elevating the performance levels of traditional materials. In the aspect of taking consideration of developments in the field, bioinspired synthesis techniques of nanomaterials hold revolutionary promise. Scientists have made great progress in using natural processes to create silica and other materials that can transform material science. For example, the review of S. V. Patwardhan et al. refers to the importance of organic molecules in silicification by demonstrating that biosilicification is possible under mild conditions, i.e., room temperature and neutral pH. These conditions of green synthesis not only reduce the energy demand but also enhance the functionality of the final materials.

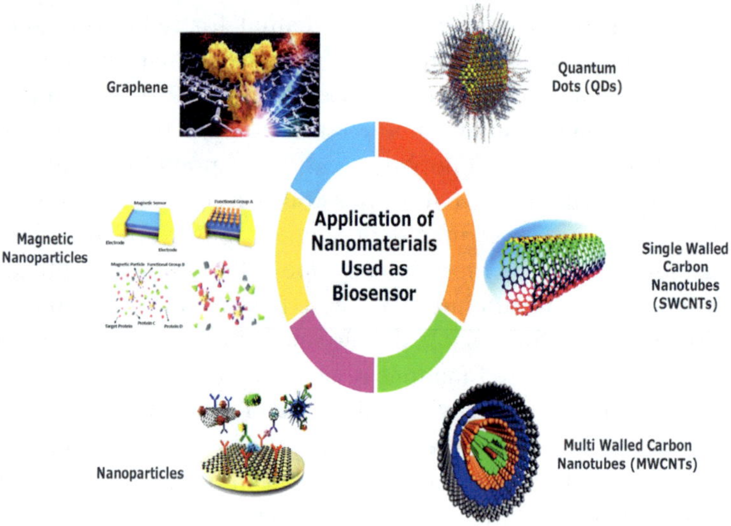

Figure 7.13. Emerging nanomaterials and technologies.

Source: https://media.springernature.com/lw1200/springer-static/image/art%3A10.1007%2Fs42247-023-00461-8/MediaObjects/42247_2023_461_Fig1_HTML.png.

The ability to control the properties of silica-based materials through biosilicifying systems underscores the versatility of nanomaterials and opens new frontiers for use from drug delivery to advanced sensors, justifying their role in the future of polymer composites and innovation (Patwardhan et al., 2011). The union of nanotechnology with polymers and composites also underscores a revolution in material design tailored to specific uses. Emerging technology allows the upscaling of new synthesis routes, thus providing larger quantities of nanomaterials for industrial uses. Scalability is crucial in that it translates basic research into usable forms that are relevant to industry. Compatibility with existing manufacturing techniques makes the lab-to-market transition as seamless as possible. From their use in coatings, hybrid composites, and catalysis, the possibility for innovation continues to expand. As shown in current literature, especially by observations reported in industry documents, ongoing discovery of nanomaterials will result in new pioneering technologies resolving principal challenges in numerous domains, driving sustainable development and high-performance technology in the polymer and composite field (Figure 7.13).

7.6.2. Research Directions in Polymer Nanocomposites

Research on polymer nanocomposites has witnessed increased interest in research directions toward enhancing the mechanical properties and performance of these materials. Of greatest concern to researchers is the ability to find a balance between the toughness and flexibility of polymers and introducing nanofillers capable of enhancing performance without compromising workability. These developments are crucial as industries seek materials that can tolerate diverse operating conditions, particularly in sectors like aerospace and automotive. Applications of nanofillers such as carbon nanotubes and nanoclays have been used to increase tensile strength and thermal stability significantly. Through the use of these materials to advance innovation, research projects aim not just to enhance existing formulations but also discover new combinations that could lead to innovations in polymer use in various fields. Aside from the development of mechanical properties, the fabrication of multifunctional polymer nanocomposites is becoming a prime focus area for research. These sophisticated materials are made with the ability to display more than one functionality, like self-healing, conductivity, or enhanced barrier properties. While companies are placing more emphasis on sustainability, there is a growing need to create sustainable nanocomposites without any trade-off on performance. Furthermore, inter-disciplinary solutions are being adopted to bridge the gap between theory and practical application. For instance, programs emphasizing collaboration among materials scientists and industry players are proving to be essential in connecting research objectives with real applications. All such cooperation is very important in the determination of the technological gaps and assured industrial adoption of the new materials in a timely basis to meet the market needs in a better manner. Future development of polymer nanocomposites will heavily rely on advancement in manufacturing technologies. Current initiatives focus on improving the scalability and efficiency in the manufacturing process of the advanced material, as evidenced with projects like 3D-LightTrans, with its focus on automating the processing of multifunctional composites. As these technologies develop, they can potentially lower costs and bring high-performance materials within easier reach for other industries, thereby creating the possibility for new applications. For instance, integrating automated fabrication with novel composite designs can lead to quicker production times, especially for lightweight transportation materials. This development can bring about a revolution in production methods and expand the market size for polymer nanocomposites, promoting a mass production trend of such advanced materials that can meet the growing demands of consumers and industrial customers (Kiekens et al., 2012a, 2012b).

7.6.3. Potential Market Growth and Economic Implications

Nanomaterial infusion into polymers and composites is likely to propagate staggering market expansion across all industries. With corporations more willing to cultivate the performance qualities of their products, the universality present in nanomaterials makes them particularly sought-after commodities. Revolutionary applications in industries such as construction, electronics, and agriculture showcase the potential offered by these materials to enhance product toughness, weight savings, and energy efficiency. In the field of agriculture, for instance, technologies derived from nanotechnology could lead to coming up with precise methods of crop disease identification and treatment, which is highly pertinent to food security in a world that is ever-changing (Claudia et al., 2014). As a result, the anticipated growth not only translates to a massive economic potential for manufacturers and suppliers but also reflects the capability of nanotechnology to transform traditional industries, indeed re-engineering their economics. Additionally, the economic relevance of incorporating nanomaterials into composites does not end at short-term market growth but instead introduces a paradigm shift towards a greener production and consumption pattern. Green economy is concerned with cutting down on resource consumption and environmental impact, yet enhancing economic resilience (Hodson et al., 2014). Nanotechnology advancements are central to achieving energy efficiency and sustainability in the production process. For example, product development using nanotech can facilitate the use of renewable sources of energy, enhance the efficiency of power delivery systems, and promote cleaner production processes. Therefore, changing paradigms of economic policies that support the integration of nanotechnology reveal not only a path toward more sustainable modes of industry but also an economic imperative to move toward greener alternatives that reflect contemporary consumer values. Lastly, in exploring the potential for market growth facilitated by nanomaterials, it is necessary to admire the attendant challenges that may impinge upon their economic deployment. Environmental health and safety issues regarding nanomaterials can hold back widespread acceptance and implementation in the market. To take an example, the knowledge of the toxic nature of these substances and answering doubts regarding exposure hazards will be critical to gaining consumer trust and regulatory acceptability. As highlighted by current studies, there is a need to strike a balance between optimum exploitation of the economic benefits of nanomaterials and ensuring the well-being of employees as well as consumers (Hodson et al., 2014). Therefore, addressing these challenges through strategic risk assessment and management measures will be critical in realizing the optimum economic potential of nanomaterials in polymers and composites while creating a safe and sustainable environment for all stakeholders.

SUMMARY

With nanomaterials examined within polymers and composites, the introduction of nanomaterials is discovered to achieve remarkable improvements in material performance. The combination of nanomaterials, such as graphite and iron oxide nanoparticles, with polymer matrices exhibits an outstanding ability to improve mechanical and functional performance. For instance, addition of exfoliated graphite to polyamide 6 (PA_6) has been evidenced to enhance the polymer by virtue of some bonding and dispersion mechanisms, thus enhancing the overall stability and durability of the composite. Such results not only enrich our understanding of material science but also point to the growing importance of nanomaterial application in most areas of application, especially engineering and technology demanding enhanced material properties (Bocchini et al., 2012). Besides, functionalization of nanoparticles is at the heart of regulating the general performance of such composites. Coating iron oxide nanoparticles with poly(acrylic acid) chains exemplifies the intricacy of stabilizing and inducing desired chemical reactivity in polymer matrices. Synthesis and characterization of such composite materials successfully illustrate how engineered surfaces of nanoparticles can be employed to enhance interaction, thus leading to amplified adsorption and functional capacity.

For instance, the excellent adsorption capacity obtained in pollutants remediation demonstrates the potential which these composites have in environmental applications. Such a multi-faceted approach is a bridge between basic research and applications, which meets a crucial gap in the study of nanomaterials in applied situations (Berret et al., 2012). Overall, the addition of nanomaterials to composites and polymers not only improves their strength but also offers avenues for new uses in various industries. Methods of synthesis and understanding of the interactions of nanoparticles in polymers are critical in developing advanced composites that can solve today's engineering problems. Optimizing processing techniques and discovering new nanomaterials should be addressed in future work in order to further advance performance and functionality. As the region evolves, possibility in nanocomposite systems will continue to advance breakthrough technologies, solidifying their status as essential building blocks of high-performance materials (Berret et al., 2012; Bocchini et al., 2012).

The integration of nanomaterials in polymers and composites introduces revolutionary material properties and applications. New research highlights the efficiencies of various nanoclays, such as halloysite, sepiolite, and laponite, as nanofillers for bioplastics, particularly Mater-Bi, applied widely in food packaging. The solvent casting method has been employed to prepare Mater-Bi/nanoclay nanocomposite films, under which it is easy to introduce nanomaterials. Of special note was the finding in tensile tests that the addition of halloysite significantly enhanced mechanical properties of resultant films to the extent of traditional petroleum-based plastics. This is an indication of the use of bio-nanocomposites not just mechanically but also in offering sustainable alternatives for traditional existing materials in food packaging and smart coatings, therefore addressing environmental problems at the expense of functional integrity (Cavallaro et al., 2018). Apart from the improvement of mechanical properties, the new uses of nanomaterials are in microelectronics and medicine. The development of biochips for high-end health monitoring exploits the versatility of nanomaterials in creating completely integrated platforms that are capable of continuously monitoring a wide range of human disease biomarkers.

The innovations point toward a direction of minimally invasive devices with real-time output, which addresses the pressing need in health monitoring. These biochips, traditionally implanted in the subcutaneous tissue, exploit the unique characteristics of nanomaterials to ensure high reliability and specificity, the secrets of successful patient care. Developments in this field not only promise improved patient outcomes but also call attention to the interdisciplinary nature of nanotechnology as it converges with medical devices and medical monitoring (Baj-Rossi et al., 2012). The study of nanomaterials in polymers and composites mirrors the need for ongoing research and development across a variety of industries. As shown, the enhanced mechanical properties through nanofillers transcend the domain of sheer performance; they create a broader environmental awareness and commitment to sustainability in materials science. The new medical uses also outline the potential for far-reaching impact on patient management and disease prevention. Future uses will necessarily be focused on the resolution of present issues such as long-term biocompatibility and safe data storage for implantable devices. Continuous innovation and cross-disciplinary convergence will be the drivers of these technologies into the future, ensuring that the entire potential of nanomaterials is realized in high-impact applications from everyday consumer products to life-saving medical interventions (Baj-Rossi et al., 2012; Cavallaro et al., 2018).

REFERENCES

1. Alison, J., Artbauer, J., Bambery, K. R., Fleming, R. J., Chapman, D. L., Crine, J. P., Dissado, L. A., et al. (2004). Internal charge behaviour of nanocomposites. https://core.ac.uk/download/9258158.pdf (accessed on 25th September 2025).
2. Baj-Rossi, C., Burleson, W., Carrara, S., Cavallini, A., et al. (2012). Fully integrated biochip platforms for advanced healthcare. https://core.ac.uk/download/286729684.pdf (accessed on 25th September 2025).
3. Barbhuiya, S., Das, A., et al. (2025). Mechanical performance of bio-based materials in structural applications: A comprehensive review. https://core.ac.uk/download/647986108.pdf (accessed on 25th September 2025).
4. Berret, J. F., Bostelmann, T., Bée, A., Courtois, J., et al. (2012). Poly(acrylic acid)-coated iron oxide nanoparticles: Quantitative evaluation of the coating properties and applications for the removal of a pollutant dye. http://arxiv.org/abs/1212.1580 (accessed on 25th September 2025).
5. Bocchini, S., Cavalleri, O., Cozza, E. S., Frache, A., et al. (2012). Simple method for the preparation of composites based on PA6 and partially exfoliated graphite. https://core.ac.uk/download/pdf/11425034.pdf (accessed on 25th September 2025).
6. Brun, V., Friedersdorf, L., Hansen, F., Keller, M., et al. (2014). Cellulose Nanomaterials–A Path Towards Commercialization Workshop Report. https://core.ac.uk/download/42901273.pdf (accessed on 25th September 2025).
7. Cavallaro, G., Lazzara, G., Lisuzzo, L., Milioto, S., & Parisi, F. (2018). Filling of Mater-Bi with nano clays to enhance the biofilm rigidity. https://core.ac.uk/download/286528402.pdf (accessed on 25th September 2025).
8. Dahiya, R., Dang, W., Khan, S., Lorenzelli, M., et al. (2015). Tuning electrical conductivity of CNT-PDMS nanocomposites for flexible electronic applications. https://core.ac.uk/download/46664171.pdf (accessed on 25th September 2025).
9. de Jong, W. H., Dekkers, S., GBO, Hagens, W. I., & Wijnhoven, S. W. P. (2014). Exposure to nanomaterials in consumer products. https://core.ac.uk/download/58768469.pdf (accessed on 25th September 2025).
10. Debackere, K., Monard, E., Raspoet, D., Ratinckx, E., et al. (2025). Technologie En Innovatie in Vlaanderen: Prioriteiten. https://core.ac.uk/download/pdf/6468460.pdf (accessed on 25th September 2025).
11. del Garrigós, M. C., Mellinas-Ciller, A. C., Ramos, M., Valdés, A., et al. (2015). New trends in beverage packaging systems: A review. https://core.ac.uk/download/32325038.pdf (accessed on 25th September 2025).

12. Fangueiro, R. (2015). Towards high performance and multi-functional structural membranes using advanced fibrous and textile materials. https://core.ac.uk/download/132529650.pdf (accessed on 25th September 2025).

13. Fratoddi, I. (2017). Hydrophobic and hydrophilic Au and Ag nanoparticles: Breakthroughs and perspectives. https://core.ac.uk/download/153366182.pdf (accessed on 25th September 2025).

14. Hodson, L. L., Hoover, M. D., Iavicoli, I., Leso, V., et al. (2014). Opportunities and challenges of nanotechnology in the green economy. https://core.ac.uk/download/55141194.pdf (accessed on 25th September 2025).

15. Hossenlopp, J., del Jimenez-Gasco, M. M., Manias, E., Manzi-Nshuti, C., et al. (2009). Polymer nanocomposites using zinc aluminum and magnesium aluminum oleate layered double hydroxides: Effects of LDH divalent metals on dispersion, thermal, mechanical and fire performance in various polymers. https://core.ac.uk/download/213079382.pdf (accessed on 25th September 2025).

16. Jeon, B., Lee, D., Park, J., Oh, K. W., Shin, T. J., Kim, T., & Tian, B. (2019). Photothermal polymer nanocomposites of tungsten bronze nanorods with enhanced tensile elongation at low filler contents. https://core.ac.uk/download/286608020.pdf (accessed on 25th September 2025).

17. Kiekens, P., & Van der Burght, E. (2012). ENMat international projects: FP7 NMP coordination action: 2BFUNTEX. https://core.ac.uk/download/55859985.pdf (accessed on 25th September 2025).

18. Kiekens, P., & Van der Burght, E. (2012). ENMat international projects: FP7 NMP large collaborative project: 3D-LightTrans. https://core.ac.uk/download/55811497.pdf (accessed on 25th September 2025).

19. Parisi, C., Rodriguez Cerezo, E., & Vigani, M. (2014). Proceedings of a Workshop on "Nanotechnology for the Agricultural Sector: From Research to the Field." https://core.ac.uk/download/38628347.pdf (accessed on 25th September 2025).

20. Patwardhan, S. (2011). Biomimetic and bioinspired silica: Recent developments and applications. https://core.ac.uk/download/9032722.pdf (accessed on 25th September 2025).

21. Samal, D. (2017). Use of nanotechnology in food industry: A review. https://media.neliti.com/media/publications/238902-use-of-nanotechnology-in-food-industry-a-b90b1cef.pdf (accessed on 25th September 2025).

CHAPTER 8

DURABILITY AND SUSTAINABILITY OF NANOMATERIALS

Contents

INTRODUCTION

The intersection of sustainability and durability marks a watershed moment in nanomaterials' research. As nanotechnology finds its way deeper into industrial processes, sweeping changes become evident across industries from construction to medicine. Not only does such change enhance the performance properties of the materials, but also satisfies the pressing need for sustainable manufacturing methods. Nanomaterials have the potential to provide more strength, corrosion resistance, and durability and hence lower overall environmental footprint due to material degradation and disposal. Greater durability translates to lower maintenance costs and product lifetimes, challenges that are mission-critical in the context of sustainability. Therefore, nanomaterials' ability to promote performance as well as environmental stewardship calls for scrupulous scrutiny in light of how their peculiar properties can be harmonized with sustainable development targets set internationally. An expansive examination of the sustainability and durability of nanomaterials requires knowledge of their extensive applications in industries.

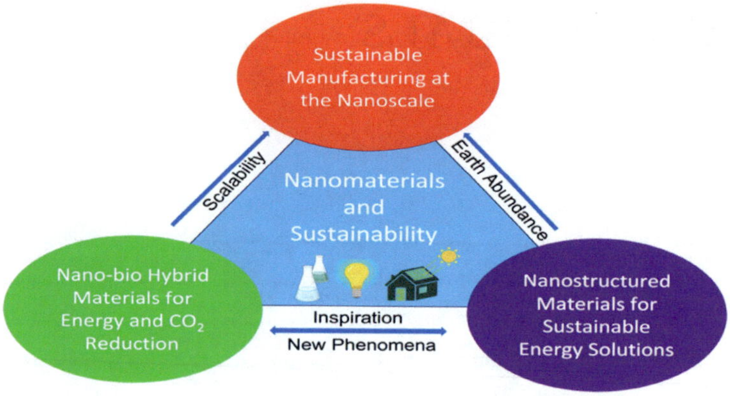

Figure 8.1. Sustainability of nanomaterials.

Source: https://encrypted-tbn0.gstatic.com/images?q=tbn:ANd9GcSBiNhhVFVlNm8o_qBXRobBX47MmBNIfe3t4g&s.

Research indicates that nanotechnology has transitioned from the lab to widespread real-world utilization, changing paradigms in manufacturing in automotive, civil engineering, and environmental management sectors. Such advances demonstrate how nanomaterials enhance existing products but also lead to new solutions that are effective and less resource intensive. Moreover, interdisciplinary collaboration between ecologists, industrialists, and scientists is called for by the integration of nanotechnology and materials science. Multilateral collaboration in this direction offers a means to develop sustainable nano-based materials that satisfy both performance excellence and environmental integrity (Malik et al., 2023). In outlining the trajectory of nanomaterials towards the future, artificial intelligence (AI) and machine learning (ML) are a desirable complement to design and deployment of long-term, sustainable materials. Through advanced modeling techniques and data treatment, the technologies enhance understanding of material properties as well as life cycle impacts. The use of smart vision and sensor technology allows for real-time monitoring and feedback across the entire nanomaterial life cycle, from design to decommissioning. This end-to-end approach allows for the development of materials that not only achieve technical performance but also adhere to circular economy and resource efficiency approaches. Hence, ongoing research is needed to overcome the challenges of designing nanomaterials that not only function to the fullest but are also regulated in a sustainable way through their lifecycle (Baduge et al., 2022). This will ultimately lead to the creation of a more sustainable and resilient future in material science (Figure 8.1).

Definition of Nanomaterials

Nanomaterials encompass generally materials with structural features in the range of nanometers, typically between 1 and 100 nanometers. It is a very significant size because this is the size range where materials are likely to exhibit new physical and chemical properties that are mostly distinct from bulk properties. Such properties occur as a result of an elevated surface area-to-volume ratio, which can lead to enhanced reactivity, strength, and conductivity. The performance of nanomaterials is very variable from each other as far as dimensions, shape, and structure are concerned, and they can be utilized for various applications in a variety of disciplines, including medical science, electronics, and ecology. Their novel behavior at the nanoscale raise a series of queries regarding their sustainability and durability, as their new properties hold promises for revolutionary progress but pose prospective risks to health and the ecosystem. Further research shows varying types of nanomaterials that have some functionalities. For instance, carbon nanomaterials such as graphene and carbon dots have attracted attention through their improved electrical and mechanical properties. Carbon dots (CDs) have been known to exhibit biocompatibility and applications through fluorescence. Their typical morphological structures like disk-like and quasi-spherical morphologies provide flexibility in synthesis processes like hydrothermal carbonization and pyrolysis. The versatility of CDs is further extended to their applications in biological systems, which enable them to serve as sensors and drug delivery systems, thereby highlighting their importance as a component of the entire branch of nanotechnology. With the continuous innovation in the field by scientists, the exploration of CDs and other nanomaterials has also introduced essential questions about not only their application but also their longevity and impact on the environment. Furthermore, the durability of nanomaterials is more and more associated with sustainability, since material composition and structure can play an important role in both performance and environmental footprint. For example, the progression of the multifunctional composite, i.e., CareGum, illustrates the potential for creating lasting materials that mimic natural systems. By blending silk fibroin and reduced graphene oxide, CareGum possesses excellent mechanical behavior and flexibility for future applications such as cyborganics and soft robotics. Such advancements suggest that through the careful consideration of material behavior and environmental impact, nanomaterials will play an important part in creating sustainable technology. Lastly, the definition of nanomaterials transcends dimensions, calling for ongoing debate about application of these materials responsibly to push durability at the cost of no environmental integrity.

Importance of Durability and Sustainability

Greater emphasis on durability in material science has deep implications for sustainability, particularly in the context of nanomaterials. As the construction industry turns towards solving ecological problems, the use of nanotechnology offers new solutions that enhance durability as well as the environmental footprint of construction materials. For example, using nanomaterials can effectively increase the durability of conventional construction materials such as concrete and steel, reducing degradation over time. All these developments show that the use of nanotechnology could lead to material replacement being less frequent, eventually saving resources and reducing wastage. Taking a policy based on the architectonic and constructive durability of each component, actors in the construction sector are better positioned to understand the consequential effects of their choice of materials, thereby inducing a culture of sustainable long-term operation in the sector (MORENO et al., 2017). Furthermore, the evaluation of nanomaterials involves both durability and being able to deliver environmental integrity. The novel applications of graphene oxide in concrete are a case in point here. Following rigorous testing, it has been demonstrated by scientists that graphene oxide-modified fly ash pervious concrete possesses enhanced stormwater infiltration, which reduces the impacts of urban runoff on nearby ecosystems. The dramatic improvement of the materials' freeze/thaw cycle and salt resistance testing properties demonstrates the promise of nanomaterials to advance both performance and sustainability concerns. This combination of durability and environmental responsibility is significant to the evolution of building practices that promote longevity with fewer environmental detriments. These advances imply that the use of advanced materials can redefine industry norms and drive construction in more sustainable ways (Shi et al., 2018). The role of sustainability and durability in developing nanomaterials cannot be overemphasized, as they offer a fundamental template for future innovation across various fields. With the world being confronted by stringent environmental issues, the ability to construct durable infrastructures that do not compromise ecological balance becomes paramount. Nanomaterials not only bring enhanced performance but also serve as agents of change in developing more sustainable practice. By focusing on products that are both durable and environmentally friendly, the sector can benefit from lower maintenance costs, reduced environmental impacts, and improved management of resources. Lastly, the use of nanotechnology in building displays a broader acknowledgment that material durability is intrinsically connected to sustainability in use, opening doors to future advancement that prioritizes human and environmental health (Figure 8.2).

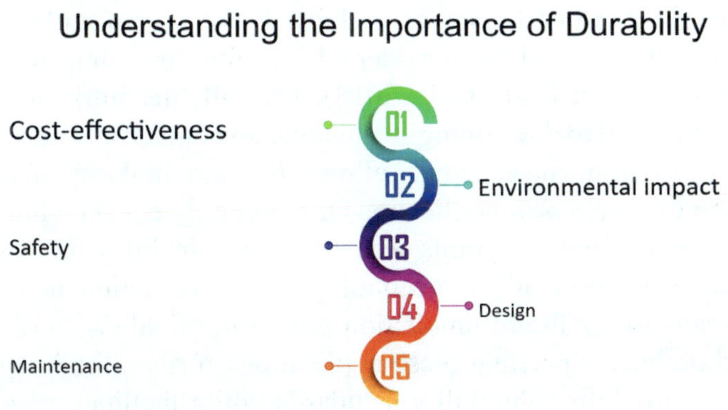

Figure 8.2. Importance of durability in material science.

Source: https://fastercapital.co/i/Durability--Choosing-Materials-for-Long-Useful-Life--Understanding-the-Importance-of-Durability.webp.

8.1. PROPERTIES OF NANOMATERIALS

The nanomaterial characteristics are critical to understanding their role in enhancing durability and sustainability across various applications, notably construction and materials science. Nanoscale materials possess varying mechanical, thermal, and electrical properties from their bulk counterparts. For instance, diminished grain size is expected to lead to increased hardness and strength, enhancing the structural integrity of construction members. This impact comes in handy when using nanomaterials to work on concrete or steel, where their application enhances exponentially the resistance to degradation and environmental stimuli, thus guaranteeing durability in structural performance. The study of nano-products indicates how they can also be utilized in architectonic and constructive durability, contributing to the tools at the disposal of current architects and builders engaged in sustainable construction activities (MORENO et al., 2017). Specifically, research has focused on the new use of nanomaterials such as graphene oxide in composite materials. Beyond conferring mechanical strength to the material, the incorporation of graphene oxide also confers chemical resistance, hence enhancing durability. By way of illustration, pervious concrete with graphene oxide modification shows dramatic enhancement in stormwater management, a reflection of its promising role in sustainable urban drainage. This changed concrete is freeze/thaw and salt attack resistant, addressing common issues experienced in harsh environments. The progress of analytical techniques, such as X-ray diffraction and nuclear magnetic resonance, better explains the behaviors at play, verifying that changes like these lead to considerable performance improvements in environmental resistance (Shi et al., 2018). Overall, the intrinsic nature of nanomaterials is central to developing sustainable and enduring solutions across industry fields. Their intrinsic nature allows for a rethinking of traditional material science to access new applications that not only enhance functionality but reduce environmental footprints too. With sustainability focusing more and more, the application of nanotechnology in construction materials and beyond is a roadmap for future innovation within the field. With researchers, engineers, and architects pushing these applications further, the integration of nanomaterials can redefine durability standards while facilitating sustainable practices in architecture and construction (MORENO et al., 2017).

8.1.1. Unique Physical and Chemical Properties

Physical and chemical properties of nanomaterials, due to their peculiarities, contribute significantly towards their sustainability and durability in various applications. One of the most typical properties of nanomaterials is their increased surface area-to-volume ratio, which makes them more reactive and stronger. By utilizing this characteristic, improved adhesion and compatibility with other materials become attainable, and in consequence of that, composites are created with mechanical properties greater than those found in their bulk states. For example, in the case of plastic surgery, nanomaterials, with size control and pH sensitivity, can successfully deliver therapeutics while maintaining the structural integrity intact under physiological conditions. Program functionality such as skin rejuvenation shows how the materials can dynamically respond to changes in the environment, thereby offering durability in medical settings, according to research indicating remarkable improvement in skin elasticity as well as anti-aging effects (Wang, 2025). The chemical composition of nanomaterials also largely results in enhanced stability and functionality, thereby making them core in various industries. For example, in dentistry, nanomaterials such as silver and titanium dioxide have shown great antibacterial properties along with improving the remineralization of dental tissues. These functionalities dual in nature show the strength of nanomaterials to improve treatment along with accelerating healing. Incorporation of such materials in restorations allows prolonged interventions, ultimately resulting in sustainable healthcare.

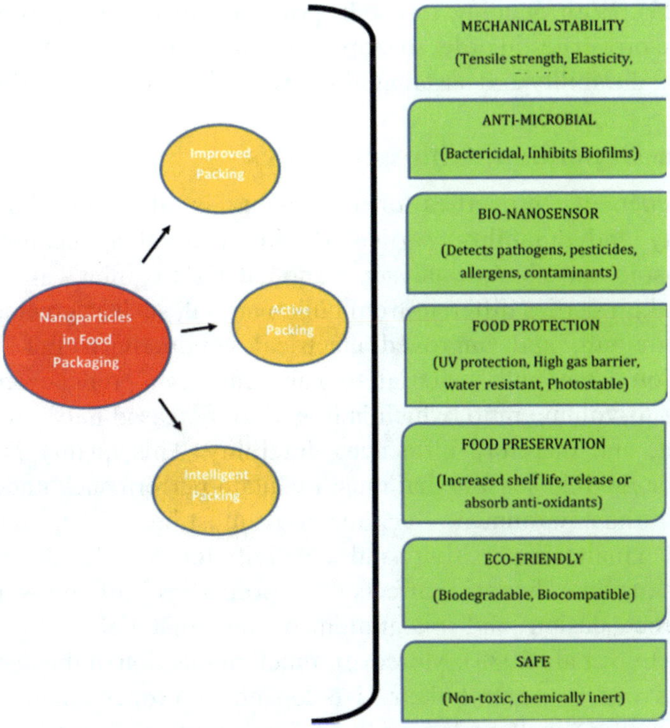

Figure 8.3. Physical and chemical properties of nanomaterials.

Source: https://www.researchgate.net/profile/Punuri-Jayasekhar-Babu/publication/358162636/figure/fig4/AS:1143128112881994@1673578337842/Different-physio-chemical-properties-of-nanomaterials.png.

Also, the inherent biocompatibility of non-metallic nanomaterials, such as chitosan, offers a new strategy to minimize ecological footprint and maximize therapeutic efficacy, as evidenced in their application towards tissue regeneration (Zeng, 2025). Moreover, nanomaterial design, with tailored physical as well as chemical attributes, extensively improves their application towards maximizing sustainability. Surface nanotexturing and dynamic gel formation are just a few instances of how such unique properties can be leveraged both aesthetically and functionally in medical treatment. The durability of these advantages, however, is expensive. Extended research has established the potential of unforeseen risks, such as the buildup of certain nanoparticles in organs, which requires a proper assessment of their safety and ecological implications. A sustainable approach necessitates a strict long-term observation system and creation of alternative production methods that minimize energy and industrial manufacturing costs (Wang, 2025). With ongoing research, priority will be given to maximizing these unique properties in order to reap the complete potential of nanomaterials for resolving durability and sustainability across the disciplines (Figure 8.3).

8.1.2. Size-Dependent Characteristics

The size-dependent properties of nanomaterials also contribute to their sustainability and durability, particularly in potential applications such as flexible sensors and electrocatalysts. Nanomaterials exhibit various physical and chemical properties different from bulk material, such as greater reactivity, increased strength, and improved electrical conductivity. For instance, a transformation from bulk material to nanoscale can produce an enhanced surface area-to-volume ratio, which is linked to increased interaction with the environment, and therefore influences durability. This quality is especially important for flexible sensors that must maintain performance under dynamic conditions of the environment. Such materials must be resistant to mechanical stress and also maintain sensitivity and selectivity for extended durations. Thus, the comprehension of how size affects these properties is of utmost importance to the correct design and integration of nanomaterials into sustainable technology (Luo et al., 2023). Moreover, functionalization of the nanomaterials also enables one to control their size-dependent properties to make them more functional as well as longer-lived. Nanomaterials may be coated and functionalized with a variety of substances to tailor surface properties towards optimizing stability against degradation and eliminating problems regarding biocompatibility and environmental worry. In the case of electrocatalysts for the oxygen evolution reaction, for example, the efficiency and stability of non-noble-metal-based materials can be significantly dependent on their size and shape. The increased active sites for reactions afforded by smaller nanoparticles can enhance efficiency, which is important in applications geared toward solving the energy crisis. But these advances should also consider seriously the lifecycle sustainability of these materials, such that their production, use, and disposal work towards fulfilling environmental goals. Thus, achieving a perfect conjunction of size, functionality, and sustainability is at the heart of leading the durability and utilitarian applications of nanomaterials (Zhi-Wu et al., 2020). Finally, sustainability and durability implications of size-dependent properties are intrinsically connected to the broader commercialization and environmental impact challenges. While nanomaterials are being integrated across various industries—ranging from consumer electronics to renewable energy—long-term sustainability needs to be measured. Size-dependent properties may have advantages; however, they may also create new challenges, e.g., toxicity and regulatory concerns. Moreover, market limitations may discourage the adoption of nanomaterials despite their attractive features. Overcoming market bottlenecks needs a coordinated approach that considers not only the material performance but also its ethical and environmental issues. By creating a coordination concept of how size influences properties and integration methods, the scientific community can make sustainable development in this sector possible and arrive at effective and sustainable nanomaterial use in society (Luo et al., 2023).

8.1.3. Influence of Surface Area on Performance

The influence of surface area on the performance of nanomaterials is one of the vital factors that define their overall durability and sustainability. Nanomaterials possess a very high surface area-to-volume ratio, which renders them more interactive with the environments surrounding them, improving catalytic and absorptive qualities. The high reactivity enables nanomaterials to perform perfectly in diverse applications, including drug delivery and environmental remediation. For instance, the greater surface area of nanocarriers will make it possible to achieve high drug loading capacity and site-specific delivery in cancer therapy, resulting in stronger and less invasive treatments for patients. Such success reveals the significance of adjusting surface area to enhance functionality and longevity of nanomaterial-based systems to make them suitable assets in sustainable operations in various fields, including health and environmental science. Besides, the specific morphology of nanomaterials plays an important role in determining their surface area characteristics, which also influence their performance. For instance, cellulose as a typical polysaccharide exists in various morphologies such as nanofibrils and microcrystalline cellulose, which have different surface area characteristics. The differing geometries and sizes of these nanomaterials based on cellulose can be optimized to the best to enhance their performance in biomedical applications such as tissue engineering and drug delivery systems. By the regulation of structure-based properties in cellulose, scientists can engineer the surface area for efficient interaction with biological tissues to boost the efficacy of therapeutic delivery and medical outcomes. The implication of surface area in determining the efficacy and sustainability of nanomaterials thus makes the imperative of future research in material science and nanotechnology all the more critical. Finally, the impact of increased surface area reaches beyond performance in specific use, even affecting lifespan and sustainability in the environment. Increased surface area assists in guaranteeing increased degradation rates in natural environments in order to confirm that nanomaterials won't be long-term environmental hazard. Degradation must, however, be carefully controlled to guarantee benefit in terms of performance rather than expense to the environment. As suggested by discussions regarding nanocarriers and anticancer therapy, manipulation of surface areas for targeted interaction can lead to materials that perform not just better but even break down in a controlled manner. In this context, increasing surface area can be one of the major drivers of nanotechnology to achieve more sustainable and durable solutions, addressing consumers' needs and environmental problems in material science today (Chehelgerdi et al., 2023; Seddiqi et al., 2021) (Figure 8.4).\

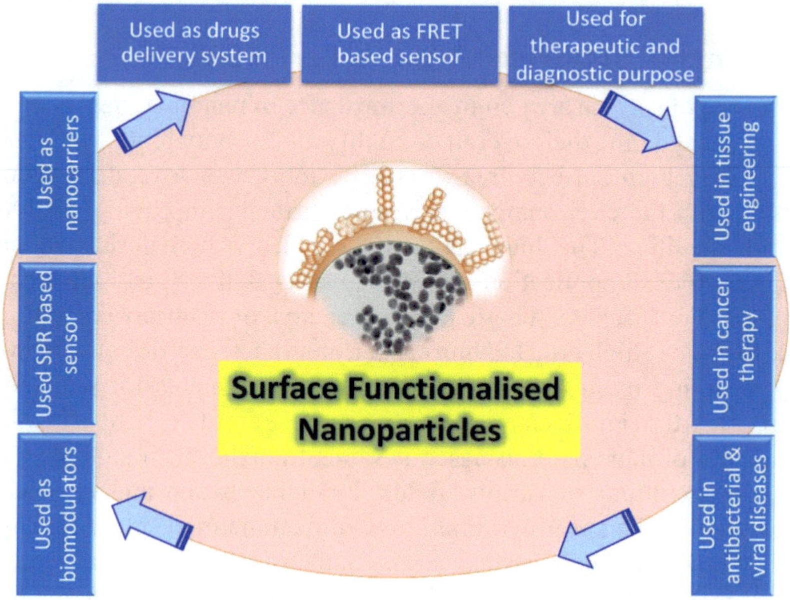

Figure 8.4. Unique properties of surface-functionalized nanoparticles.

Source: https://www.mdpi.com/nanomaterials/nanomaterials-12-01333/article_deploy/html/images/nanomaterials-12-01333-g001.png.

8.2. APPLICATIONS OF NANOMATERIALS

Applications of nanomaterials in building and construction reflect a significant intersection of sustainability and longevity. By taking advantage of the unique properties of nanomaterials, construction experts can enhance the performance and longevity of building materials. Nanomaterials, for instance, improve the durability against degradation of steel and concrete so that these materials continue to be effective with environmental stress over many years. This integration not only lengthens the lives of buildings but also minimizes the number of repairs or replacements, opening the way for a sustainable construction methodology. Moreover, the use of specialty nano-products such as water-repellent coatings and high-performance sealants lowers water intrusion and mold growth, further extending the durability of building components. Hence, the use of nanotechnology in construction operations can potentially achieve a more sustainable built environment, the foundation of which is set by efficiency and long-term resilience, as reported in MORENO et al. (2017). Apart from construction, nanomaterials also offer promising advances in other fields, particularly energy and environmental sectors. For instance, the use of nanomaterials in photovoltaic cells and batteries demonstrates their ability to improve energy efficiency and storage. These developments are significant since societies are trying to transition towards cleaner energy sources and reduced carbon emissions. Secondly, nanomaterials can be utilized to design complicated water purification systems so that contaminants may be removed on a molecular level effectively. This is not only a consideration in accessing clean water but also in designing an environmentally friendly future. The integration of nanotechnology and green technologies foreshadows the wider degree of impact of nanomaterials in bringing forth sustainable practices for industries at large, ultimately resulting in more environmentally friendly options and resource conservation, as mentioned in Brun et al. (2014). Nanomaterial commercialization, with cellulose nanomaterials being especially prominent, is a very enticing option for enhancing product development sustainability, as elaborated upon by the U.S. Forest Service, biodegradable cellulose nanomaterials have the ability to replace conventional plastics and other non-renewable products in a multitude of applications. Not only is it a step towards global sustainability, but economic growth also occurs by creating jobs in rural areas where these materials are obtained. The collaboration between the U.S. Department of Agriculture and the National Nanotechnology Initiative indicates a concerted effort to ignite innovation and bridge commercialization barriers. In providing heavy emphasis on research and development according to market needs, stakeholders can ensure that nanomaterials sufficiently meet performance levels while promoting environmental stewardship. This strategic emphasis underscores the focal position of nanotechnology in the evolution of sustainable and durable solutions to address current societal demands, thereby testifying to the significance of sustained investment in the field, as highlighted in Brun et al. (2014).

8.2.1. Use in Construction and Building Materials

The construction industry has increasingly appreciated the value of nanomaterials to act as a means of enhancing the sustainability and durability of building materials. Nanotechnology presents innovative solutions to central problems such as material degradation, resource utilization, and long-term sustainable performance in response to varying environmental conditions. For example, nanoscale additives have the potential to significantly increase the mechanical strength and environmental stress resistance of commonly employed building materials such as steel and concrete. Inclusion of such advanced materials not only increases the structural rigidity but also enhances the durability of constructions, thereby reducing repair and replacement rates. This minimization of material fatigue and failure is congruent with green building principles, aiming at use of material with minimal environmental footprint throughout their life. Thus, nanomaterial application in construction paves the way for developing more sustainable infrastructure that will endure the test of time and weather adversity (MORENO et al., 2017). Among the thousands of nanomaterials available, graphene oxide is a strong candidate in enhancing the properties of pervious concrete, which previously has been utilized for its capacity to provide infiltration of stormwater. Studies prove that combining graphene oxide with coal fly ash-based pervious concrete not only extends its durability but also its environmental sustainability by reducing the concentration of harmful pollutants in stormwater runoff. Specifically, testing proves enhanced resistance to freeze/thaw exposure and salt-induced degradation, which are paramount to maintaining structural integrity in diverse climates. The chemical reactions initiated by graphene oxide in the binder system lead to highly improved mechanical properties and durability characteristics of the concretes. Thus, through the use of nanomaterials like graphene, not only is building process sustainability promoted, but innovative solutions are also initiated to address urban flooding and water quality issues (Shi et al., 2018). Furthermore, the application of nanomaterials is not mere physical enhancement but a systems-thinking method of sustainability in building. The combination of nanotechnology with traditional materials provides significant gains in energy efficiency and material use. For instance, nanoscale treatments and coatings lead to thermal insulation and energy efficiency improvements in buildings, lessening heating and cooling energy requirements. Furthermore, this method can lower the entire carbon footprint of the construction industry by reducing reliance on energy-hungry materials. By establishing a situation where the application of nanomaterials into the construction process is done systematically, not only does the industry move closer to building resilient and durable structures but also to an environmentally cleaner future. Generally, the strategic use of nanomaterials in building construction is a significant step towards meeting the double objective of sustainability and durability within the built environment (MORENO et al., 2017) (Figure 8.5).

Figure 8.5. Role of nanotechnology in construction.

Source: https://external-content.duckduckgo.com/iu/?u=https%3A%2F%2Fwww.constructioncost.co%2Fimages%2Fimg%2Fnanoparticles-construction.jpg&f=1&nofb=1&ipt=44cbb9b4ba7791b5fb228797cbe7a0006d2f2756453f7a6d147d0c6ebee1fffd.

8.2.2. Role in Electronics and Energy Storage

The use of nanomaterials in electronics and energy storage is a significant intersection of innovation and sustainability. A number of nanomaterials, such as carbon nanotubes and graphene, are being widely utilized for enhancing the efficiency and performance of electronic devices. Their high conductivity and structural characteristics make them ideal for miniaturization and increased device lifespan. This increased lifespan is particularly advantageous for consumer electronics, where thinner, lighter, and more efficient products are desired. By integrating nanomaterials into the fabrication of capacitors and batteries, researchers are discovering ways to move energy density forward while reducing the total environmental impact of such technology. Such progress makes it increasingly necessary to consider beyond functionality when measuring material application in electronics with the movement towards greener technological methods (MORENO et al., 2017). In energy storage systems, the introduction of nanomaterials has revolutionized battery and supercapacitor technologies. For instance, employing nanostructured materials in lithium-ion batteries significantly enhances charge capacity and cycle stability, both of which are essentially related to longevity. Nanomaterials enable the development of a better electrode structure with high electron and ion conductivity rates, resulting in enhanced overall battery performance. Moreover, these innovations are indispensable for renewable energy, where the storage systems of energy are imperative in order to achieve optimal energy generation from volatile resources like solar and wind. The long-term application of these innovative batteries is instrumental in developing a more reliable energy grid, thus sustaining the ultimate goal of energy technology sustainability (Fangueiro et al., 2015).

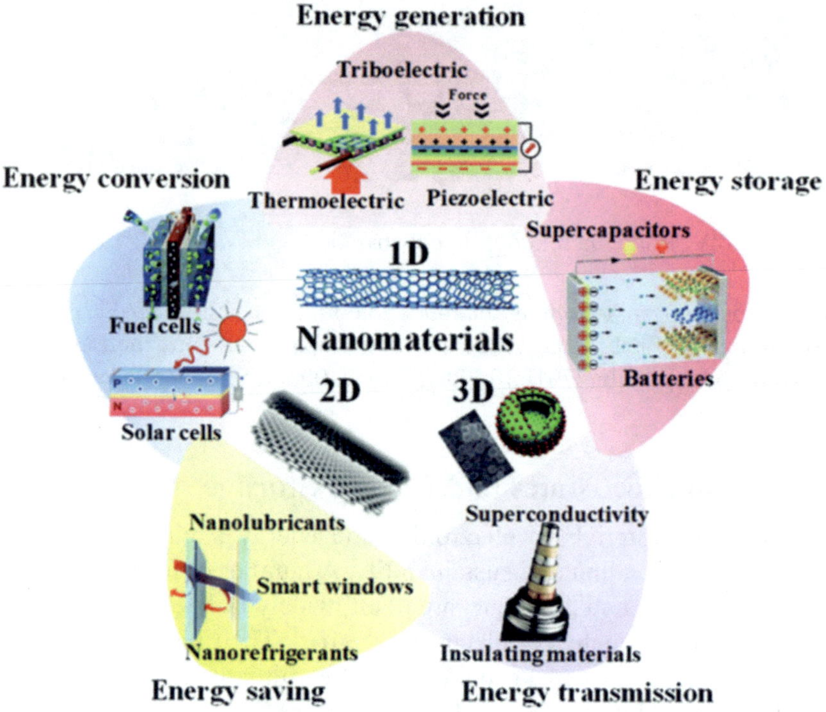

Figure 8.6. Nanomaterials for energy storage applications.

Source: https://media.springernature.com/lw685/springer-static/image/chp%3A10.1007%2F978-981-16-1888-8_7/MediaObjects/511627_1_En_7_Fig3_HTML.png.

In addition, nanomaterials' versatility is beyond normal utilization, paving the way for new applications in energy storage devices. As an example, the development of multifunctional structural membranes from fibrous and textile nanomaterials has offered light, tough, and energy-efficient components for clean energy systems. Such membranes are able to be designed with self-cleaning property and temperature regulation, greatly enhancing their efficiency of work and durability. These developments demonstrate the way in which nanotechnology integration into the energy infrastructure can introduce materials that are not only efficient but also sustainable by cutting down wastage and resource consumption. With rising requirements for energy storage to evolve higher-performing and multifunctional material requirements, exploration of new nanomaterials becomes increasingly crucial in order to achieve sustainable and durable technological outcomes for the energy sector (Fangueiro et al., 2015) (Figure 8.6).

8.2.3. Applications in Medicine and Healthcare

Nanotechnology has been much touted for its revolutionary medical and healthcare applications, with the promise of a new therapeutics and diagnostics era. At the forefront is the development of nanocarriers for targeted drug delivery, with the spotlight shining brightest in oncology. These nanoscale vehicles improve the precision of therapy by targeting the therapeutic agents specifically to the cancer cells, thus reducing the collateral injury to normal tissues. This precision not only maximizes the efficacy of the drugs but also reduces many of the harsh side effects normally encountered with traditional chemotherapy. As attested by studies conducted, many nanocarrier preparations have advanced to human trials, but only a few have been approved by the regulatory bodies for clinical application. This gap emphasizes the need for ongoing research to fill the gaps in translating laboratory success to effective clinical therapies, thus making nanotechnology a promising future direction in cancer treatment (Chehelgerdi et al., 2023). Outside of oncologic practice, the utility of nanomaterials pervades other areas of medicine, such as diagnostics and regenerative medicine. Nanoscale systems are also at the core of advancing imaging techniques and biomarker discovery, both of which are crucial in early diagnosis of diseases. In increasing the sensitivity of diagnostic tools, nanotechnology can facilitate timely interventions, which ultimately lead to enhanced patient outcomes. Additionally, its uses in regenerative medicine, especially in tissue engineering, have been promising in repairing injured tissues through scaffolding that replicates the natural extracellular matrix. This innovative capability of diagnostic and therapy modalities underscores the wonderful potential of nanotechnology as a foundation of modern medicine, driving the evolution of medical procedures and introducing a new era of precision medicine (Malik et al., 2023). Despite all the celebrated advances, however, the application of nanotechnology in medicine is not without problems, with concerns regarding safety, environmental friendliness, and ethics. With the expanding application of nanomaterials in clinical settings, clinicians and researchers must collaborate in multidisciplinary consortia to address these problems proactively. This collaboration is crucial in creating standard protocols for the use of nanotechnology such that innovations are effective and safe. In addition, a proper understanding of the long-term effects of nanomaterial exposure on human health and the environment must have priority. Solving these complicated challenges not only promotes the stability of nanomaterials in medicine but also enhances people's trust in such new technology, thereby making it easier to have broader application and acceptance in hospitals (Malik et al., 2023).

8.3. DURABILITY OF NANOMATERIALS

The durability of nanomaterials is also one major factor in determining their viability for various applications, particularly in building and architecture. These materials, composed of nanoscale particles, possess distinct traits that enhance functionality and longevity compared to their traditional equivalents. For instance, nanomaterials can significantly increase the strength of steel and concrete structures since they could stop deterioration while retaining the ability to structurally exist. Notably, existing literature presents a range of existing nano-products used in construction, which may have multiple roles from enhancing the durability of building components to service life extension. These developments mark a groundbreaking shift in the manner in which designers and constructors respond to the design and construction process, steering it toward sustainable and extended practices with a view to reversing challenges posed by the built environment today (MORENO et al., 2017). The integration of next-generation nanomaterials like graphene oxide equally illustrates startling improvements in durability performance. In the area of novel concrete solutions, graphene oxide-modified pervious concrete is an example of the potential of nanotechnology to enhance material properties. Material performance can be measured through rigorous testing, exhibiting outstanding resistance towards environmental conditions such as freeze/thaw and salt exposure. These are key characteristics for applications in weather-prone regions. Furthermore, adding graphene oxide not only meets durability standards but also enhances the capacity for removing levels of pollutants from stormwater, promoting environmental sustainability. Such a synergy of advantages points to the transformative potential of nanomaterials to transform the construction industry through an integration of structural integrity and environmental stewardship (Shi et al., 2018). Aside from uses in construction, nanomaterials' longevity is discovered in numerous other fields, including electronics, coatings, and healthcare. As an example, nanomaterials would extend the lifespan of electronic components by improving thermal and electrical conductivity, thereby sustaining performance for longer durations. Similarly, in the protective and coatings industry, nanomaterials possess the capability to generate water repellence and abrasion resistance, thereby extending the lifespan of coated surfaces. In medicine, nanoscale drug delivery system stability can improve therapeutic effectiveness with reduced accompanying side effects. These multi-tasking applications not only highlight the versatility of nanomaterials, but further highlight the urgent necessity for more investigation to clearly define their long-term behavior under different conditions. Lastly, proper evaluation of nanomaterial durability contributes significantly to sustainable development and inspires future developments in many areas (Figure 8.7).

Figure 8.7. Durability of nanomaterials.

Source: https://external-content.duckduckgo.com/iu/?u=https%3A%2F%2Fthumbs.dreamstime.com%2Fb%2Fenhancing-durability-nanomaterials-science-labs-experimenting-innovations-techniques-explore-how-innovative-methods-354666605.jpg&f=1&nofb=1&ipt=42bea520bfa5cb0b81fdd5eb2de33eee9d37751e4163b059613c10fa6a3aa32c.

8.3.1. Factors Affecting Durability

The durability of the nanomaterials depends on a multitude of factors that contribute crucially to ensuring their effectiveness in various uses. One such critical factor is the environmental conditions to which these materials are exposed. For instance, extreme environments characterized by dominance of chemical exposure, humidity, and temperature fluctuations can deteriorate the strength of nanomaterials. These conditions induce degradation processes such as oxidation or corrosion, particularly for composite materials that are mixed with metals. Having knowledge about the specific vulnerabilities of nanomaterials in these conditions allows for more optimal design choices and material selection, ensuring higher durability and longer lifespan. Further, the non-symmetric exposure to the environmental factors can induce the unbalanced degradation of the system as a whole. Solving these environmental concerns through wise material selection and defensive strategies will be needed to optimize the longevity of nanomaterials in real implementation (Fomin et al., 2025). Mechanical stresses are critical parameters for nanomaterial durability as well, particularly in structural applications under loads. Redistribution of loads, whether due to dynamic loads like wind and seismic forces or as a result of uneven weight distributions, can create concentrated stress points. The effect can lead to premature failure because materials cannot endure imposed mechanical stresses during long-term service. Moreover, reinforcing components such as nanofibers or nanoparticles in composites control the overall toughness of the nanomaterial. If the reinforcing components themselves corrode or weaken, the resulting compromise can significantly reduce the load-carrying capacity of the structure. Hence, considering mechanical stresses and material design to resist these loads is important in the durability of nanomaterials under varying conditions (Fomin et al., 2025). Lastly, the intrinsic properties of the nanomaterials, including the composition and microstructure, form the foundation of their durability and longevity. Changes in synthesis processes lead to differences in the arrangement and bonding of the nanoparticles, which bear significant implications on durability. For example, geopolymer concrete, an alternative binder material, has been under the spotlight due to its sustainability and enhanced durability. The unique chemistry of geopolymer concrete renders it resistant in harsh conditions, thereby providing a promising path towards sustainable construction. Evaluation and enhancement of these physical properties during the design and manufacturing process can significantly improve nanomaterials' durability and dependability. Therefore, an understanding of these intrastructural factors is required to further advance the field of nanomaterials and utilize them in sustainable technologies with success.

8.3.2. Testing Methods for Durability

Durability test methods for nanomaterials must be developed to validate their durability for long-term use in various industries. Accelerated aging tests are one of the commonly applied methods, which simulate long-term exposure to environmental factors such as temperature fluctuations, humidity, and UV radiation. These tests help in understanding nanomaterial degradation over time and can even predict the working life of nanomaterials. In addition, mechanical testing methodologies like tensile strength and fatigue tests identify how nanomaterials behave when they withstand stress and strain. An understanding of these is of immense importance, particularly for application in building and aeronautical engineering, where material failure will prove deadly. The designs of these tests must account for the specific nanostructure and its interactions with the surrounding materials, so that test results are an accurate representation of actual conditions. Dynamic mechanical analysis (DMA) is another valuable way to test nanomaterial durability. DMA is the measurement of the viscoelastic behavior of materials by measuring the mechanical response of the material to varying frequencies and temperatures.

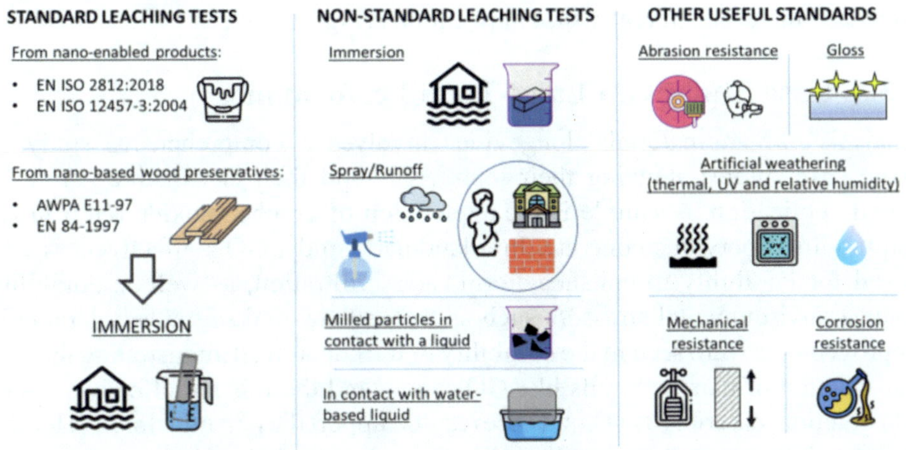

Figure 8.8. Testing methods for durability of nanomaterials.

Source: https://media.springernature.com/lw685/springer-static/image/art%3A10.1186%2Fs12302-021-00493-z/MediaObjects/12302_2021_493_Fig1_HTML.png.

This technique is highly efficient in the characterization of nanocomposites, as it provides a clear notion of how the incorporation of nanoscale fillers influences the overall performance and durability of the material. Moreover, water absorption tests are vital in determining the aqueous resistance of nanomaterials, crucial in applications exposed to water or humidity. With careful analysis of these parameters, researchers are able to come up with ways through which the performance of nanomaterials can be enhanced. Significantly, the addition of new test methodologies allows for a better understanding of durability and allows for the formulation of durable materials that are sustainable. Scalability of durability testing protocols is another important component involved in nanomaterial evaluation. As the material transitions from research in the laboratory to commercial sustainability, there is a need to have assurance that protocols can be scaled up. For instance, the application of standardized test procedures for nano-enhanced asphalt within Warm Mix Asphalt technology, according to the most recent literature, highlights the importance of robust methods for measuring moisture susceptibility and mechanical properties (Hasib et al., 2024). Similarly, applications in wind turbines necessitate stringent, physics-driven test procedures that quantify their effects on structural performance and mechanical behavior (Furi et al., 2025). Therefore, establishing standard testing procedures is crucial in the construction of confidence and acceptance of the use of nanomaterials in the different industries. These activities not only serve to ensure compliance with the regulations but also to the broader sustainability narrative of the use of advanced materials in real-world applications (Figure 8.8).

8.3.3. Case Studies on Long-Term Performance

Analysis of nanomaterials' longevities involves a comprehensive study of some case studies showing their longevities and their sustainable nature in actual application. A sample is the application of graphene oxide (GO) in fly ash-modified pervious concrete. In a landmark study, GO's effectiveness was tested for its ability to enhance stormwater infiltration, as well as durability against environmental stressors such as freeze/thaw and salt attack. Dramatic improvements were seen in the durability of the concrete, demonstrating that the introduction of nanomaterials like GO can extend the life and effectiveness of construction materials by years. These results support the potential for developing stronger infrastructure that still performs over extended periods of time, able to sustain the increasing demand for green construction practices (Shi et al., 2018). Another excellent case study is the addition of plant-based nanofibers, in this instance, nanocellulose, to cement-based composite materials. Plant-based fibers are increasingly being accepted by the civil construction sector due to their affordability and environmental friendliness. Experiments have proven that the mechanical properties of cement composites are significantly improved through the addition of nanocellulose. The current case study has a double advantage: apart from it producing stronger materials, it is sustainable as it uses renewable resources. However, the study also identifies challenges, such as processing plant fibers and achieving a good fiber/matrix interface, that are critical in making the end composite material last long. Overcoming these challenges will be critical in facilitating extensive application of plant fibers in the construction sector (Ahmad et al., 2017). The longevity of nanomaterials is also shown through their application in engineered cement composites, for example, reducing the environmental impact of traditional materials. Tests to quantify durability through a series of stress tests such as freeze/thaw and salt scaling resistance indicate that nanomaterials are not only more resistant but also possess improved physical characteristics such as water permeability and chemical attack resistance. These advances suggest that the application of nanomaterials can facilitate sustainable construction methods that reduce maintenance and increase building lifespan. The application of nanomaterials presents a visionary approach to civil engineering, meeting both the needs of sustainable and long-lasting building solutions. Through such case studies, the potential of nanomaterials to redefine the standards of sustainability and longevity in construction comes into wider view (Ahmad et al., 2017; Shi et al., 2018).

8.4. SUSTAINABILITY CONSIDERATIONS

Concerns regarding sustainability are becoming increasingly significant in the argument over the application of nanomaterials. Incorporating sustainable usage into the production and use of the materials can provide immense benefits to various industries. Handling nanomaterials appropriately should account for their effects on the environment, most importantly their life cycle and potential influence on natural surroundings. Methods such as the application of green nanotechnology, an approach that emphasizes efficiency in energy and least utilization of resources, can be useful in forming a sustainable economy. Certain practices that get developed should seek to optimize material efficiency, reduce cost annually, and reduce environmental degradation, thus ensuring that advancements in nanomaterials benefit overall sustainability goals. Such a strategy not only enhances the utility of nanomaterials but also places them at the pinnacle of sustainable technological progress, and it is, therefore, a point of prime necessity to evolve innovation that minimizes environmental footprints. The role played by policy frameworks and regulations towards the adoption of nanomaterials' sustainability cannot be overstated. Effective governance can contribute to the transition towards sustainable practice through the establishment of standards that prioritize health, security, and environmental considerations. It is up to policymakers to balance innovation and development on one side and risks associated with nanomaterials, particularly regarding worker safety and environmental integrity on the other. Moreover, ongoing research and collaboration between industries and regulatory bodies are essential in terms of best practices sharing and assisting stakeholders in adopting sustainable processes. This synergy not only enhances the occupational health of the workers but also maintains innovations with long-term ecological benefits. Through the application of effective regulatory mechanisms, the journey toward sustainability in nanotechnology can be effectively undertaken, minimizing risks and maximizing the diverse advantages of nanomaterials.

Sustainable approaches to nanotechnology can have deep implications beyond direct environmental returns, impacting economic and social facets as well. The transition to greener production methods will drive job creation in new green industries and enhance community resilience to environmental shocks. Specifically, the application of construction practices that adopt the utilization of nano-enabled materials could guarantee optimal resource utilization, while enhancing the durability of infrastructure with the lesser expense of maintenance in the long run. However, this transition also calls for obstacles related to consumer acceptance as well as the moral implications of nanomaterial utilization to be overcome. As highlighted, public opinion and market forces are central to the success of such initiatives. Dialogue and education on the benefits and risks of nanotechnology are important, with sustainability looming above investment choices in nanomaterials by the public and private sector, leading towards a sustainable future.

8.4.1. Environmental Impact of Nanomaterials

The environmental consequences of nanomaterials are one of the largest topics in current research, especially with their increasing application in various industries, including construction and water management. Their introduction into pervious concrete, for instance, has been very promising. Not only does the new material enhance the structural performance and lifespan of the concrete but also includes environmental benefits by improving stormwater management. Studies have established that GO-modified pervious concrete effectively removes pollutant concentrations in stormwater like copper, zinc, and phosphates, thereby preventing subsequent ecological damage. Such double-duty function of enhancing material performance as well as promoting environmental sustainability presents the potential of nanomaterials to help bring about beneficial inputs to ecological well-being to broader agendas of sustainable development within the construction industry, as the evidence in Shi et al. (2018) indicates. Besides, the sustainability in the use of nanomaterials is furthered by the aspect that they help improve the long-term durability of building materials, thus reducing the environmental footprint associated with replacements and maintenance. A number of nanotechnology applications have been developed with the aim of furthering the functionality of traditional building materials such as steel and concrete through the improvement of their resistance against deterioration and environmental degradation. Nanomaterial products have been observed to exhibit greater durability, i.e., buildings will be more resistant to climatic and chemical attacks for an extended period. Building materials with greater durability must be replaced less often, so the outcome is a consequent reduced consumption of resources, generation of wastes, and environmental degradation. This mutual dependency between sustainability and durability highlights the potential of nanomaterials to significantly redefine the construction industry's impact on the preservation of the environment, as developed in the comprehensive review presented in MORENO et al. (2017). Finally, there is a need to consider the long-term implications of the introduction of nanomaterials in environments. While the benefits of increased performance and environmental cleaning are great, the effects of nanomaterials on ecosystems and biodiversity need to be evaluated carefully. The potential for nano-scale particles to interact adversely with biological systems is problematic and needs to be addressed by scientists so that safe utilization can take place. Additional cross-disciplinary research needs to occur in order to fully assess these interactions and allow nanotechnology to be placed into sustainable practices responsibly. Regulatory framework establishment and long-term effect studies will be required for protecting the integrity of the environment while capitalizing on the extensive advantages presented by nanomaterials in terms of durability and sustainability in the construction sector and others (Figure 8.9).

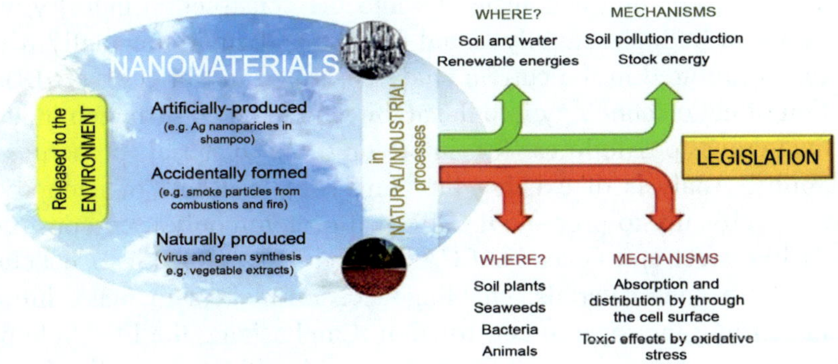

Figure 8.9. Environmental impact of nanoparticles.

Source: https://pub.mdpi-res.com/materials/materials-14-00166/article_deploy/html/images/materials-14-00166-ag-550.jpg?1621230952.

8.4.2. Life Cycle Assessment (LCA) of Nanomaterials

Understanding the LCA of nanomaterials is pivotal to determining their overall sustainability and durability in the long term. LCA provides a holistic platform that estimates the environmental impacts associated with each phase of a material's life, from raw material extraction to processing, use, and disposal.

Through the application of LCA in nanomaterials, researchers can identify potential pathways through which ecologically focused footprints can be minimized while optimizing the performance of materials. This approach is especially relevant considering the unique features of nanomaterials, which could influence their performance and interactions with the environment in a manner distinct from conventional materials. Besides, the integration of LCA in developing nanomaterials would simplify compliance with regulations and stimulate innovation for creating more sustainable applications of nanomaterials. An excellent example of how LCA translates to nanomaterials is applying graphene oxide (GO) for use in pervious concrete production. Here, LCA contrasts the environmental benefits of adding GO, such as improved stormwater management and reduced levels of pollutants, with the sustainability of fly ash as a replacement binder. Outcomes of studies comparing GO-modified fly ash pervious concrete demonstrate enhanced material durability, particularly freeze/thaw cycles and salt attack resistance. Consequently, these results alert architects and engineers to the prospect of utilizing GO in increasing infrastructure sustainability. Such advances capture the multifaceted benefits of incorporating nanomaterials into the construction industry, which finally result in both structurally sound designs and environmentally friendly practices, as captured in the current study described in Shi et al. (2018). The integration of nanotechnology into the architecture and construction industries offers numerous possibilities for improving the material properties and sustainability. Analysis of existing literature reveals a host of nanoproducts designed specifically to promote the lifespan of commonly used construction materials like steel and concrete. LCA methodology serves as a benchmark to assess these nanomaterials, enabling decision-makers to make informed choices regarding their use in construction. Emphasizing the life cycle of the nanomaterials, construction practitioners can identify new applications that significantly improve energy efficiency and longevity. Besides, the studies on nanoproducts continue to increase with applications ranging from photovoltaic cells to advanced water filtration systems, as enumerated in recent literature. Emphasizing LCA ensures that the integration of such technology promotes sustainability alongside organizing resilience and durability, enhancing the entire environmental responsibility and efficiency blueprint described in MORENO et al. (2017).

8.4.3. Regulatory Frameworks and Guidelines

The regulatory systems and guidelines around nanomaterials are critical to their sustainability and long lifespan across different applications. In the last few years, the regulatory systems have been evolving in line with the novel properties and challenges of nanomaterials, demanding a multi-dimensional approach to environment and safety. Some of the organizations responsible for providing such frameworks and guidelines include the European Chemicals Agency (ECHA) and the U.S. Environmental Protection Agency (EPA) have established particular guidelines that highlight the need for thorough risk analysis as well as accounting for lifecycle effects. These frameworks promote regulation in line with nanomaterial usage and offer necessary guidance for manufacturers as well as researchers. The arrival of concepts like Safe and Sustainable by Design (SSbD) further strengthens these legislations by integrating sustainability standards in the chemical life cycle in order to promote the development of materials that are not just powerful but also sustainable. The EU Chemical Strategy for Sustainability (CSS) embodies the growing need to align regulatory approaches to nanomaterials by introducing a new framework for establishing criteria for SSbD chemicals and materials.

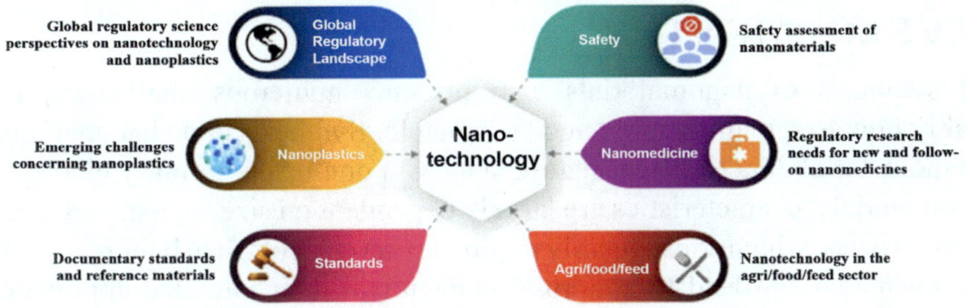

Figure 8.10. Regulatory frameworks and guidelines of nanomaterials.

Source: https://ars.els-cdn.com/content/image/1-s2.0-S0273230021000258-gr1.jpg.

This push highlights the integration of safety and sustainability considerations into the evaluation processes of alternative and incumbent chemicals. The operationalization of the SSbD principle encompasses a variety of sustainability dimensions when taking into account parameters and indicators used to quantify the safety of materials and environmental impacts along their life cycle. Through the integration of such a comprehensive approach, the CSS aims to improve the creation and use of innovative nanomaterials with the capability of serving societal needs without compromising the integrity of the environment. As nanotechnology advances further, proper regulatory mechanisms remain at the core of promoting responsible research and industrial operations. Besides the efforts in Europe, a number of initiatives are in progress around the globe to enhance regulatory governance of nanomaterials. Countries begin to realize the potential threats and benefits of the materials and redesign their regulatory mechanisms accordingly. For instance, industry players and environmental organizations are already collaborating to establish standards for transparency and accountability in the production and utilization of nanomaterials. The dilemma is to balance such various regulatory approaches to formulate a single international standard that can efficiently cover the issues of nanomaterial research and application. Also, the need for constant surveillance and adjustment of these frameworks is essential in order to keep them relevant and effective in light of changing technological advancements in the field in order to safeguard public health and environmental sustainability in the long term (Figure 8.10).

8.5. CHALLENGES IN NANOMATERIAL DEVELOPMENT

The synthesis of nanomaterials also presents numerous challenges, most notably how to remain durable and sustainable. Synthesis and characterization of nanomaterials' inconsistent nature stands as one of the primary challenges. Nanomaterials' characteristics are largely dependent on size, shape, and surface characteristics, which consequently requires strict control when being processed. Any mismatch can lead to abnormal behavior in their intended applications, which could compromise their structural integrity and performance. Most studies have also concentrated on individual applications, such as in cementitious composites, without real understanding of their performance under varying environments and scales. This lack of rigorous research limits the potential to develop standard protocols that would offer reliability and durability in applied uses, becoming a massive hindrance towards broader application. It is the same attribute that makes nanomaterials valuable that makes it difficult to produce them with environmental and health concerns. Their small size enables them to penetrate biological membranes and have harmful interactions with living systems. Systemic risk evaluation is key to identifying the long-term impacts of exposure to nanomaterials on human health and the environment. This work is key to providing insight into safe production processes, regulation, and sustainable development goals. As the technology advances, preventing the risk of toxicity and ecological impacts becomes increasingly important. This need highlights safe innovation, ensuring that nanotechnology advances, such as the application of graphene oxide in ultra-high-performance concrete Lambert et al. (2020), comply with safety and sustainability protocols to minimize the risk of adverse impact. Economic factors also play a pivotal role in the challenges of nanomaterial development. The initial cost of research and production often deters investment, and a gap exists between possible research advances and their possible application to industries. For instance, though nanomaterials' use has been postulated to make various infrastructures like concrete blends stronger, the cost of applying these materials on a mass scale is unclear. Thus, there is a strong need for investment and cross-sectoral collaboration for additional research and development. Such endeavors could supply innovative solutions that not only address durability concerns in buildings but also towards achieving global problems in sustainability. As researchers draw upon nanotechnology to solve grand challenges of development Daar et al. (2005), they must also be mindful of the economic implications and strive to use cost-effective techniques that promote sustainable practices (Figure 8.11).

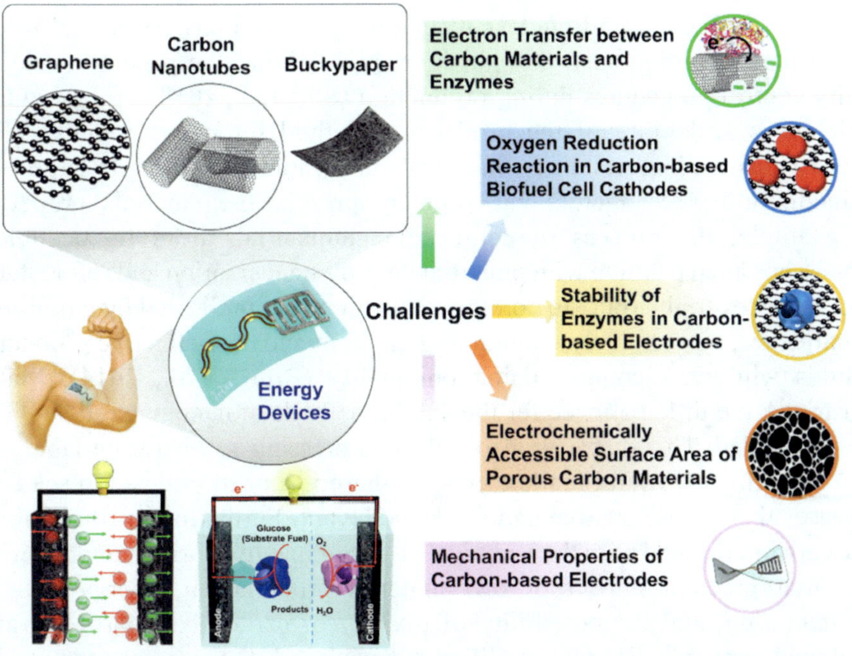

Figure 8.11. Challenges in nanomaterial development.

Source: https://external-content.duckduckgo.com/iu/?u=https%3A%2F%2Fwww.mdpi.com%2Fcarbon%2Fcarbon-05-00062%2Farticle_deploy%2Fhtml%2Fimages%2Fcarbon-05-00062-g001.png&f=1&nofb=1&ipt=cc06066e647bc329bccb069caab3674193466bbf8d15fefb8587bed87c00e790.

8.5.1. Technical Challenges in Production

Production of nanomaterials is confronted with several technical challenges that hinder their widespread application, most prominently durability and sustainability. Among the chief challenges is material raw processing to achieve desired nanoscale properties without losing structural integrity. The example of synthesis of graphene oxide (GO)-functionalized pervious concrete shows the role of addition of nanoscale improvements in toughening of the material. However, success in modifying production requires tight control of the processes involved, both physical and chemical. Faults in this control lead to inconsistencies in the final products properties, and hence, both its performance and lifespan are compromised.

As seen in the GO-modified fly ash concrete research, advanced techniques like X-ray diffraction and nuclear magnetic resonance have a pivotal role to play in understanding such complexities, highlighting the need for stringent quality control procedures during production (Shi et al., 2018). Another major challenge is scalability of the production method for nanomaterials. While there are positive impacts from small-scale fabrication, upscaling to large-scale production introduces factors that could compromise the materials' properties. For example, the process of cellulose nanomaterials from lignocellulosic biomass has been positive with potential employment creation and reforestation effects. However, the route to commercialization is complicated by a multitude of challenges, most notably maintaining efficient and cost-effective production methods while not compromising on quality (Brun et al., 2014). Mature industries have little patience for the subtle methods of nanomaterial synthesis that are needed. This disparity also reflects a pressing gap between laboratory breakthroughs and useful market uses, so there is a need to develop scaleable processes that are innovative and market-relevant. Furthermore, sustainability issues are intertwined with the technical issues of manufacturing nanomaterials. The environmental impact of raw material procurement, energy-intensive manufacturing, and the possibility of toxicity of certain nanoparticles raises significant concerns. For GO-modified materials, while these enhance product durability like pervious concrete, sustainability during graphene production is a cause for concern. Of significance is to look at the entire lifecycle of nanomaterials from production through waste disposal so that sustainability is achieved in each phase. Overcoming such problems demands a multidisciplinary approach using chemistry, engineering, and environmental science to produce responsibly and sustainably. This will not just make nanomaterials more durable but also align their production processes with the general goals of protecting the environment and ensuring sustainable resource use.

8.5.2. Economic Factors Influencing Sustainability

The economic drivers of sustainability are at the core of consideration of the long-term longevity of nanomaterials. As industries across the world increasingly focus on sustainability, the economic burden of producing and utilizing nanomaterials is of significant concern. The production of nanomaterials is often dependent on novel but expensive processes, which dissuades their mass application. The material saving advantages and extended product lifespan, however, can recoup themselves in the long run. Economic feasibility of utilizing nanomaterials, therefore, must account for not only short-term costs but also potential yields in terms of enhanced performance and sustainability (Lahiri et al., 2023). This holistic economic perspective is needed by stakeholders who want to make sense of the transition to more sustainable material technologies in sectors like healthcare, consumer products, and construction. Besides production cost, market demand for sustainability also contributes significantly to the economic viability of nanomaterials. Consumers and entities are also focusing more on green products, thereby pushing producers toward greener processes and materials, including nanomaterials. This change can lead to greater competition among producers, ultimately driving innovation and price reductions. Furthermore, regulatory frameworks that promote sustainability offer economic benefits to companies that invest in sustainable technologies. A case in point is the creation of geopolymer concrete, produced from lower energy demand than traditional materials, which illustrates how economic incentives can speed up the uptake of new sustainable solutions in construction (Wong, 2022). The dynamic interaction among economic support, consumer behavior, and cost competitiveness underscores the imperative of an enabling economic environment for the dissemination of sustainable nanomaterials. Moreover, integration of nanomaterials into current economic systems entails the restructuring of lifecycle studies and environmental analyses. The long-term lifespan and durability of such materials will need to weigh their economic advantages against possible ecological effects throughout their life cycles. For instance, durability-design nanomaterials can avoid degradation and the subsequent environmental footprint of pollution and waste. Yet, their production might include substantial energy consumption as well as resource extraction, which can erode their overall sustainability record (Wong, 2022). Their intricacies necessitate that they are solved with a robust economic framework which captures not only direct monetary cost but also broader ecological reach. Ultimately, it is important for industry leaders and policy makers to be aware of these economic considerations in order to facilitate a sustainable future where the benefits of nanomaterials are optimized and their detriments minimized (Figure 8.12).

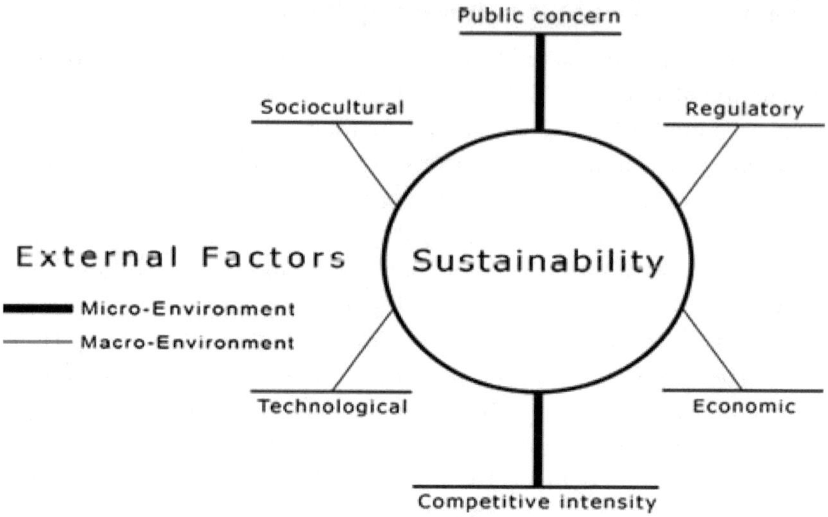

Figure 8.12. Economic factors influencing sustainability.

Source: https://external-content.duckduckgo.com/iu/?u=https%3A%2F%2Fthf.bing.com%2Fth%2Fid%2FOIP._tT9za23FF5h_QbQwWZuZgHaEo%3Fc-b%3Dthfc1%26pid%3DApi&f=1&ipt=f7aa21672c3bfc9b6e9b9c637e1e7bd9e3368618746958de1337cff1248febef.

8.5.3. Public Perception and Acceptance

Public perception and acceptance of nanomaterials are crucial to their adoption in various sectors, as these define a great deal of how they are used sustainably and for how long. Although the potential of nanotechnology to transform medicine, agriculture, and the environment is limitless, it is commonly tainted by public fear of potential health and environmental impacts. The issues at the community level are that there is reduced understanding of nanomaterials that leads to skepticism about their efficiency and safety. As such, the absence of intersubstance between scientific facts and public information demands efficient communication strategies to build confidence and enhance perception. By addressing these issues and increasing the public awareness of the risks and advantages of nanomaterials, stakeholders can ensure greater acceptance and develop more informed dialogue about their sustainable application. In addition, public acceptance is strongly affected by nanomaterials' sustainability assessment. Literature indicates that such approaches as life cycle assessment, risk assessment, and multi-criteria decision analysis are vital in the evaluation of the potential impacts and advantages of nanomaterials on health and the environment. Though the tools enable a structured framework for informing stakeholders, they also depict areas of uncertainty further generating public concern. For instance, multi-criteria decision analysis is the most suitable tool because it can facilitate diverse evaluation criteria as well as enable intense decision-making processes. Through the use of open and participatory sustainability appraisal, academics and policymakers can easily communicate outcomes to the public, finally resulting in higher acceptance and the sustainable development of nanotechnology (Cinelli et al., 2018). Finally, with changing consumption patterns of the world and the demand for sustainable choices growing, public sentiment with regard to food safety and nanotechnology merge incredibly. Observing how socio-economic and regulatory forces, more specifically in the management of foodborne diseases, affect public sentiment indicates how trends can contribute towards shaping public opinion (Cole et al., 2010). Issues such as increasing food demand and globalization underscore a pressing call for emerging technologies, including nanomaterials, in ensuring food safety. Public perception of the deployment of such applications is, however, generally inclined by previous food safety scandals and visible impacts of technological advancement. Establishing trust, therefore, through efficient regulation and effective communication capabilities is imperative. Instilling trust in the role of nanotechnologies to enhance sustainability can pave the way for broader acceptance and implementation in everyday practice, ultimately contributing to long-term and sustainable benefits.

8.6. FUTURE TRENDS IN NANOMATERIALS

The study on nanomaterials has great potential to enhance the durability and sustainability of various applications across numerous industries. As environmental sustainability awareness among the public keeps growing, the integration of nanotechnology into building materials becomes increasingly imperative. Traditional methods of construction have preferred to prioritize mechanical performance over environmental concerns, resulting in unsustainable development that consumes gigantic amounts of resources. To correct this imbalance, future trends are bound to aim at optimizing the growth of nanomaterials that not only deliver enhanced physical performance but also minimize the environmental load generated by construction materials. Research work under the guidance of paradigms such as the Millennium Development Goals is vital in transforming priorities around sustainability, with the focus on more sophisticated materials that meet strength with environmental priorities (Labrincha et al., 2013). In looking at directions for nanomaterials in the future, one particularly significant direction is the development of eco-friendly alternatives such as geopolymer concrete. This new construction material, produced from industrial waste like fly ash, is a move toward more sustainable methods. The use of nanotechnology in geopolymer materials can facilitate the removal of contaminants from wastewater, thereby solving serious environmental issues using materials otherwise wasted. Nanomaterials are efficient agents for contaminant removal processes with their multifunctionality demonstrated. Using materials that are good for durability and sustainability, such as those in the case of geopolymer concrete reinforced with zeolite or titanium dioxide, companies can minimize their environmental impact while enhancing resource efficiency (Aris et al., 2015).

Figure 8.13. Future trends in nanomaterials.

Source: https://external-content.duckduckgo.com/iu/?u=https%3A%2F%2Fmiro.medium.com%2Fv2%2Fresize%3Afit%3A1080%2F1*4jNwGuPQGLWcvOQUgUfYpA.png&f=1&nofb=1&ipt=513242db516aff4fd5b11de1bf9b80c33e36ecb6986b8e05aaea40ffc7caf743.

As the world continues to advance into the 21st century, it is increasingly important to study the lifecycle impacts of nanomaterials. Future research will likely involve detailed examination of how the materials endure over time, particularly longevity and recyclability. Researchers will need to consider the potential for nanomaterials to contribute to circular economy principles, where resources are used more effectively and waste is minimized. This entails not only determining the physical and durability properties but also understanding how nanomaterials can be repurposed or recycled once their initial purposes have ended. Ultimately, the future of nanomaterials will likely be set by ongoing technological advancement and one focus toward establishing environmentally sound practices on the basis of societal requirements (Labrincha et al., 2013) (Figure 8.13).

8.6.1. Innovations in Sustainable Nanomaterial Design

The recent application of nanotechnology to sustainable materials has been a revolutionary approach in environmental management and construction. One of the innovations is graphene oxide-modified pervious concrete, which indicates the capability of nanomaterials to generate more advanced enhancements for traditional construction materials. With coal fly ash as a binder and in combination with graphene oxide (GO), scientists have demonstrated improved resistance to severe weather conditions, including freeze/thaw cycles and salt attack, which are vital for buildings exposed to environmental aggressions. Furthermore, this structure of nanomaterial effectively reduces amounts of common stormwater pollutants such as copper and zinc, delineating the double benefits of increased structural performance and ecological sustainability in cities as described in research findings on GO-modified concrete more recently (Shi et al., 2018). While material science continues to leap ever further forward, research innovation is still far removed from a practical realization in construction. The construction sector has, by tradition, given more emphasis to mechanical properties as compared to environmental effects, and this has led to the misinterpretation of practices required for the construction of future infrastructures in a sustainable way. The focus on enhancing materials' mechanical properties without their environmental effects has promoted the use of resources in an unsustainable manner. Highlighting the work of the Millennium Development Goals (MDGs), that is, the seventh goal that focuses on ensuring environmental sustainability, it is certain that the intersection of sustainability and nanotechnology needs more emphasis and urgency in the area of construction materials (Labrincha et al., 2013). Sustainable innovation in nanomaterial design can be used to align with overall environmental efforts, promising smaller paths towards lower ecological footprints. Furthermore, the future nanomaterial design is where cutting-edge material science and good environmental policy intersect. Researchers are tasked not only with considering the mechanical advantages of nanomaterials but to examine the ecological implications that accompany their utilization in detail. This convergent approach will require collaboration on multiple fronts, including material science, environmental engineering, and regulatory mechanisms, such that innovations are translated into meaningful tools of sustainability. Also, an understanding of the life cycle of nanomaterials, from production to disposal, will be crucial to developing sustainable methods that minimize ecological impacts and maximize material performance durability. Therefore, continued research and development investments on sustainable nanomaterials will not only fulfill scientific curiosity but also usher in a future for an environmentally friendly and more responsible building industry (Labrincha et al., 2013).

8.6.2. Potential for Recycling and Reusability

The recycling and reusability potential of nanomaterials presents a critical aspect of their sustainability, particularly in light of increasing environmental concerns and diminishing resources. Nanomaterials, like Metal-Organic Frameworks (MOFs), have been found to be highly flexible and adaptable, whose property can be leveraged to maximize their lifecycle management. MOFs, due to their structural versatility and surface area, can be designed specifically for reuse by utilizing sophisticated synthesis techniques emphasizing eco-friendliness as much as efficiency. As highlighted in new studies, the incorporation of Safe and Sustainable by Design (SSbD) principles during the development of these materials not only makes them more stable but also lowers potential environmental impacts throughout their life cycle, favoring a circular economy system in nanomaterial use (Bhadane et al., 2025a). The use of several nanomaterials, particularly cross-material synergies, has enormous potential for increased recycling and reuse. For instance, the hybridization of carbon nanomaterials with MOFs and Covalent Organic Frameworks (COFs) can yield multi-functional composite systems with general-purpose enhancement and resolution of problems related to structural integrity and selectivity. Integration offers a platform for designing hybrid systems that not only improve pollutant removal and resource recovery but can also be recycled more efficiently (Bhadane et al., 2025b). This approach promotes sustainability goals by transforming waste materials into valuable resources and representing the responsible consumption and production culture. Through reusability at the start of nanomaterial design, scientists have the ability to revolutionize the sustainability narrative in the material science arena. Nanomaterial recyclability goes beyond enhancing their environmental impact; it involves generating economic value for industries. Process innovation that facilitates the reuse and recycling of nanomaterials makes industry save on raw material procurement costs at the same time it complies with more stringent environmental regulations. Furthermore, the recycling of nanomaterials has the potential to drive innovations for use in such areas as cleaning up the environment and advanced drug delivery systems. Through the evaluation of the lifecycle of such materials and taking steps for their recycling and refresh, stakeholders are able to release the full potential of nanotechnology without compromising the resilience of ecological systems and advancing sustainable development goals (Bhadane et al., 2025a, 2025b). Hence, addressing the reusability and recycling of nanomaterials is not an environmental compulsion alone but also a way towards economic resilience.

8.6.3. Emerging Research Areas and Technologies

Research into nanomaterials has seen a wide range of new fields of study dedicated to their potential sustainability and longevity for application in architecture and construction. Efforts to incorporate nanotechnology into traditional building materials are accelerating as builders and architects work to enhance the lifespan and performance of buildings. Recent literature confirms an intense correlation between nanomaterials and improved durability performance, especially for steel and concrete, the fundamental materials in all building activities. Methods such as nanocoatings and self-healing concrete, besides prolonging the life of such materials, also contribute to the decrease in maintenance requirements and in environmental damage. These advancements point to the necessity of bringing scientific research into practice through applications that show a paradigm shift in instituting more sustainable building processes without sacrificing stability (MORENO et al., 2017). Outside the traditional material of concrete, the field of fibrous and textile materials has become a premier front for instituting durability and sustainability through guaranteeing cutting-edge applications. The incorporation of nanotechnology into textiles has resulted in the development of new functional materials employed in different industries, including architecture and building construction. The advancements in structural membranes with characteristics such as self-cleaning, thermoregulation, and energy harvesting capability show the contribution of textiles to developing energy-efficient buildings and sustainable architecture. These materials are flexible and lightweight in addition to having design flexibility and mechanical flexibility. As science progresses, the projected use of these new materials in construction will be higher, further solidifying nanotechnology as the foundation of sustainable architecture (Fangueiro et al., 2015). New nanomaterial technologies are also broadening the range of conventional construction methods, making it easier to provide more emphasis on resource conservation as well as environmental sustainability. Scientists are investigating the applications of nanostructured materials in renewable energy technology devices, water purification, and waste management, critical for sustainable development. Nanomaterials applied to solar panels and power storage devices enhance efficiency while lowering the overall environmental footprint. As the technologies evolve, they promise great potential to revolutionize building processes and material usage. Therefore, ongoing research on the durability and sustainability of nanomaterials is crucial, not only for meeting current construction needs but also for addressing future needs in the wake of global environmental issues (Fangueiro et al., 2015; Moreno et al., 2017).

SUMMARY

The study on the durability and sustainability of nanomaterials, it is evident that the new materials have enormous potential to enhance various sectors, particularly in relation to environmental sustainability. Their new properties — high strength-to-weight ratios and enhanced electrical conductivity, to name a few — enable innovations that have the ability to radically reduce use of resources and waste. The successful penetration of industries with nanomaterials, however, depends importantly on the removal of existing obstacles, especially as far as environmental impacts and life-cycle studies are concerned. With increasing research, it is crucial for researchers and practitioners to assign top priority to sustainability indicators in the development processes of nanomaterials so that benefits may not be compromised at the cost of ecological integrity and health. Commercialization of nanomaterials will also have to be handled with a strategic eye on technical as well as non-technical issues. It must be pointed out that while monumental advances have been made in the lab environment, practical translation to the marketplace is limited by several factors, including regulatory constraints, economic feasibility, and social acceptability. Notably, discussion of ethical issues in nanotechnology is increasingly becoming essential.

By promoting interdisciplinary partnerships that examine these variables, stakeholders can bring about the opportunity for more robust frameworks that not only sustain the growth of the nanomaterial sector but also its adoption and integration into society. It is important that these variables are addressed to ensure the scalability of nanomaterials aligns with broader environmental goals. Moreover, the research future of nanomaterials needs to focus on developing smart systems with adaptive properties to their surroundings, advancing their functionality and sustainability. For instance, the generation of bendable sensors and incorporating them into a great variety of applications will transform many sectors, from healthcare to environmental monitoring. However, significant challenges lie in delivering these technologies that are usable and feasible in actual conditions and ensuring sustainable practice. By addressing and resolving the performance, compatibility, and commercialization bottlenecks identified in current literature, the industry of nanomaterials can play significant roles in a sustainable future. Additional investments in new solutions will ultimately enable researchers and industry specialists to harness the full potential of nanomaterials and hence play their part in an emission-lowered sustainable carbon cycle envisioned by current research activities (Luo et al., 2023; Stephens et al., 2022).

The nanomaterial discovery, particularly graphene oxide (GO) and cellulose nanomaterials, demonstrated great advancements in durability and sustainability. Research regarding GO-modified pervious concrete indicates how materials used for urban engineering infrastructure can be enhanced. Preliminary findings from Phase I determined that using the application of coal fly ash as the sole binder had the potential for sustainable outcomes. Phase II follow-up studies also highlighted the increased stormwater infiltration rate of this nanocomposite material, increased freeze/thaw stability, and salt resistance. These characteristics not only point to a more durable performance of building materials but also to appreciable reduction of environmental contaminants, such as metals such as copper and zinc, pointing to the two-part benefit of nanomaterials application in terms of sustainability and durability (Shi et al., 2018). Commercialization of cellulose nanomaterials is another important area ventured in the broader backdrop of nanotechnology. Defined in a U.S. Forest Service report, these naturally derived materials present promising avenues in developing biodegradable and renewable products. Their capacity to create employment with enhanced product development sustainability across various industries is highly relevant. A U.S. Department of Agriculture and U.S. National Nanotechnology Initiative convened the researchers to share commercialization options and challenges. Industry-led feedback gathered was meant to coordinate research and development with industry needs, and it was a strategic move towards embracing cellulose nanomaterials into existing markets with a win (Brun et al., 2014). Embedding nanomaterials into various applications is a paradigm shift in consumer products and building materials. The developments in GO in cellulose nanomaterials and concrete identify not just technological advancements in material science but the integration with sustainability as well. With companies being asked to become more sustainable, the incorporation of nanomaterials is an important initiative. Not only are the products more resistant, but the utilization of these materials also presents the potential for waste reduction and pollution. Furthermore, the development and application of these novel materials also place emphasis on interdisciplinary research and smart research investment, thereby positioning nanomaterials at the forefront of sustainability initiatives in academia and industry.

REFERENCES

1. Ahmad, M., Almeida, M., Angellier, N., Angles, M. N., Araki, J., Ardanuy, M., Arsene, M. A., et al. (2017). Macro and nano dimensional plant fiber reinforcements for cementitious composites. https://core.ac.uk/download/154277082.pdf (accessed on 25th September 2025).
2. Aris, A., Berhanuddin, M. S., & Shamsudin, S. (2015). An overview of geopolymer materials-based zeolite as an antifouling layer in improving water quality. https://core.ac.uk/download/83533062.pdf (accessed on 25th September 2025).
3. Baduge, S. K., Thilakarathna, P. S. M., Perera, J. S., Arashpour, M., Sharafi, P., Teodosio, B., Shringi, A., et al. (2022). Artificial intelligence and smart vision for building and construction 4.0: Machine and deep learning methods and applications. Automation in Construction, 141, 104440–104440. https://doi.org/10.1016/j.autcon.2022.104440 (accessed on 25th September 2025).
4. Bhadane, P., & Chakraborty, S. (2025). Cross-material synergies of carbon nanomaterials, MOFs, and COFs: Innovative approaches for sustainable environmental remediation and resource recovery. https://core.ac.uk/download/648323994.pdf (accessed on 25th September 2025).
5. Bhadane, P., Chakraborty, S., Dhumal, P., Ibrahim, M., et al. (2025). Evaluating the path to sustainability: SWOT analysis of safe and sustainable by design approaches for metal–organic frameworks. https://core.ac.uk/download/647248970.pdf (accessed on 25th September 2025).
6. Brun, V., Friedersdorf, L., Hansen, F., Keller, M., et al. (2014). Cellulose Nanomaterials–A Path Towards Commercialization Workshop Report. https://core.ac.uk/download/42901273.pdf (accessed on 25th September 2025).
7. Chehelgerdi, M., Chehelgerdi, M., Allela, O. Q. B., Cosme Pecho, R. D., Narayanan, J., Rao, D. P., Thamaraikani, T., et al. (2023). Progressing nanotechnology to improve targeted cancer treatment: Overcoming hurdles in its clinical implementation. Molecular Cancer, 22. https://doi.org/10.1186/s12943-023-01865-0 (accessed on 25th September 2025).
8. Cinelli, M., Coles, S. R., & Windsor, R. (2018). Comparison of tools for the sustainability assessment of nanomaterials. https://core.ac.uk/download/158370331.pdf (accessed on 25th September 2025).
9. Cole, M. B., Cook, P. E., Gorris, L. G. M., Quested, T. E., et al. (2010). Trends in technology, trade and consumption likely to impact on microbial food safety. https://core.ac.uk/download/pdf/29244276.pdf (accessed on 25th September 2025).

10. Daar, A. S., Bapsy, S., Persad, D. L., Martin, D. K., Court, E. B., Salamanca-Buentello, F., et al. (2005). Nanotechnology and the developing world. https://core.ac.uk/download/pdf/3637206.pdf (accessed on 25th September 2025).

11. Ezuldin, N. Y., Raoof, S. M., Alattar, A., & Hamada, H. (2024). Factors affecting durability properties of GPC: A review. Research on Engineering Structures and Materials. https://www.semanticscholar.org/paper/5f67e7ea2069536aa15a9e718229b33866c84ae7 (accessed on 25th September 2025).

12. Fangueiro, R. (2015). Towards high performance and multi-functional structural membranes using advanced fibrous and textile materials. https://core.ac.uk/download/132529650.pdf (accessed on 25th September 2025).

13. Farcal, R., Mancini, L., Moretti, C., Patinha Caldeira, C., Rasmussen, K., Rauscher, H., Riego Sintes, J., et al. (2022). Safe and sustainable by design chemicals and materials: Review of safety and sustainability dimensions, aspects, methods, indicators, and tools. https://core.ac.uk/download/642866432.pdf (accessed on 25th September 2025).

14. Fomin, V., & Fomina, I. (2025). Analysis of the sustainability of a spatial building. Mechanics and Mathematical Methods. https://www.semanticscholar.org/paper/f388f80b1ad60936908c1988c2c2fdf6dd4b52ef (accessed on 25th September 2025).

15. Furi, A. T., & Gonfa, D. G. (2025). The role of advanced materials in the optimization of wind energy systems: A physics-based approach. https://core.ac.uk/download/636390634.pdf (accessed on 25th September 2025).

16. Hasib, S. N. R., Phiri, M., & Raihan, P. M. S. (2024). Integration of nano-additives in warm mix asphalt: Enhancing moisture resistance and long-term durability. https://core.ac.uk/download/636358661.pdf (accessed on 25th September 2025).

17. Hernandez Moreno, S., Solache De La Torre, S. C., et al. (2017). Nano-technological products in architecture and construction. https://core.ac.uk/download/154796653.pdf (accessed on 25th September 2025).

18. Hodson, L. L., Hoover, M. D., Iavicoli, I., et al. (2014). Opportunities and challenges of nanotechnology in the green economy. https://core.ac.uk/download/55141194.pdf (accessed on 25th September 2025).

19. Jorns, M., & Pappas, D. (2021). A review of fluorescent carbon dots, their synthesis, physical and chemical characteristics, and applications. Nanomaterials, 11, 1448–1448. https://doi.org/10.3390/nano11061448 (accessed on 25th September 2025).

20. Kadumudi, F. B., Hasany, M., Pierchała, M. K., Jahanshahi, M., Taebnia, N., Mehrali, M., Mitu, C. F., et al. (2021). The manufacture of unbreakable bionics via multifunctional and self-healing silk–graphene hydrogels. Advanced Materials, 33. https://doi.org/10.1002/adma.202100047 (accessed on 25th September 2025).
21. Labrincha, J. A., & Pacheco-Torgal, F. (2013). The future of construction materials research and the seventh UN Millennium Development Goal: A few insights. https://core.ac.uk/download/55622767.pdf (accessed on 25th September 2025).
22. Lahiri, A., Maji, A., Potdar, P. D., Singh, N., Parikh, P. M., Bisht, B., Mukherjee, A., et al. (2023). Lung cancer immunotherapy: Progress, pitfalls, and promises. Molecular Cancer, 22. https://doi.org/10.1186/s12943-023-01740-y (accessed on 25th September 2025).
23. Lambert, P., Liu, C., Wu, Y. Y., et al. (2020). Effect of graphene oxide nanosheets on physical properties of ultra-high-performance concrete with high volume supplementary cementitious materials. https://core.ac.uk/download/305116797.pdf (accessed on 25th September 2025).
24. Luo, Y., Abidian, M. R., Ahn, J. H., Akinwande, D., Andrews, A. M., Antonietti, M., Bao, Z., et al. (2023). Technology roadmap for flexible sensors. ACS Nano, 17, 5211–5295. https://doi.org/10.1021/acsnano.2c12606 (accessed on 25th September 2025).
25. Malik, S., Muhammad, K., & Waheed, Y. (2023). Emerging applications of nanotechnology in healthcare and medicine. Molecules, 28, 6624–6624. https://doi.org/10.3390/molecules28186624 (accessed on 25th September 2025).
26. Malik, S., Muhammad, K., & Waheed, Y. (2023). Nanotechnology: A revolution in modern industry. Molecules, 28, 661–661. https://doi.org/10.3390/molecules28020661 (accessed on 25th September 2025).
27. Seddiqi, H., Oliaei, E., Honarkar, H., Jin, J., Geonzon, L. C., Bacabac, R. G., & Klein-Nulend, J. (2021). Cellulose and its derivatives: Towards biomedical applications. Cellulose, 28, 1893–1931. https://doi.org/10.1007/s10570-020-03674-w (accessed on 25th September 2025).
28. Shi, X. (2017). Recent Advances in Sustainable Winter Road Operations–A Book Proposal. https://core.ac.uk/download/286999895.pdf (accessed on 25th September 2025).
29. Shi, X., & Xu, G. (2018). Environmentally Friendly Pervious Concrete for Treating Deicer-Laden Stormwater: Phase II. https://core.ac.uk/download/185623460.pdf (accessed on 25th September 2025).

30. Stephens, I. E. L., Chan, K., Bagger, A., Boettcher, S. W., Bonin, J., Boutin, E., Buckley, A. K., et al. (2022). 2022 roadmap on low temperature electrochemical CO_2 reduction. Journal of Physics Energy, 4, 042003–042003. https://doi.org/10.1088/2515-7655/ac7823 (accessed on 25th September 2025).
31. Wang, Z. (2025). Functional applications of nanomaterials in plastic surgery: Strategies, practice and evaluation. Highlights in Science, Engineering and Technology. https://www.semanticscholar.org/paper/51a7c6c2cd143ec4d732fabeb3f2b496d778c551 (accessed on 25th September 2025).
32. Wong, L. S. (2022). Durability performance of geopolymer concrete: A review. Polymers, 14, 868–868. https://doi.org/10.3390/polym14050868 (accessed on 25th September 2025).
33. Wu, Z. P., Lu, X. F., Zang, S. Q., & Lou, X. W. (2020). Non-noble-metal-based electrocatalysts toward the oxygen evolution reaction. Advanced Functional Materials, 30. https://doi.org/10.1002/adfm.201910274 (accessed on 25th September 2025).
34. Zeng, A. (2025). Advances in nanotechnology: Applications of nanomaterials in oral dentistry. Advances in Engineering Technology Research. https://www.semanticscholar.org/paper/d82a9721873f25845abe6af759b4aebdbd33777e (accessed on 25th September 2025).

CHAPTER 9

SMART NANOMATERIALS IN STRUCTURAL ENGINEERING

Contents

INTRODUCTION

The application of smart nanomaterials in structural engineering is a vital step forward in the pursuit of more efficient, durable, and responsive building materials. Smart nanomaterials possess unique properties at the nanoscale that enhance functionalities lacking in normal materials. Smart nanomaterials are designed to be dynamically responsive to environmental conditions such as temperature, light, and pressure, making buildings adaptive to changing circumstances.

This flexibility not only increases material performance and durability but also facilitates construction sustainability. The implications extend beyond mere material property improvements; these advances have the capacity to redefine how structures behave and respond to their environment, hence challenging contemporary urban design and management problems. The development of such materials requires an intrinsic knowledge of their potential, setting a new paradigm for engineering practices. With advancements in research, the scope of smart nanomaterial applications in structural engineering keeps increasing, covering various functionalities that enhance both performance and robustness. For instance, novel developments in nanocarrier design and application for the delivery of active ingredients are significant. Such nanomaterials can be designed through various engineering methods, including surface engineering to render them capable of being targeted to specific environments or stimuli (Grenha et al., 2020). Such customization is particularly beneficial in structural applications where specific performance results, such as enhanced durability or environmental sensitivity, are desirable. Moreover, applications of nanostructured materials have been related to an increased resistance against environmental stress factors, thereby improving the durability and safety of construction. Examining the synthesis and application of such intelligent nanomaterials reflects a path forward for addressing critical issues regarding urban infrastructure. For instance, the employment of advanced fabrication techniques, such as chemical vapor deposition, has facilitated the creation of advanced nanostructures with complex functions (Balbuena et al., 2018). Such materials have been highly promising in the guise of photocatalytic activity and self-cleaning properties and are suitable candidates for smart windows and exterior building surfaces capable of air cleaning and reducing maintenance costs. The use of such technology makes the urban landscape more sustainable, where buildings serve not only the overt purpose of serving as homes but also actively contribute to the improvement of urban air quality and energy efficiency. The evolution of smart nanomaterials heralds the beginning of a revolutionary era in structural engineering, where innovation and sustainability coalesce to serve humanity.

Definition of Smart Nanomaterials

Smart nanomaterials is the term used to refer to nanomaterials capable of responding dynamically to external stimuli, such as temperature, humidity, pH, or magnetic fields. These materials, due to their nanoscale size, often possess unusual physical, chemical, and biological properties that are not seen in bulk materials. Their responsivity provides the opportunity to control functionalities in a manner that they can adapt themselves in response to varying conditions in real time. Smart nanomaterials are also important in structural engineering for making structures perform and live longer. Smartness here can result in structures that not only monitor their own health through self-sensing capability but also self-heal after being damaged, thereby extending the life of engineering structures. All these developments amount to a revolution toward a shift in how building materials are made to be stronger and more efficient at a time when sustainability is the priority.

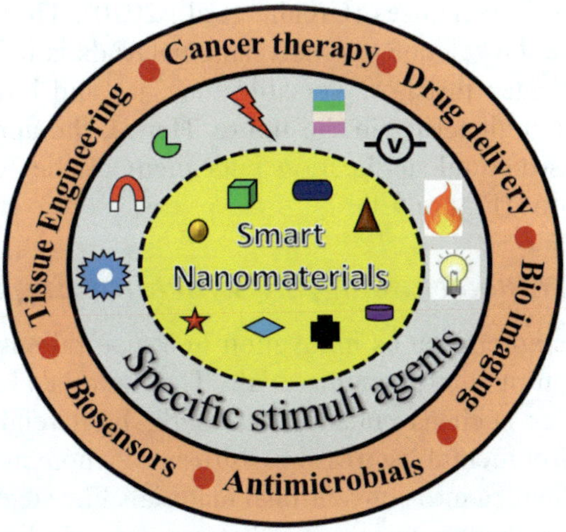

Figure 9.1. Smart nanomaterials.

Source: https://media.springernature.com/lw685/springer-static/image/chp%3A10.1007%2F978-981-15-4802-4_15/MediaObjects/487299_1_En_15_Figa_HTML.png.

Other than the innate capacity of smart nanomaterials to modify, their multifunctionality also reinforces their place in modern engineering practice. Smart nanomaterials are capable of being employed to carry out several functions simultaneously, for example, as a sensor and a structural component. For example, by incorporating nanoparticles into ordinary concrete, the composite material that results not only can support loads but also provide real-time feedback about stress and strain conditions. These sensing capabilities are critical for guaranteeing the integrity and safety of structures during their lifetime. Furthermore, smart nanomaterials can be designed to trigger upon specific environmental fluctuations, such as humidity or temperature changes, so that there is a complex control over structural behavior. This amalgamation of properties shows how smart nanomaterials are revolutionizing the landscape of structural engineering, ushering in new design methodologies. Advancement in materials science and nanotechnology fuels the innovation of smart nanomaterials significantly, motivating researchers to come up with innovative applications in structural engineering. The majority of the potential for smart nanomaterials stems from their ability to incorporate numerous functionalities into a single platform, which can promote insensitivity to mechanical and environmental loading. In drug delivery through lungs, a similar approach is being utilized to design nanocarriers that are multifunctional in order to achieve maximum therapeutic responses (Grenha et al., 2020). This is indicative of a broader trend since the science of smart nanomaterials is influencing multiple arenas, which indicates just how versatile they are and how much they can potentially change engineering in the future. Hence, the application of smart nanomaterials in structural application puts them at the forefront of green engineering solutions (Figure 9.1).

Importance of Nanotechnology in Structural Engineering

Nanotechnology is an enormous innovation in structural engineering, namely through the creation of smart nanomaterials. The materials have extraordinary mechanical qualities such as enhanced strength, lightweight, and improved resistance to environmental aggressors. Infusing nanomaterials like carbon nanotubes and nanosilica into conventional materials like steel and concrete has the possibility of producing composites that are not only lightweight but also stronger and longer lasting. This assumes more importance as infrastructures age and require materials that can withstand rough conditions but maintain structure integrity. Therefore, the addition of nanotechnology will revolutionize the designing, constructing, and maintaining of structures and also set new industry standards for safety and performance. In addition, smart nanomaterials are not just viable for simple mechanical reinforcement but also for real-time monitoring and self-healing capability in structures. The incorporation of nanosensors enables structural health to be continuously monitored, allowing engineers to pinpoint strains and expected failures before they become major threats. Additionally, intelligent nanomaterials may be created with self-healing properties that activate whenever cracks or damages occur. This aspect is especially important in main structures such as bridges and skyscrapers, where safety becomes a topmost concern. By integrating these technologies, structural engineers not only extend the life of buildings but also reduce maintenance expenses to a great degree, thereby leading to more sustainable construction. Furthermore, the environmental impact of conventional building materials can be reduced by adopting nanotechnology in structural engineering. Through material optimization at the nanoscale, one can reduce the environmental impact of the construction and demolition processes. For instance, the production of nanocomposites with recycled content or lower manufacturing energy requirements falls within the modern definition of sustainability. Incorporating nanotechnology not only addresses the structural performance but also allows for more sustainable building practices, closing the gap between environmental responsibility and engineering. As more is learned about sustainability, nanotechnology's role in structural engineering will increasingly become the basis for future advances, dictating not only strong infrastructure but a sustainable built environment too (Debackere et al., 2025; Grenha et al., 2020) (Figure 9.2).

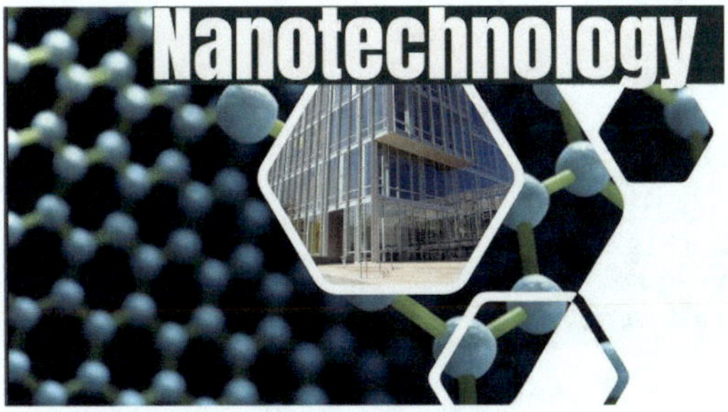

Figure 9.2. Nanotechnology in structural engineering.

Source: https://i.ytimg.com/vi/5fE6awEBV9c/hq720.jpg?sqp=-oaymwEhCK-4FEIIDSFryq4qpAxMIARUAAAAAGAElAADIQj0AgKJD&rs=AOn4CLB1ndkQG45qxI1iuc8CXQgdukJobg.

9.1. PROPERTIES OF SMART NANOMATERIALS

The new properties of smart nanomaterials have rendered them transformative mediators in structural engineering, particularly by their potential to respond dynamically to external stimuli. The materials often exhibit mind-boggling properties such as self-healing, enhanced strength, and light weight, rendering them an ideal choice for futuristic construction needs. Self-repairing ability, one of the characteristic features of some smart nanomaterials, allows for the repair of damages and microcracks independently, thus prolonging the life of structural members. It not only increases the durability but also saves cost in the procurement of maintenance over time. Moreover, the tensile strength of traditional materials can be increased using nanoparticles such that buildings become strong and durable as well, solving the problem of durability that normally plagues civil infrastructure in harsh environments. Additionally, the multi-functionality of smart nanomaterials extends their application beyond the improvement of structures. For instance, the incorporation of nanosensors in concrete allows the measurement of stress, temperature, and strain in real-time, providing the potential for the structural health to be sustained throughout the life of a building. The nanosensors leverage the advances in nanotechnology to provide data capable of proactively diagnosing potential failure, thus allowing timely intervention. With over 290 research works analyzed, it becomes evident that the diverse applications of smart nanomaterials, including carbon allotrope-based and inorganic materials, create a new paradigm in healthcare as well as engineering diagnostics (Altintas et al., 2019).

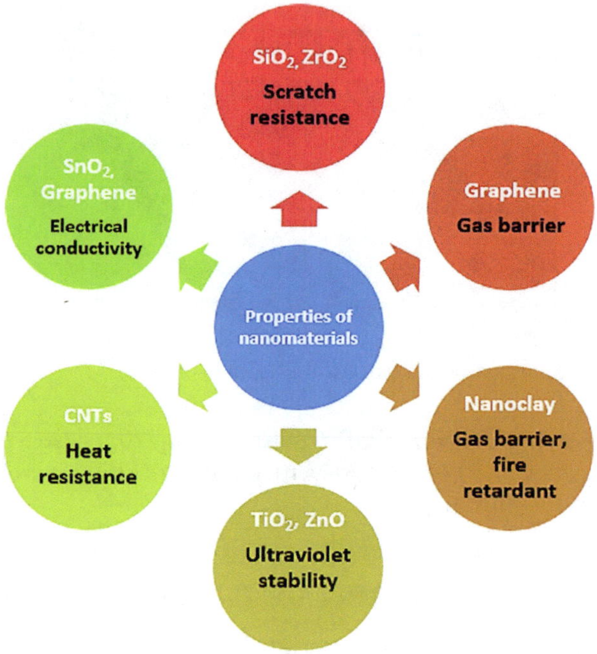

Figure 9.3. Properties of smart nanomaterials.

Source: https://www.researchgate.net/publication/349441468/figure/fig3/AS:1012692588453888@1618456136143/Properties-of-nanomaterials-22.png.

This level of interactivity not only enhances human safety but also optimizes resource usage in construction, paving the way for smarter urban environments. Besides sensing and durability, smart nanomaterials are increasingly known for the ability to contribute to energy-efficient constructions. Smart nanomaterial response to environmental stimuli can lead to adjustments in thermal characteristics, which can unlock more effective insulation and energy saving in buildings. As an example, smart nanomaterials capable of changing thermal conductivity based on the temperature variation can help to better control indoor climates. Besides that, their low weights help in reducing total transportation and installation energy. As we proceed deeper into the evolving era of smart nanomaterials, we can see that their unique properties not only ensure innovations in structure design but also establish greener building practices. By fitting with the needs of modern engineering, these materials underscore the pivotal role played by forward-looking technology to redesign the building future (Figure 9.3).

9.1.1. Mechanical Properties and Strength

The development of smart nanomaterials has brought superior developments in mechanical properties and strength, revolutionizing the topology of structural engineering at its core. The incorporation of nanostructures, such as graphene and carbon nanotubes (CNTs), enhances the natural mechanical properties of composite materials, leading to immense enhancements in toughness and durability. These are attributed to the improved aspect ratios and natural strength of nanomaterials, which are able to distribute the stress over a broader area. Consequently, structures are less prone to increased loads and deformation due to stress. As reported, experimental addition of trace amounts of graphene or CNTs in mineral oils resulted in a coarse 70%–80% increase in the breakdown voltage of AC and, in turn, improved the materials reliability under extreme conditions, which is significant for power system applications (Chattopadhyay et al., 2018). This shows how the addition of nanomaterials brings the performance of traditional materials to new heights. In addition, smart nanomaterials possess an exceptional ability for adaptation and responsiveness to environmental changes, thus improving their mechanical strength. Textile and fibrous materials, for instance, have been designed to impose mechanical flexibility with residual strength appropriate for construction and transportation sectors. New fibrous structures of structural membranes allow for tailored mechanical properties that correspond to specific engineering needs. These membranes not only save weight but also confer properties like self-cleaning along with energy efficiency through integrated solar cells, thereby maximizing their use in structural applications (Fangueiro et al., 2015). This multifunctionality places emphasis on the imperative role of smart nanomaterials in the creation of new solutions for modern engineering challenges, both performance and sustainability. The application of intelligent nanomaterials therefore encourages a paradigm change in designing and using structural members, pushing the boundaries of mechanical performance. The materials enable engineers to pursue novel avenues for maximizing load-carrying structures with less material use and embodied energy, resulting in sustainable construction practices. Understanding the mechanical properties of the materials, including their hardness, tensile strength, and resistance to fatigue, becomes important in order to predict their behavior under different loads and conditions. With continuous research work, the co-use of sophisticated computation models along with experimental methods will ensure the accuracy of knowledge about the mechanical properties. Such efforts will ultimately lead to the development of even more sophisticated structural solutions that not only provide greater strength but also better adaptability to the dynamic requirements of various applications of engineering (Chattopadhyay et al., 2018; Fangueiro et al., 2015).

9.1.2. Thermal and Electrical Conductivity

One of the most distinctive features of smart nanomaterials applied to structural engineering is their thermal conductivity, which also predominantly influences the efficiency and performance of materials in terms of applications. Titanium dioxide (TiO_2) and aluminum oxide (Al_2O_3) nanomaterials have been utilized to synthesize nanofluids, showing unprecedented improvements of thermal conductivity compared to traditional fluids. The research has indicated that the incorporation of 1.5 vol.% TiO_2-Al_2O_3 nanoparticles resulted in an increase of thermal conductivity by around 25%, which indicates the thermal nanofluids' potential in thermal management systems. This enhanced thermal performance results from the higher Brownian motion of particles and their ability to conduct heat at a molecular level, thus better heat dissipation properties suitable for structural members under varying thermal loads (Ahmadi et al., 2020). In addition to thermal conductivity, electrical conductivity of nanomaterials has also attracted vast attention, particularly in insulation and transmission systems of energy. The addition of carbon nanotubes (CNTs) and graphene to insulating oils has been found to be with improved electrical characteristics, which increase the breakdown strength and reliability of the material when subjected to high-voltage conditions. It has been found from research that these nano-enhanced oils can achieve up to 70%–80% enhancement in breakdown voltage compared to conventional oils. The process in play is electron scavenging by the nanostructures, which leads to a slowed growth rate of streamers in ionized liquids and hence better overall electrical performance. Such amplified enhancement is key to the design of safe and efficient power systems, demonstrating the capability of intelligent nanomaterials to transform structural engineering paradigms where electrical integrity matters (Chattopadhyay et al., 2018). The synergy of electrical and thermal conductivity in smart nanomaterials not only enhances their performance but also has the significant advantage of being applied to multifunctional structures within structural engineering. The ability of such materials to conduct heat effectively while maintaining greater electrical properties means structures can be created that are not only electrically and thermally resistant but also functionally integrated with smart technology. For example, structural materials fabricated from these nanomaterials can improve energy efficiency, safety, and reliability of performance. Therefore, the exploration of new nanomaterials continues to be the core interest in accelerating the development of the field of structural engineering with innovation harmonized with the ever-growing demand for adaptive and sustainable building technologies. The synergistic response achieved due to the integration of enhanced thermal and electric conductivity showcases the revolutionary potential of smart nanomaterials in such use cases (Figure 9.4).

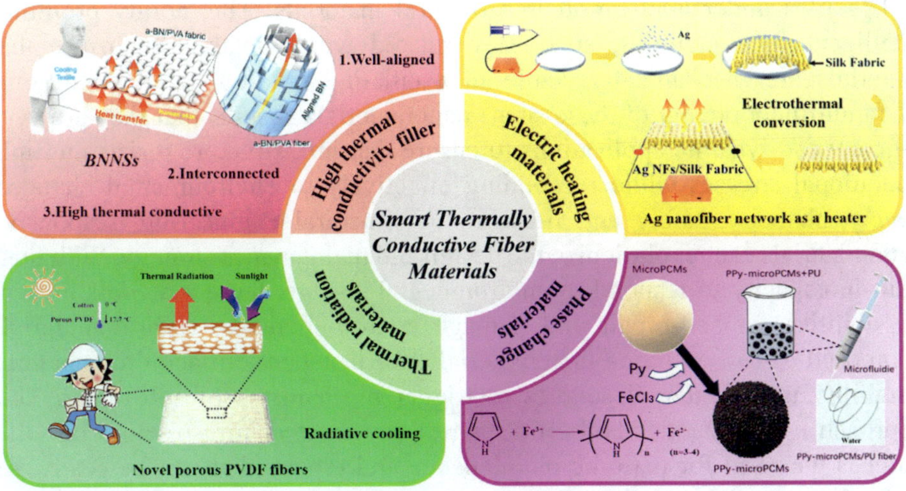

Figure 9.4. Thermal and electrical conductivity of smart nanomaterials.

Source: https://www.mdpi.com/nanomaterials/nanomaterials-14-00154/article_deploy/html/images/nanomaterials-14-00154-g001.png.

9.1.3. Responsiveness to Environmental Stimuli

Responsiveness to environmental stimuli is key to the use of smart nanomaterials, particularly in the domain of structural engineering. The ability of these materials to respond dynamically to fluctuations in their environment enhances the performance in most applications. By integrating temperature, pH, and electromagnetic sensitivity with other properties, engineers can create materials that not only sense environmental conditions but also improve durability and strength of structure. For instance, nanomaterials like carbon nanotubes can be functionalized with molecular switches that exhibit reversible changes upon the action of some stimuli. This property enables their optical and electrical properties to be regulated and therefore the best performance of the materials when applied structurally. This adaptability is useful in systems whose structural components are exposed to varying stresses and environmental conditions and therefore this constitutes a very basic consideration in designing modern infrastructure. Moreover, incorporation of stimuli-responsive smart nanomaterials has the capability to achieve gigantic revolution in construction technology. For example, use of photo-sensitive hydrogels and polymers can result in materials that can change their properties on exposure to light, heat, or pH conditions. Synthetic formation of new copolymers like poly-NiPAAm-chitosan using microgel assemblies makes incorporation into fabrics and textiles under application in structural materials. This ability promotes intelligent functionalities, such as regulated water uptake, to ensure that materials are mechanically stable in oscillating environmental conditions. The multifunctionality of these smart nanomaterials ensures that they are very efficient for dynamic applications to ensure that infrastructure resists multiple operational stresses while responding effectively to environmental changes, hence enhancing safety and lifespan. In structural engineering, responsive nanomaterials can revolutionize the design and maintenance strategy. With their built-in capacity to respond to environmental change, these materials present possibilities for self-repair and damage control methods. Hybrid materials composed of carbon-based nanomaterials, for instance, can acquire enhanced durability using external sources to trigger mechanisms of self-repair. This approach not only extends the lifespan of constructed infrastructure but also reduces the need for periodic maintenance and repairs that can be costly. As technology further unveils new responsive properties in nanomaterials, future developments can lead to the development of completely smart infrastructure that is self-sustaining regardless of human intervention. This change is a paradigm shift in the thought and realization of resilient infrastructures that are not only strong but also smart in their reaction to their operational states, fostering sustainability and security to future generations.

9.2. TYPES OF SMART NANOMATERIALS

Structural engineering has seen smart nanomaterials as pioneering building blocks which enhance mechanical performance and operating functionality. One of the shining types includes shape memory alloys (SMAs), whose key characteristic is to revert to their initial shape under temperature changes. SMAs are particularly beneficial in dynamic response applications under environmental conditions, such as in smart bridges and buildings. By integrating SMAs into structural elements, engineers can enhance seismic resistance and thermal integrity under varied conditions. Such kinds of innovations mark the shift from fixed engineering solutions to more adaptive ones, with structures that can not only survive but actually respond beneficially to changing stresses and loads over their lifespan. Self-healing concrete is yet another significant category of smart nanomaterials that exploit microcapsules or vascular networks of healing agents. Once cracks are developed, these agents are activated, initiating a chemical reaction that heals material integrity.

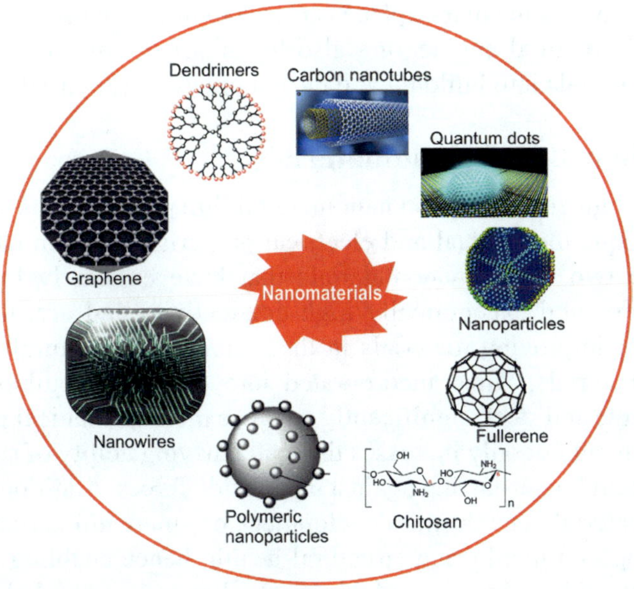

Figure 9.5. Types of smart nanomaterials.

Source: https://external-content.duckduckgo.com/iu/?u=https%3A%2F%2Ftse3.mm.bing.net%2Fth%2Fid%2FOIP.UxxRUSlxu_C_jj6EUxe9-QHaG9%3Fr%3D0%26pid%3DApi&f=1&ipt=a2199384e695391f619b1411b320cddf6a13a659e4132dbc6af1d21de63b1951.

The use of self-healing concrete has vast impacts on infrastructure durability and maintenance, with consequences of reduced repair costs and enhanced safety. This technology addresses one of the biggest challenges in civil engineering: material degradation with time due to exposure to the environment and mechanical forces. Through the integration of self-healing capabilities into concrete, engineers can vastly prolong service life, facilitate sustainable building, and reduce the environmental impact of recurrent repair and reconstructions. Furthermore, stimuli-responsive nanomaterials represent a large family of smart materials possessing the capability to respond to external stimulation such as temperature, pH, or electromagnetic fields. These materials are capable of changing their properties and form with a change in environment, and hence are quite apt for application in adaptive structural members. Scientists are studying, for instance, the potential application of nanocomposites whose stiffness or damping characteristics are responsive to externally applied forces and thereby optimize structural performance in a scenario like an earthquake or gust of wind. This flexibility allows increased safety and comfort, in harmony with the contemporary goals of structural engineering for high-performance and dynamic systems. As the marketplace is racing to integrate these new materials, the future of structural engineering also looks strong with the integration of smart nanomaterials into building and design practices (Figure 9.5).

9.2.1. Carbon-Based Nanomaterials

In structural engineering, carbon nanomaterials are proving to be revolutionary with their unique mechanical and electrical properties. Carbon nanotubes and graphene are two of the nanomaterials that have extremely high strength-to-weight ratios and exceptionally high conductivity and are, therefore, top contenders for improving materials in the majority of structural applications. These nanomaterials, when incorporated into traditional building materials such as concrete and steel, significantly enhance their mechanical performance. This enhancement not only increases the load-carrying ability of structures, but also makes them longer-lasting against the outside forces. Additionally, carbon-based nanomaterials enable the development of smart infrastructure capable of self-sensing and monitoring structural health, hence enabling engineers to actively fix possible failures ahead of time. Utilizing the remarkable properties of carbon-based nanomaterials, the practice of structural engineering will benefit from more efficient and secure design. Also, the applications of carbon nanomaterials go beyond the simple enhancement of mechanical properties but extend to advancements in environmental sustainability. As the construction industry grapples with reducing its carbon footprint, the use of such nanomaterials can lead to the production of lighter, more energy-efficient structures needing less raw material. The reduction in the use of materials has a direct correlate in lower emissions during manufacturing and transportation. Moreover, carbon nanomaterials are also studied with the intent to integrate them in energy-saving technologies, such as piezoelectric composite materials utilized in energy harvesting. As researchers keep studying new applications of these materials in structural components, sustainable construction becomes more promising with time. The shift towards carbon-based nanomaterials not only optimizes the performance of existing structures but also paves the way for green engineering solutions that fulfill global sustainability goals. The future of structural engineering is poised to experience gigantic change through ongoing exploration of carbon-based nanomaterials. With emerging nanotechnology, researchers are continuously discovering novel applications that increase the benefits of these materials in intelligent infrastructures. For instance, incorporation of carbon nanotubes and graphene into insulation materials has proved effective at promoting thermal and electrical conductivity while concurrently improving strength. Such technologies play a critical role in developing buildings that will be able to respond to shifting environmental conditions as well as user demands. Moreover, recent research on the dielectric properties of carbon-based nanomaterials, their use in electrical systems, proves them to have a capability to enhance the structural material's reliability under high-stress conditions. As reflected in recent studies, for instance, the employment of carbon nanomaterials to enhance breakdown strength in insulating oils, the utilizations of these nanomaterials in practice towards structural integrity and safety monitoring continue to expand Altintas et al. (2019) and Chattopadhyay et al. (2018). The convergence of nanotechnology and structural engineering will certainly contribute to the construction of durable and intelligent building systems.

9.2.2. Metal-Based Nanomaterials

The use of metal-based nanomaterials in structural construction is a revolutionary development because the materials possess some unique properties that optimize the functionality and durability of building materials. The highest strength-to-weight ratio is one of the most desirable properties of metal nanomaterials that not only leads to the optimization of total structural load but also optimizes material sustainability by optimizing resource utilization. Besides, nanomaterials can also facilitate multifunctional properties such as self-healing, resistance to corrosion, and enhanced thermal conductivity. Finc-scale control of metal particles can provide some control in engineering that is unprecedented, to facilitate design innovation and application innovation in new construction technologies. In addition, characterization of metal-based nanomaterials has significantly increased such that engineers are able to tailor their properties for individual applications in structural engineering. For instance, adding nanoparticles can make concrete and steel have improved mechanical properties, in effect tightening their tensile resistance and durability. Such improvement is particularly important in retrofitting existing infrastructure and integrating smart technologies in urban planning. The fact that these nanomaterials can dynamically respond to applied forces highlights the role of such materials in smart materials. With advancements in the science of materials, the interactions achieved by metal nanoparticle joining from diverse metal particles could lead to revolutionary engineering solutions tailored to solve specific environmental challenges or performance needs in structures. Besides their mechanical enhancement, use of the metal-based nanomaterials would also significantly affect sensing roles of the structure elements. As highlighted in recent studies, the integration of nanomaterials with sensors provides real-time analysis and monitoring of structural condition. For instance, metal nanoparticles may be engineered to respond to specific stress, strain, or environment, thus facilitating preventive maintenance and security control. These advancements reflect a critical intersection of nanotechnology and structural engineering, in which technological advancement allows not just the production of robust materials but also the development of an intelligent built environment, ultimately leading towards increased security and operational efficiency within infrastructure systems. Further exploration into these applications will further establish the role of metal-based nanomaterials within the future of structural engineering (Figure 9.6).

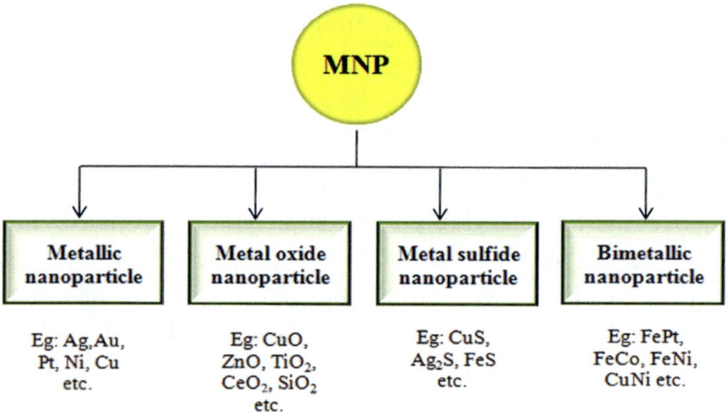

Figure 9.6. Metal-based nanomaterials.

Source: https://external-content.duckduckgo.com/iu/?u=https%3A%2F%2Fthf.bing.com%2Fth%2Fid%2FOIP.a_l_sPQMvBEKDw4ngi7dJAHaFt%3Fcb%3Dt-hfc1%26pid%3DApi&f=1&ipt=17449b87d3a316dcac91ef758980889e63b676cbc1509f6c0922e8438dd66208.

9.2.3. Polymer-Based Nanomaterials

Applications of polymer-based nanomaterials are transforming structural engineering through their unique features, such as enhanced mechanical strength, flexibility, and lightness. These nanomaterials have many benefits compared to traditional materials through which innovative designs and improved performance may be realized in construction processes. Applications of polymers at the nano-scale may lead to the development of very tough and resilient materials, thus extending the life of structures. As a result, engineers can develop more robust and secure infrastructure, particularly crucial in regions exposed to rigorous conditions. Moreover, functionalization possibilities enable the adaptation of polymer nanocomposites to specific engineering needs, hence making them more versatile across various structural engineering applications. Advances in production techniques of polymer-based nanomaterials have also facilitated extensive usage in structural applications. Techniques such as thermoplastic processing and electrospinning facilitate economically viable mass scaling of nanomaterials with no sacrifice in desirable properties. The development of sustainable manufacturing methods also facilitated production with the smallest possible environmental impact, placing the construction industry in accordance with current sustainability goals.

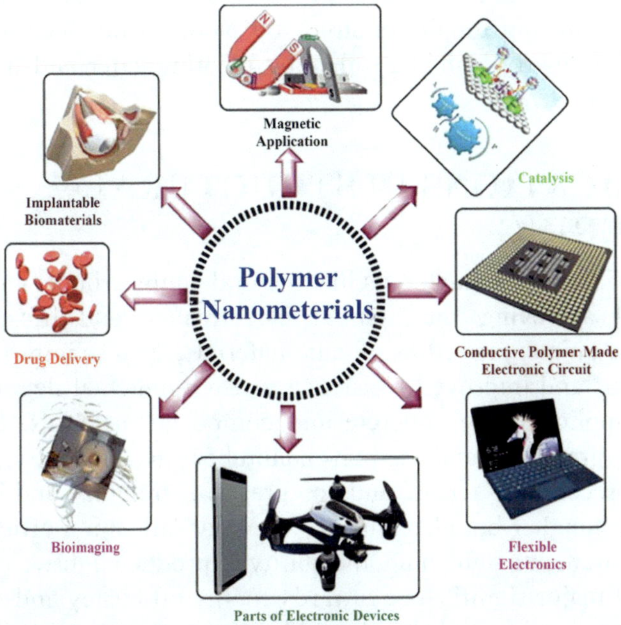

Figure 9.7. Polymer-based nanomaterials.

Source: https://external-content.duckduckgo.com/iu/?u=https%3A%2F%2Fwww.pharmaexcipients.com%2Fwp-content%2Fuploads%2F2021%2F07%2FPolymer-Nanomaterials.jpg&f=1&nofb=1&ipt=a722c4aa518cd72c95e578ba94d3c91502c352a2a073ed7579cff9a2892b682d.

These advances enable structural engineers to have a variety of options in order to develop materials that not only fulfill their purposes but also help drive sustainability. The integration of these groundbreaking polymer-based materials enables the development of smart structures that are able to self-sense and adapt in dynamic conditions, further enhancing the operational efficiency of engineering projects. In addition, the multifunctional nature of polymer-based nanomaterials renders them fundamental constituents in the emerging field of smart nanomaterials. Through the ability to sense exterior stimulus—temperature, pH, or stress—the design of materials that can modify their mechanical properties in real time is feasible. Such a feature becomes particularly valuable in applications related to adaptive response to changing conditions, i.e., earthquakes or harsh climatic conditions. The integration of sensors into polymer nanocomposites can lead to intelligent materials that dynamically monitor structural integrity, providing vital information for maintenance and safety studies. These innovations supplement ongoing work on the nano-optimization of materials for infrastructural use—a tendency that is extremely promising for future structural engineering since it brings forth the potential of smart technology, thereby creating safer and more effective structures (Figure 9.7).

9.3. APPLICATIONS IN STRUCTURAL ENGINEERING

Application of smart nanomaterials in structural engineering is a revolutionary stride toward achieving the optimal performance and lifespan of civil structures. The unique properties of nanomaterials, such as improved strength-to-weight ratios and improved resistance to environmental degradation, have led to their application in concrete and composite materials. For instance, incorporating graphene and carbon nanotubes in concrete mixtures has displayed unprecedented tensile and compressive strengths, and buildings not only are made tougher but also enhance material utilization efficiency. While structural requirements tighten up, the ability to produce lightweight yet robust materials is of major significance towards energy efficiency and sustainability in building construction and thereby addresses economic and environmental concerns in structural engineering. Furthermore, the multifunctionality of smart nanomaterials opens up new avenues in the field of health monitoring of infrastructure. Implantation of nanoscale sensors in the materials employed to build structures enables real-time monitoring of structural health via data on stress, strain, and regions of probable failure. This technology enables preventive maintenance, allowing engineers to respond to dangers linked with structural failure prior to their ability to create significant risks. For example, nanosensors embedded in concrete can detect micro-cracks and transmit data, conserving maintenance costs as well as greatly improving safety standards.

Such technological advancements ensure that intelligent nanomaterials not only enhance the physical properties of engineering materials but also revolutionize the way engineers analyze and monitor structural condition over the life cycle of a building. Finally, smart nanomaterials' ability to sense environmental stimuli and adapt correspondingly is one reason self-healing building materials are innovative. This self-healing ability results from the incorporation of certain healing nanostructures that repair cracks autonomously upon formation. Through the use of encapsulated healing agents within the material matrix which react and solidify when in contact with air or water, there is maintenance of structural integrity over time without a requirement for extensive repair. The development of such materials from idea to use can result in dramatic savings in repair and downtime, commonly associated with conventional construction. These advancements not only symbolize innovative research but also have great potential to shape the future of sustainable engineering practices in an age where resilience and longevity are of utmost priority in structural design (Figure 9.8).

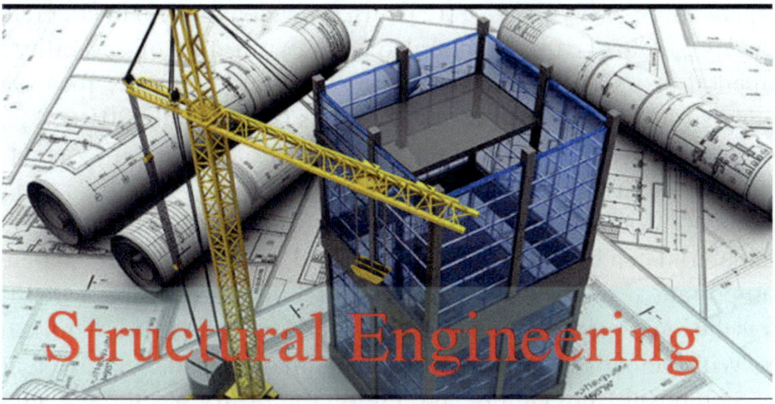

Figure 9.8. Structural engineering.

Source: https://external-content.duckduckgo.com/iu/?u=https%3A%2F%2Fwww.careerindia.com%2Fimg%2F2015%2F07%2F03-1435913359-structuralengineering.jpg&f=1&nofb=1&ipt=d83ae26578d019c2d3a993818c38eeec2f913ddac718c86d88e5c1dc4900f440.

9.3.1. Use in Concrete and Composites

The use of smart nanomaterials in concrete as well as composite materials is a breakthrough in structural engineering, as these materials enhance mechanical characteristics as well as longevity. Concrete is one of the most widely used construction materials, and it is plagued by disadvantages such as brittleness and environmental sensitivity. Through the addition of nanomaterials such as multi-walled carbon nanotubes (MWCNT) and calcium silicate hydrate (C-S-H), the quality of the concrete has been noticed to increase exponentially. The nanoscale enhancements not only change the microstructure of the concrete but also its strength and toughness, extending the lifespan of structures without compromising security. This new idea is one step towards more durable materials that can withstand the adversities of varying environmental conditions, and hence extending the lifespan of concrete structures. Nanocomposites in cement-based materials are particularly promising since it could impact significantly on mechanical strengths. Present studies indicate that adding MWCNTs and C-S-H of specific ratios of calcium and silicon yields greater compressive and tensile strengths. For instance, nanocomposites with 0.8 Ca/Si C-S-H have exhibited strength gains within the range of as much as 30% above normal controls. The optimal performance results from the reactivity of the C-S-H, enabling reduction in porosity and improvement in structural integrity. These developments point toward the strategic use of nanomaterials in concrete as not just improving its load capacity but also its maximization of durability against environmental degradation, which is an area of highest priority for research in present-day structural engineering applications (Duart et al., 2018). Moreover, the environmental factor for the use of smart nanomaterials in concrete plays an important role in the fact that the utilization of such materials can lead to more sustainable building. The improvement of concrete properties indicates that less material can be used to obtain required strength properties, potentially saving important resources. In addition, durability created by nanomaterials minimizes the long-term maintenance problems usually burdening infrastructure projects. With scientists still studying the full range of benefits associated with nanocomposite applications, the building industry has the potential to be enhanced in terms of performance as well as conservation of resources, hence adding value to further research on the application of intelligent nanomaterials in sustainable development in structural engineering.

9.3.2. Integration in Smart Sensors and Monitoring Systems

The use of intelligent sensors and surveillance systems is a paradigm shift in structural engineering, particularly if integrated with smart nanomaterials. Such sensors are designed to provide real-time information on the strength and performance of structures by monitoring parameters such as strain, temperature, and environmental conditions continuously. By incorporating such advanced sensing technology into building materials, engineers enhance the ability to detect precursor indications of failure or deterioration. Smart nanomaterials, because of their sensitivity, enhance the efficiency of such sensors with better sensitivity and accuracy. With this cooperation not only guaranteeing durability and security of buildings but also minimizing costly repairs through predictive maintenance strategies, the fusion of smart sensors and nanomaterials ends up yielding more sustainable and durable structural designs, which means there is a need to integrate these technologies into current engineering practices. Within structural components, the integration of smart sensors into sensing systems is employed to enhance the functionality of traditional materials.

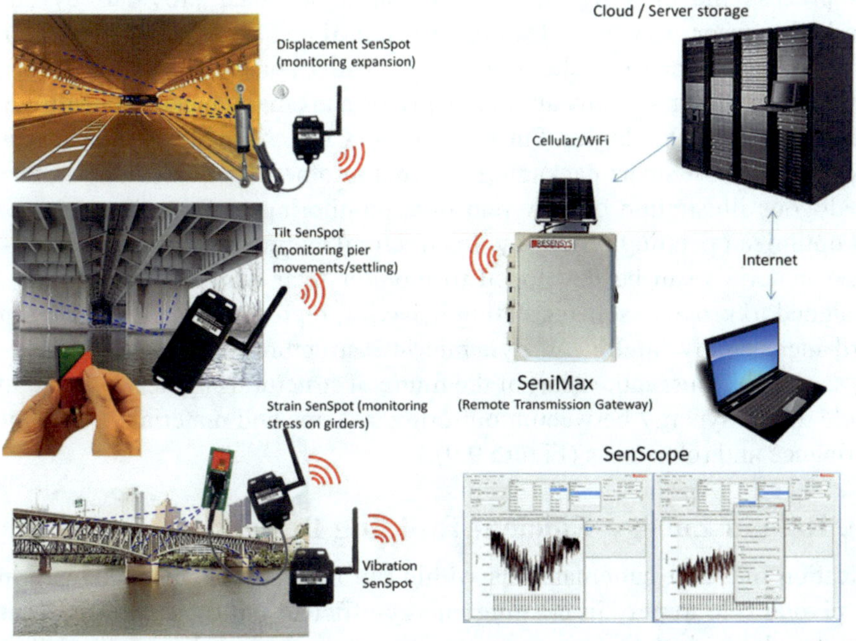

Figure 9.9. Structural engineering integration in smart sensors and monitoring systems.

Source: https://external-content.duckduckgo.com/iu/?u=https%3A%2F%2Fwww.resensys.com%2Fimages%2Fproducts%2FAppNote%2FResensysAllSolutions.jpg&f=1&nofb=1&ipt=540e3093b622d7152618de926484f6fcbc61ddb5047b5e7541708ee7d7926df5.

By integrating nanofibers or nanocomposites with intrinsic sensing capabilities, engineers can create structures that respond actively to stimuli. For example, piezoelectric nanomaterials can be used to generate electrical signals under stress and feed these directly into monitoring systems that analyze load capacities. Moreover, structural health monitoring is enhanced through the utilization of such materials since they provide vital information for describing structural behavior under various conditions, such as seismic events or harsh weather conditions. Structural material development through intelligent nanotechnology allows for a shift towards proactive rather than reactive management systems in engineering, ultimately raising levels of safety and reliability across the built environment. Uses of integration of smart sensors are more varied than mere monitoring and have the potential to significantly enhance automated response systems for smart buildings. Advanced algorithms could process the ongoing streams of continuous data provided by such networked sensor systems, leading to real-time assessments informing decision-making. For example, a building sensing abnormal vibrations could modulate supporting systems automatically or alert maintenance staff through networked smart technologies. Furthermore, as Carrasco et al. (2016) suggests, the development of smart packaging systems in other disciplines can serve as an analogous illustration of how real-time monitoring and adaptive response could optimize operating efficiency. In structural engineering, this implies that not just materials can be developed to monitor their own status, but also to be designed to provoke self-regulating behavior, representing a visionary path toward increasingly intelligent, dynamic infrastructures. The convergence of smart sensors is, thus, at the heart of the future of structural engineering, making possible a new synergy between monitoring systems and materials to enhance performance and robustness (Figure 9.9).

9.3.3. Role in Energy-Efficient Building Designs

Application of smart nanomaterials within the field of structural engineering has a pivotal role to play in creating energy-efficient building designs. These nanomaterials, with their unique properties on the nanoscale, are engineered specifically to optimize thermal performance and reduce energy consumption considerably. Phase change materials (PCMs), for instance, are used in building envelopes to control indoor temperature using optimized heat storage and release. Not only can this capability deliver increased occupant comfort but also minimize use of heating and cooling systems and, ultimately, reduce greenhouse gas emissions. At a time when sustainability is the highest priority, strategic deployment of such advanced materials positions them as leaders in bringing about energy-efficient buildings. Additionally, the smart nanomaterials serve a purpose that goes beyond thermal management; they are applied to enhance the mechanical properties of traditional building materials as well. This integration gives rise to the creation of more robust yet lighter composites with lower energy requirements for processing and transportation. As such, nano-clay composites and carbon nanotubes enhance the tensile strength of steel and concrete, reducing the overall amount of material required for buildings. This reduction, in turn, not only reduces the environmental impact of raw material extraction and processing but also encourages durability in building design. As cities continue to grapple with issues from urbanization, smart nanomaterials mark a major breakthrough in building durable and efficient infrastructures. Moreover, their merging must be founded on a successful understanding of how they behave in the long term and what effects they have on the environment. While scientists and engineers continue to challenge the limits and refine the properties of smart nanomaterials, it is the moment to balance their life cycle, from construction to dismantling, to guarantee that they contribute to sustainable building. Interdisciplinary cooperation between various engineering disciplines, highlighted in the engineering division news reports, is crucial in determining the design protocols that effectively integrate these materials into existing paradigms. Beyond negating potential disadvantages, interdisciplinarity fosters innovations in energy-efficient building designs that conform to existing standards for sustainability. In conclusion, the use of smart nanomaterials in structural engineering is an indication of the influence that technology has in shaping a more sustainable tomorrow.

9.4. BENEFITS OF USING SMART NANOMATERIALS

Use of smart nanomaterials in construction engineering has a number of decisive benefits, which essentially determine the durability and performance of the materials used in construction. Their greatest benefit lies in their enhanced mechanical properties. These nanomaterials increase the strength, durability, and pliability of regular materials like concrete and steel, which ultimately results in buildings capable of resisting stressors like earthquakes and natural disasters. For instance, some nanoparticles can make the structural matrix more robust, hence improving load-carrying capacity and reducing the likelihood of catastrophic collapse of the structure. Therefore, the application of smart nanomaterials leads to safer, more resilient infrastructure that is appropriately able to satisfy the demands of modern engineering practices. Moreover, the application of smart nanomaterials promotes sustainability in structural engineering through efficient resource utilization. They are usually self-healing due to their nanoscale structure, with the ability to heal automatically from damage on small scales, enhancing the lifespan of structures and diminishing the amount of regular maintenance and repairs needed. This, in turn, conserves resources and minimizes the carbon footprint associated with manufacture of new material. Also, smart nanomaterials can be designed to respond to environmental signals, including temperature or humidity, making it possible for structures to adapt in a dynamic manner, maximizing their comfort and energy efficiency.

This double advantage of increasing durability and minimizing ecological footprint makes smart nanomaterials central players in the innovation of green construction processes (Grenha et al., 2020). Smart nanomaterials' versatility is also exhibited in their multifunctional nature. Through the incorporation of nanoscale actuators and sensors, the materials can monitor structural health in real time, providing critical information to inform maintenance actions and safety protocols. Detection of the first signs of structural fatigue or failure allows for proactive intervention and minimizes risks while enhancing the structure's lifespan. Moreover, smart nanomaterials are capable of increasing energy efficiency through the improvement of energy harvesting and thermal insulation properties, with the aim of overall sustainability in buildings. Thus, the application of smart nanomaterials not only enhances the structural function and strength of engineering structures but also encourages new approaches to solving problems in modern construction challenges as well as architectural design.

9.4.1. Enhanced Durability and Longevity of Structures

Application of smart nanomaterials in structural engineering has been significantly responsible for enhanced durability and longevity of structures. Nanocomposites and nanoparticles are a few materials that possess improved mechanical behavior with increased tensile strength and improved resistance against environmental degradation. By adding nanomaterials to conventional building materials, engineers can create structures that not only withstand extreme conditions but also are easier to maintain over a prolonged period of time. For instance, the use of nanocoatings will protect concrete from water infiltration, and thus reduce corrosion risk in reinforcing materials. This preventive action for material design is critical in prolonging the lifespan of structures and infrastructure, ultimately lowering costs and enhancing safety for residents and the environment, further validating the application of smart nanomaterials in contemporary construction practices. In addition to their protective capabilities, smart nanomaterials also facilitate real-time monitoring and assessment of structural well-being, contributing to longevity. The addition of micro-electromechanical systems (MEMS) in building materials has introduced the possibility of sophisticated autonomous structural health monitoring (SHM) systems that can identify future failures before they become full-fledged issues. According to recent research, MEMS devices incorporated into structures are capable of providing real-time data for parameters such as temperature, humidity, and stress levels, allowing continuous performance evaluation without human intervention. In addition to enhancing system safety and reliability against natural hazards, these innovations allow for predictive maintenance strategies that help towards drastically reduced lifecycle costs—both during construction and over the operational life of structures. Such coordination of intelligent nanomaterials and advanced monitoring systems is a giant leap toward the establishment of hard infrastructure. Also, the phenomenon of renewable power sources has emphasized the importance of greater durability in support structures, particularly in wind power systems. With the help of high-tech materials like carbon fiber reinforced polymers (CFRPs) and nanomaterials such as graphene, structural engineers are enhancing wind turbine components' mechanical properties and longevity. These materials are not only lightweight but also show improved fatigue resistance and stress tolerance, thereby providing a major improvement in the operational life of turbines. As the chapter suggests, the application of smart materials also allows for real-time monitoring, which is crucial for effective maintenance and ensures optimization of performance in the long run. Finally, smart nanomaterials are not just revolutionizing structural engineering in terms of enhanced durability and longevity but are also pushing the boundaries of sustainable practice in the energy sector for a more resilient future (Figure 9.10).

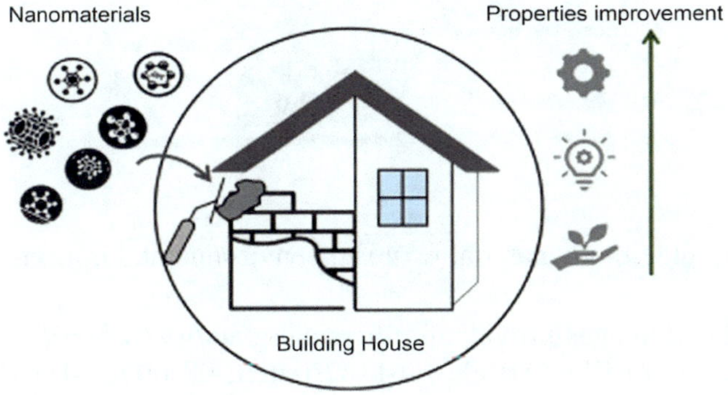

Figure 9.10. Enhanced durability and longevity of structures.

Source: https://ars.els-cdn.com/content/image/1-s2.0-S2666016424002573-ga1.jpg.

9.4.2. Reduction in Material Waste and Environmental Impact

Smart nanomaterials have now emerged as a key factor in reducing wastage of materials in structural engineering, bringing about a fundamental transformation of conventional methods. These materials have enhanced strength and durability characteristics, allowing the manufacture of structures that require less raw material input but maintain performance. By integrating nanotechnology with construction methods, engineers can create materials that are light yet stronger, which not only minimizes material volume needed but also minimizes waste materials used during production and construction phases. Besides stifling wastage of materials, using smart nanomaterials in recycling processes allows reusing existing material, creating a closed-loop process that greatly stunts wastage and fosters sustainability in structural engineering, aligning construction practices with environmentally friendly standards. The use of smart nanomaterials in structural engineering also helps greatly in reducing environmental impact.

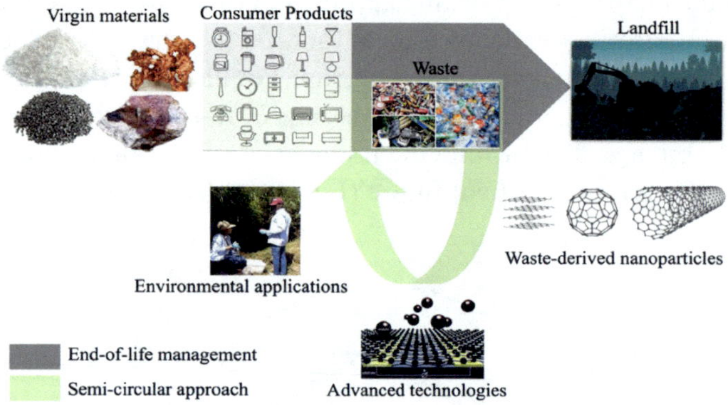

Figure 9.11. Reduction in material waste and environmental impact.

Source: https://images-provider.frontiersin.org/api/ipx/w=480&f=webp/https://www.frontiersin.org/files/Articles/561187/fchem-08-00782-HTML/image_m/fchem-08-00782-g001.jpg.

The efficiency of such products typically results in less energy being required to make them and similarly in the working phase of construction life cycles. For instance, nanomaterials can be designed to function as better thermal insulators, implying reduced heating and cooling energy demand for buildings. This will translate into a lower carbon footprint since lower energy use automatically implies lower greenhouse gas emissions. Also, the life and resilience of structures built from smart nanomaterials are such that they require few repairs and replacements over a long period of time, thus bringing long-term environmental benefits much in excess of the period for initial construction, thus initiating a new trend towards sustainable construction. Large-scale adoption of smart nanomaterials is also essential towards addressing the immediate needs of the environment, such as posed by the construction sector. As highlighted by recent studies, the need for sustainable development has provoked new responses in material sciences, directing towards intercoalition between scientists, engineers, and industry stakeholders. Issues of transition to such sophisticated materials include the scale of production, regulatory requirement issues, and environmental concerns—domains addressed comprehensively by recent studies aiming to make intelligent materials usage an everyday practice (Luo et al., 2023). As industries across the globe continue to explore and implement nanotechnology, it becomes increasingly evident that intelligent nanomaterials not only promise material efficiency but also a way forward for environmentally friendly engineering solutions, as debated in efforts to enhance various industries through new procedures (Malik et al., 2023) (Figure 9.11).

9.4.3. Improved Safety and Performance Under Stress

Integration of smart nanomaterials into structural engineering has significantly enhanced the field, particularly in terms of improved safety and performance under stress conditions. Such nanomaterials, such as carbon nanotubes (CNTs) and graphene, strengthen the mechanical properties of construction materials to support greater loads and resist failure better. For instance, structures with such nanostructures within them have a greater tensile strength and elasticity, benefiting the performance under extreme weather or unexpected load conditions. With increased material structural integrity at a microscale, engineers can create buildings and infrastructure that are not only stronger but longer lasting, contributing to improved safety for inhabitants and users. The role of smart nanomaterials to improve the electrical properties of insulating materials again raises the security of structural applications. Research finds that the addition of graphene or CNTs to insulating oils has the potential to significantly improve the dielectric breakdown strength, making electrical systems integrated into infrastructure more reliable. More specifically, introduction of such nanostructures has demonstrated impressive enhancements of about 70%–80% in breakdown voltage, and this is particularly vital for systems under high voltage stress. Such enhancements ensure that electrical devices remain functional under harsh conditions, thereby precluding the potential for catastrophic failures that are capable of jeopardizing safety and incurring massive losses of resources in structural engineering applications (Chattopadhyay et al., 2018). This improved performance enables safer constructions that can better protect against electrical hazard. Additionally, the application of smart nanomaterials transcends mechanical and electrical benefit, promising the advent of a new generation of multifunctional material delivery systems in structural engineering. Such systems may be designed to respond to specific environmental inputs, which can in turn be applied to enhance safety features. For instance, targeted protection and repair material delivery nanocarriers are activated under pre-defined states of stress, thereby responding to weakness before it leads to failure and structural collapse. In addition to the active monitoring of providing consistent monitoring of structure condition, it also enhances the reliability and longevity of the engineering asset. With advancing technology, utilization of such tactics has the capability of sustaining and enhancing safety standards in construction and infrastructure management (Grenha et al., 2020).

9.5. CHALLENGES AND LIMITATIONS

Application of smart nanomaterials in structural engineering is not free of a number of challenges and limitations that might slow their widespread application. A key challenge is scalability of manufacturing processes for the materials. While laboratory-scale synthesis generally yields high-quality nanomaterials with novel properties, upscaling to an industrial level may cause material quality and performance variations. These variations are key to the dependability and longevity of structures that incorporate smart nanomaterials. Furthermore, the complexity of synthesizing nanostructures that are uniformly of high engineering quality raises issues of cost-effectiveness and manufacturing feasibility and therefore creates a bottleneck to their viable application. An additional important limitation concerns understanding long-term performance and environmental effects of nanomaterials used in structural applications. Though investigations pinpoint prospective advantages of nanomaterials toward the improvement of performance, material longevity under real-world conditions is not clearly understood. Exposure to water, temperature fluctuations, and UV radiation, for instance, could lead to degradation and thus compromise structural integrity. Moreover, the potential of nanomaterials to induce unexpected environmental impacts requires extensive research to examine their life cycle and sustainability. Addressing these issues will require substantial R&D costs to ensure that smart nanomaterials are able to achieve desired benefits while complying with environmental safety standards. Finally, regulatory and standardization concerns add further complexity to the adoption of smart nanomaterials within the field of structural engineering. The unique properties of nanomaterials usually fall outside the borders defined by prevailing industry standards, and this leads to confusion among producers, engineers, and constructors about approved methodology and safety protocols. Further, the requirement of rigorous testing procedures unique to nanomaterials raises the cost and time involved in certification and compliance. The same applies to issues that are faced when dealing with plasma technology application, in which development not only needs to meet engineering standards but also compliance with environmental standards and production regulations (Arias et al., 2019; Grenha et al., 2020). With the progress in technology, it will be crucial to have a clear regulation and promotion of collaboration among stakeholders to overcome such regulatory hurdles and unlock the full potential of smart nanomaterials for structural engineering.

9.5.1. High Production Costs and Scalability Issues

The integration of smart nanomaterials with structural engineering can revolutionize the discipline, but significant barriers remain in the form of high production costs and issues of scalability. The processing of nanomaterials often involves high-technology, low-volume processes that drive costs up, making them unavailable for mass applications. For instance, processes like chemical vapor deposition or sol-gel processing can be efficient in adding functionality but are resource-intensive and economically impractical for mass production. As such, the cost of installing such smart nanomaterials in construction plans can exceed budgetary capacities, discouraging engineers and developers from adopting intelligent nanotechnology solutions. Addressing these monetary challenges is necessary in the successful integration of smart nanomaterials into structural applications. Scale-up issues also heighten challenges that accompany high production cost in the nanomaterials sector. While encouraging advances in the performance properties of nanostructured materials have been made, scaling small-scale experimental data to large-scale production is especially demanding. Among some of the issues that are accompanied by this are maintaining uniform quality and properties in large quantities, which can introduce variability that undermines the reliability of the materials in engineering structures. The transition from laboratory-scale synthesis to commercial production involves significant capital investment in technology and facilities, which has led the manufacturers to shy away from such developments in the past. These problems will need to be solved by further research and new ideas for scaling materials to enhance constructability of nanomaterials. In order to address these issues, it is critical to seek out other forms of manufacturing that not only serve the function of cost saving but also environmentally friendly function. For example, plasma processing technology advancement has brought potential applications in plastic and textile alteration and production, related to the construction industry (Arias et al., 2019). These technologies are emphasizing environmentally sustainable processes at lower costs that can assist in cutting down on some of the cost implications of traditional nanomaterials production. The energy sector has also experienced similar developments where composite materials development for wind turbine blades has significantly minimized cost while enhancing performance (Sapra et al., 2024). By adopting such novel technologies, the structural engineering sector can further overcome costs of production and scalability issues, thereby facilitating the broader application of smart nanomaterials (Figure 9.12).

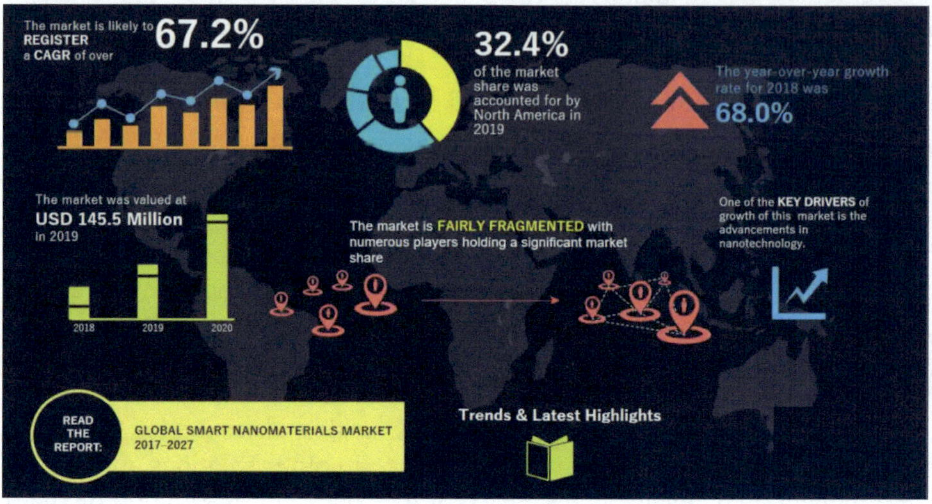

Figure 9.12. Smart nanomaterials market size.

Source: https://www.emergenresearch.com/images/reportImages/smart-nanomaterials-market.webp.

9.5.2. Regulatory and Safety Concerns

The application of smart nanomaterials in structural engineering has numerous new opportunities; however, it also poses significant regulatory and safety concerns that must be addressed seriously. The nanotechnology and unique characteristics of nanomaterials necessitate reconsidering existing regulatory frameworks traditionally designed for bulk materials. As the properties of these materials vary when they reach the nanoscale, the regulatory bodies have to change their approach to identifying the likely hazards in the right way. It encompasses the identification of the health hazards resulting from exposure to nanomaterials in their created form and upon environmental degradation. It is crucial that robust guidelines be in place to guarantee responsible development and utilization of these substances in engineering practice and thus promote public health and safety in the environment as highlighted in arguments made in Stander et al. (2013). Health risk assessments form a central element in the regulatory framework for smart nanomaterials. Traditional risk assessment methods may fail to identify the novel risks introduced by nanomaterials due to their increased reactivity, bioavailability, and the possibility of unforeseen toxicological consequences. The authorities are free to design standard test protocols that would correctly predict the safety of such materials. Moreover, in formulating such standards, one must also consider the life cycle of nanomaterials, e.g., their manufacturing, application, and ultimate disposal or degradation. This integrated design is required so that any likely risks can be assessed in detail, allowing scientists and engineers to use necessary precautions during the entire life of the material. Therefore, it is essential to evolve our information through research and regulatory modification in order to neutralize any adverse effects, as the observations unveiled in Stander et al. (2013). Finally, an important aspect of addressing regulatory and safety concerns regarding smart nanomaterials is promoting coordination between researchers, the regulatory bodies, and industry stakeholders. With the application of nanotechnology becoming more prevalent in the field of structural engineering, sharing knowledge and experience can facilitate best practice development and guidelines that reflect actual implementation. It is also important to have constant discourse regarding public opinion and ethics. By discussing the implications to society and demonstrating the possible benefits as well as hazards, stakeholders are able to collaborate towards developing a system that assists in propelling innovation while remaining safe. The process of developing nanocarriers, particularly drug delivery systems, is a case in point for the need for ongoing evaluation, as presented in. In the future, the development of a balance approach emphasizing innovation as well as regulation will be imperative in making the potential of smart nanomaterials productive.

9.5.3. Technical Challenges in Material Integration

Implementation of smart nanomaterials in structural engineering is beset with an array of technical challenges, most notably material compatibility and performance under fluctuating environmental conditions. The biggest challenge amongst them is how to achieve a smooth interface between dissimilar materials to facilitate maximum mechanical properties for composite structures. The thermal expansion coefficient and elastic modulus differences can cause stress concentrations in joints and potentially compromise structural integrity. Additionally, precise control of nanomaterial dispersions within matrices is of great concern in order to generate consistent performance. Compatibility solutions to these challenges require detailed experimental investigations for the development of protocols that prevent delamination and other types of failure. Even after the advances made in nanotechnology, the quest for reliable integration techniques is a significant roadblock to the full utilization of the potential of smart materials for engineering applications. Scalability and reproducibility of the synthesis of nanomaterials are another crucial challenge that impacts directly on the feasibility of their integration into structural systems.

Nanomaterials of high quality can be synthesized at a laboratory scale, but mass production by factories tends to create variation in material properties. This variation has the potential to significantly change the intended performance results, rendering the dependence on such sophisticated materials in practical applications questionable.

Figure 9.13. Technical challenges in material integration.

Source: https://external-content.duckduckgo.com/iu/?u=https%3A%2F%2Ftse2.mm.bing.net%2Fth%2Fid%2FOIP.q5rVDjG2pXvS-cqTXuNsrgHaHF%3Fpid%3DApi&f=1&ipt=5233a4d5d19f2cb8d5d7b25ddd1a89eefe865b070f8899b854e3a3ca50d6cae8.

In structural engineering, where reliability and safety are of prime importance, the absence of homogeneously produced materials can be a hindrance to large-scale implementation. It would be essential that researchers and industry stakeholders establish standard manufacturing techniques and quality management processes to guarantee the incorporation of smart nanomaterials maintains desired performance criteria in numerous applications (Debackere et al., 2025). Establishing these standards would be fundamental in overcoming technical hurdles in the implementation of new materials within the industry. In addition, long-term stability and long-term behavior of smart nanomaterials under varied loading conditions pose significant technical challenges in integration.

Among the primary concerns are material resistance to fatigue, corrosion, and other forms of degradation upon the elapse of time. These can lead to premature structural failure, raising questions on durability as well as maintenance needs of such materials. Establishing robust test protocols that reflect real-world usage is critical to verifying these smart nanomaterials can perform as expected in the long term. Further, an understanding of nanomaterial interactions with environmental conditions such as moisture, temperature extremes, and chemical exposure will be critical for the development of effective integration schemes that optimize overall structural performance. These considerations emphasize the need for implementing a holistic approach of material integration taking into account not just short-term performance but also long-term sustainability and resilience in structural engineering applications (Figure 9.13).

9.6. FUTURE TRENDS IN SMART NANOMATERIALS

The rapid innovation in smart nanomaterials is set to transform the field of structural engineering through enhanced performance, durability, and sustainability of the built environment. One of the most likely applications is in integrating multifunctionality in nanomaterials. The ability to manage material structure through chemical modification is a strong way of modifying their responsiveness to external stimulus so that they can customize to a range of environmental conditions. Nanomaterials, for instance, can be designed to self-heal, increasing the lifespan of structures by healing micro-defects before they can turn into catastrophic failure. With the growing demand for green construction methods, development of sustainable nanomaterials that can reduce energy consumption and emissions involved in the construction process is becoming more and more important. These are proofs of the significant role that smart nanomaterials are going to play in future resilient and adaptive infrastructure development. In addition to creating structural strength, smart nanomaterials are paving the path for the creation of intelligent buildings that can perceive and respond to their surroundings in real time. Sensors based on nanomaterial embedded in structures can provide real-time feedback regarding structural integrity, allowing active maintenance and minimizing the likelihood of failure. The integration of these technologies enables one to monitor stress, temperature, and humidity in real-time inside buildings so that engineers can make logical decisions regarding repairs and maintenance. The potential of smart nanomaterials for enhancing energy efficiency is also significant, as they can be tailored to provide maximum thermal insulation or help in the harvesting of solar power. All these developments point toward a robust shift to synergistic coexistence between structural engineering and smart technology in which nanomaterials form the basis for advanced, green building technologies. The future of smart nanomaterials is their application in the new field of adaptive architecture, where buildings adapt continuously in response to their context. This adaptability is of utmost significance in various scenarios, such as earthquake-prone regions, where nanomaterial-reinforced buildings could potentially take in seismic energy and release it, thereby reducing damage. Nanomaterials being tailored to respond to environmental changes, such as fluctuating temperatures and humidity levels, will also enhance occupant comfort and energy consumption. With growing interest in tailor-made building solutions, the development of smart nanomaterials holds the promise of revolutionizing architecture. As investment and research in smart nanomaterials continue to develop, so does their potential to transform the practice of structural engineering and ultimately alter our constructed world, heralding a future world where resilience and efficiency are at the forefront of building concerns (Figure 9.14).

Figure 9.14. Future trends in smart nanomaterials.

Source: https://external-content.duckduckgo.com/iu/?u=https%3A%2F%2 Fball science.net%2Fwp-content%2Fuploads%2F2024%2F01%2FThe-Future-of-Smart-Technology-in-Balls-Evolving-Trends.jpg&f=1&nofb=1&ipt=71f798f6257 681a8256c9bb30e92826cc6ef687ec07d5c9710a1075e19d5f3c2.

9.6.1. Innovations in Material Synthesis and Design

The field of structural engineering has been significantly enhanced with advancements in material design and synthesis, particularly through the employment of smart nanomaterials. These nanomaterials, synthesized at the nanoscale, exhibit unique characteristics that enhance the performance and longevity of structural components. Advances in synthesis methods, such as sol-gel processes and electrospinning, have enabled engineers to synthesize nanocomposites with increased strength, elasticity, and weather durability. This change in material design not only increases structural integrity but also impacts building practice sustainability.

The ability to tailor material properties at the molecular level characterizes the changing nature of building materials and signals the complete potential of growth in this area of research and practice. Incoporating nanotechnology in structural engineering is not only pertinent to the physical properties of the material, but also to their design functionality, which has seen a complete transformation. Adaptive nanomaterials can be designed to respond dynamically to external stimuli, such as temperature or stress, and this makes adaptive architecture a real possibility. They possess self-healing properties, therefore significantly enhancing the life of structures and reducing maintenance costs. Furthermore, the integration of nanomaterials provides the ability to produce lightweight, high-performance structures that are resilient and reduce the consumption of resources.

A nanomaterial systematic review of their use in construction reveals that new nano-products can potentially enhance the life of essential building materials, such as concrete and steel, thereby promoting the efficiency of the engineering process and its outcome (MORENO et al., 2017). Application of smart nanomaterials in structural engineering provides a groundbreaking answer to addressing the exigencies of climate change and shortage. As sustainable and green construction continues to be a growing need, the role of sophisticated material design in ensuring they are met continues to grow more significant. Not only do smart nanomaterials increase the mechanical properties of traditional building materials, but they also offer functionalities lacking in traditional materials. Through enhanced reflection of light or insolation, and through energy efficiency of machinery, these materials can go a long way in sustainable architecture. As has been put forth in recent literature, the widespread availability of many nano-products that can be applied successfully within the construction industry is such that architects and builders are now able to design buildings which are not merely long-lasting but also environmentally friendly.

9.6.2. Potential for Self-Healing Materials

The discovery of self-healing materials is a breakthrough in structural engineering, primarily by the adoption of smart nanomaterials. Such next-generation materials are designed to heal automatically, thereby extending the life span and enhancing the safety of a structure. By incorporating nanoparticles into traditional building materials, researchers have made composites that can respond to the formation of microcracks or any other flaws in the structure. Such a self-healing mechanism also finds application in structural membranes, which must be preserved. Such developments can revolutionize the use of fibrous and textile materials in construction, with efficient deployment of design combined with long life and durability. Incorporation of such self-healing properties aligns with growing demands for strong infrastructure, demonstrating the capability of intelligent nanomaterials to provide solutions for advanced engineering problems in an environmentally friendly way. In addition, the integration of advanced fibrous structures into self-healing materials holds potential for multifaceted applications beyond mere structural robustness. For instance, structural membranes based on smart nanomaterials may offer such functionalities as thermoregulation, energy harvesting, and even self-cleaning properties.

By applying the properties of nanostructured materials, engineers can design membranes not only to react to environmental stimuli but also to actively contribute to energy efficiency and sustainability. As evidenced Fangueiro et al. (2015), emerging textile materials can enhance the multifunctionality of structural elements to make them indispensable in modern architectural applications. Hence, the potential of self-healing materials can lead to smarter, more responsive built space that adapts to both occupants' needs and the environmental conditions where they happen. In medicine, merging smart nanomaterials and self-healing in innovative hydrogels also presents a twin with potential. Hydrogels with nanostructured carbon exhibit superior mechanical and electrical properties with high potential for numerous applications, notably in tissue engineering and drug delivery. By incorporating the fundamental aspects of self-healing with very advanced materials, researchers can develop smart systems that adapt appropriately under varying conditions. This capability for on-demand response and healing can enhance the value of hydrogels in structural and biomedical application significantly, closing the gap between the two fields. As highlighted Bosi et al. (2016), the scope for self-healing materials is without bound, with the potential to have a revolutionary impact not just on engineering, but healthcare and environmental sustainability in general. The integration of these disciplines holds the promise for a future age where intelligent nanomaterials can revolutionize the boundaries of material science.

9.6.3. Advancements in Smart Infrastructure and Urban Development

Application of smart nanomaterials in infrastructure design represents the bold leap towards sustainable urban growth. With the fast rate of urban growth, having paradigm-breaking construction solutions that leave minimal footprints on the environment is increasingly crucial. These innovations not only enhance the physical functionality of buildings but also bring new solutions to energy efficiency issues. Research indicates that the building industry is a major source of greenhouse gases globally and buildings account for a vast majority of energy consumed. Therefore, the creation of new technologies, such as nanostructured materials, is needed to obtain meaningful energy savings and minimize the carbon footprint involved in urban environments, as discussed in Pacheco-Torgal et al. (2017). The capacity to design adaptive building materials that adjust their behavior according to changing environmental conditions is a clear example of how smart infrastructure can transform urban landscapes while responding to critical environmental issues. Augmented functionality with intelligent nanomaterials can also be used to make the structural integrity and longevity of urban infrastructure better. For instance, materials such as multifunctional cementitious geocomposites have been found to be abnormally effective, combining self-sensing and self-heating capabilities. These materials effectively monitor structural condition and environmental conditions, allowing for proactive maintenance strategies that ensure long-term performance. The inclusion of graphene nanoplatelets and carbon nanotubes enhances mechanical properties to endure higher resilience under varied loads, as seen in Abedi et al. (2024). The self-sensing capability also gives real-time feedback regarding stress and strain, which can alter the approach city planners and engineers take in maintaining and rehabilitating transportation infrastructure. This proactive approach will not only reduce costs but assist in extending the lifespan of urban infrastructure so that cities are able to accommodate growing numbers and increasing transportation requirements efficiently. Moreover, as the population of the world keeps increasing, building smart infrastructure is essential to designing sustainable cities. Urban cities are expected to house close to 68% of the world's population by the year 2050, and with them come varied challenges to energy consumption and resource utilization. Smart nanomaterials are poised to contribute to building sustainable infrastructure capable of supporting such growth without compromising on resource efficiency. The development of energy-efficient building structures through the implementation of these materials can significantly minimize energy needs, moving towards a sustainable built environment. As the earlier Pacheco-Torgal et al. (2017) noted, the increasing demand for energy-efficient buildings necessitates materials that facilitate this transition. Smart nanomaterials can facilitate urban planners and civil engineers in accomplishing this by enabling them to create infrastructure that not only meets existing needs but also paves the way for future resilient and responsive cities.

SUMMARY

This research into smart nanomaterials in structural engineering has revealed how they have the potential to revolutionize material design and application. As illustrated throughout this essay, the incorporation of nanostructures such as ZnO and WO_3 not only enhances physical characteristics but also enables multifunctionality. For instance, the fabrication of ZnO-WO_3 nanoheterostructures, characterized by the nanoscale porous morphology and their heterojunction density, reflects a significant advancement in photocatalytic applications to clean the environment, that is, to destroy NOX gases. This is reflective of a general trend in structural engineering, where the intersection of new materials and sustainable practices is front and center. These innovations offer very promising pathways for reducing the environmental impact of traditional building materials, as well as enhancing their performance (Balbuena et al., 2018). Apart from environmental factors, the presence of nanocomposite oils doped with material such as graphene or carbon nanotubes also shows spectacular improvements in electrical performance, which is of significant concern for structural applications involving electrical insulation. The results from experiments indicate that the nano-oils increase the AC breakdown strength by approximately 70%–80%, marking their high dependability and efficiency in high-voltage applications.

The ability to harness nanomaterials for the enhancement of dielectric characteristics in insulating liquids is a synergy of electrical engineering and structural resilience. As structural needs keep evolving, the use of such high-tech materials can render power systems safer and more efficient and hence play a part in having a solid infrastructure (Chattopadhyay et al., 2018). In summary, the current research and application of smart nanomaterials are a paradigm shift in structural engineering principles.

The examples presented during this discussion demonstrate not just the technical advantages of these materials but also the influence on sustainability and efficiency. As the world keeps developing to meet growing environmental problems and the need for long-term infrastructure, nanomaterials are predestined to spearhead innovation. Future studies ought to work towards fully optimizing these materials for additional uses, examining synergistic combinations, and creating mass production techniques to meet the demands of industry. The future for smart nanomaterials is good, and potential to transform the practice of structural engineering is on the horizon (Balbuena et al., 2018; Chattopadhyay et al., 2018).

Smart nanomaterials are now a game-changer in structural engineering, bringing a lot of advantages enhancing the design as well as performance. One of the most contentious issues is how they can dynamically respond to environmental changes such as temperatures, humidity, and structural stress. The dynamic response allows such materials to adapt as well as maintain structural integrity and hence longer lifetime with reduced maintenance costs. Various research has determined that incorporating smart nanomaterials in structural components can enhance significantly towards environmental pressures, ultimately contributing to safer and more sustainable structures. The advancements in this field are indicative of future applications, opening up possibilities for more robust construction techniques that make use of the unique features of nanomaterials, making them a necessary topic in the field of structural engineering (Debackere et al., 2025). The multifunctional nature of intelligent nanomaterials is yet another area that also needs recognition. Researchers have explored how one intelligent material can simultaneously exhibit characteristics such as self-sensing, self-healing, and increased mechanical strength, thereby performing multiple roles in construction operations. For instance, integrating piezoelectric nanoparticles into concrete can enable self-sensing properties, allowing monitoring of the structural status in real time. This technology not only enhances safety but also facilitates predictive maintenance practices. In addition, smart nanomaterials are also able to optimize energy efficiency since they can act in their environment in a way that ordinary materials cannot. Such a multifunctional capability expands the range of potential applications, and it opens enormous opportunities for innovation in safe and efficient design of structures, particularly in those infrastructure environments (The Boston University Photonics Center, 2016). Finally, problems with the use of smart nanomaterials in structural engineering cannot be overemphasized. Scalability, cost, and long-term performance under extreme conditions are some of the main concerns. While much progress has been made in research, most smart nanomaterials require intense testing and validation before they can be deemed fit for widespread use. In addition, the monetary effect of integrating such smart materials into usual construction processes raises eyebrows over percentage investment and adoption globally. Resolution of these challenges is paramount to realizing the full potential of smart nanomaterials so that their value may finally be realized in actual application in the field of structural engineering (Debackere et al., 2025).

REFERENCES

1. Abedi, M., Adresi, M., Correia, A. G., et al. (2024). An advanced cement-based geocomposite with autonomous sensing and heating capabilities for enhanced intelligent transportation infrastructure. https://core.ac.uk/download/624081563.pdf (accessed on 25th September 2025).
2. Ahmadi, M. H., Ghazvini, M., Khadang, A., et al. (2020). Prediction of thermo-physical properties of TiO_2-Al_2O_3/water nanoparticles by using artificial neural network. https://core.ac.uk/download/322853318.pdf (accessed on 25th September 2025).
3. Altintas, Z., & Pirzada, M. (2019). Nanomaterials for healthcare biosensing applications. https://core.ac.uk/download/287885382.pdf (accessed on 25th September 2025).
4. Balbuena, J., Barreca, D., Carraro, G., et al. (2018). WO_3-decorated ZnO nanostructures for light-activated applications. https://core.ac.uk/download/154329384.pdf (accessed on 25th September 2025).
5. Bosi, S., Da Ros, T., Iglesias, D., et al. (2016). The glitter of carbon nanostructures in hybrid/composite hydrogels for medicinal use. https://core.ac.uk/download/53748661.pdf (accessed on 25th September 2025).
6. Buss, A., Ceylan, H., Ceylan, H., et al. (2013). Highway infrastructure health monitoring using micro-electromechanical sensors and systems (MEMS). https://core.ac.uk/download/38934593.pdf (accessed on 25th September 2025).
7. Canal Arias, J. M., Canal Barnils, C., Cernák, M., et al. (2019). White paper on the future of plasma science and technology in plastics and textiles. https://core.ac.uk/download/187265494.pdf (accessed on 25th September 2025).
8. Cardano, F., Frasconi, M., & Giordani, S. (2018). Photo-responsive graphene and carbon nanotubes to control and tackle biological systems. https://core.ac.uk/download/168406510.pdf (accessed on 25th September 2025).
9. Carrasco, R., Fuertes, G., Lagos, C., et al. (2016). Intelligent packaging systems: Sensors and nano sensors to monitor food quality and safety. https://core.ac.uk/download/288914520.pdf (accessed on 25th September 2025).
10. Chattopadhyay, A., Dhar, P., Harikrishnan, A. R., et al. (2018). Streamer evolution arrest governed amplified AC breakdown strength of graphene and CNT colloids. http://arxiv.org/abs/1807.02698 (accessed on 25th September 2025).

11. Chen, K. S., Hersam, M. C., Kang, J., et al. (2015). Solvent exfoliation of electronic-grade, two-dimensional black phosphorus. http://arxiv.org/abs/1505.00878 (accessed on 25th September 2025).
12. Debackere, K., Monard, E., Raspoet, D., et al. (2025). Technologie en innovatie in Vlaanderen: Prioriteiten. https://core.ac.uk/download/pdf/6468460.pdf (accessed on 25th September 2025).
13. Duart, M. A., Mortari, S. R., & Rodrigues, O. E. (2018). Carbonic nanoparticles and C-S-H insertion into cementitious nanocomposite. https://media.neliti.com/media/publications/237458-carbonic-nanoparticles-and-c-s-h-inserti-0cd4709e.pdf (accessed on 25th September 2025).
14. Fangueiro, R. (2015). Towards high performance and multi-functional structural membranes using advanced fibrous and textile materials. https://core.ac.uk/download/132529650.pdf (accessed on 25th September 2025).
15. Furi, A. T., & Tolasa, D. G. (2025). The role of advanced materials in the optimization of wind energy systems: A physics-based approach. https://core.ac.uk/download/636390634.pdf (accessed on 25th September 2025).
16. Grenha, A., & Pontes, J. F. (2020). Multifunctional nanocarriers for lung drug delivery. https://core.ac.uk/download/304375181.pdf (accessed on 25th September 2025).
17. Hernandez Moreno, S., Solache De La Torre, S. C., et al. (2017). Nano-technological products in architecture and construction. https://core.ac.uk/download/154796653.pdf (accessed on 25th September 2025).
18. Jocic, D., Kulkarni, A., Tourrette, A., et al. (2010). Microgel-based surface modifying system for stimuli-responsive functional finishing of cotton. https://core.ac.uk/download/12042432.pdf (accessed on 25th September 2025).
19. Luo, Y., Abidian, M. R., Ahn, J. H., Akinwande, D., Andrews, A. M., Antonietti, M., Bao, Z., et al. (2023). Technology roadmap for flexible sensors. ACS Nano, 17, 5211–5295. https://doi.org/10.1021/acsnano.2c12606 (accessed on 25th September 2025).
20. Malik, S., Muhammad, K., & Waheed, Y. (2023). Nanotechnology: A revolution in modern industry. Molecules, 28, 661–661. https://doi.org/10.3390/molecules28020661 (accessed on 25th September 2025).
21. N/A. (2016). The Boston University Photonics Center Annual Report 2015–2016. https://open.bu.edu/bitstream/2144/29754/1/Photonics-Center-Annual-Report-2016-final.pdf (accessed on 25th September 2025).
22. Pacheco-Torgal, F. (2017). High tech startup creation for energy efficient built environment. https://core.ac.uk/download/76177771.pdf (accessed on 25th September 2025).

23. Sapra, V. K., et al. (2024). Highlight materials innovation, performance optimization, and applications in renewable energy systems. https://core.ac.uk/download/660321479.pdf (accessed on 25th September 2025).

24. Stander, L., & Theodore, L. (2013). Regulatory concerns and health/hazard risks associated with nanotechnology. https://core.ac.uk/download/46712612.pdf (accessed on 25th September 2025).

CHAPTER 10

FUTURE TRENDS AND APPLICATIONS OF NANOMATERIALS

Contents

INTRODUCTION

Nanotechnology, through the manipulation of materials at a nanolevel, has revolutionized various industries and holds enormous future potential. The new science encompasses a vast array of sciences like engineering, chemistry, and material science, thus leading to the development of novel nanomaterials that exhibit novel properties and functionality. Unlike single-bulk materials, nanomaterials can exhibit distinguishing optical, electrical, and thermal properties, which make them invaluable in medicine, electronics, and environmental science. While scientists continue pushing the boundaries of single-bulk materials to understand nanoscale processes, it is increasingly important to study the use and implications of these materials for addressing the world's challenges, from clean energy technologies to medicine and environmental remediation. In addition, the incorporation of nanomaterials in everyday products has already begun to reshape industries, with the potential for greater performance and efficiency.

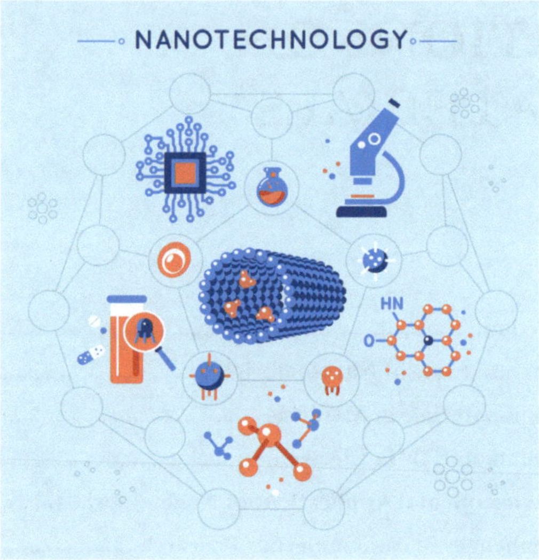

Figure 10.1. Future trends and applications of nanomaterials.

Source: https://external-content.duckduckgo.com/iu/?u=https%3A%2F%2Frevolutionized.com%2Fwp-content%2Fuploads%2Fsites%2F5%2F2020%2F03%2Fnanotechnology.png&f=1&nofb=1&ipt=aa7021c0f6697da6c89cf56fe8c12385363dad78fd9b2ecb7f8d06d50455b912.

To illustrate, in electronics, nanomaterials allow the miniaturization of devices with increased conductivity and less energy demand. In medicine, nanoparticles are being designed for targeted drug delivery, with the possibility of more effective treatment with fewer side effects. The versatility of nanomaterials also stems from their application in the environment, where they are employed in water filtration and pollution remediation. With increasing demand for sustainable and innovative solutions, the versatility and adaptability of nanomaterials keep them at the heart of the development of emerging technologies. It is, thus, important to study future trends and applications of nanomaterials in order to unlock their full potential in a rapidly changing world.

As we advance the future applications and technology of nanomaterials, it is vital to address the ethical and social issues that accompany their application. The unusual characteristics of nanomaterials, as useful as they are, give rise to safety issues, environmental issues, and concomitant issues over human health. State control and public opinion are also issues to be considered as nanotechnology advances. Institutions and scientists must tread a fine line between innovation and prudent research practice to ensure that society reaps the benefits of these advances while not compromising on security or environmental integrity. The ongoing clash between scientists, policymakers, and the public will be in the business of charting the future direction of nanomaterials, influencing policy and practice in this innovative and potentially lucrative sector. Hence, an in-depth understanding of nanomaterials contributes not only to scientific progress but also assists in developing guidelines to facilitate responsible use and application in diverse domains (Figure 10.1).

Definition of Nanomaterials

Nanomaterials are materials with structural features at the nanoscale, between 1 and 100 nanometers. In such a scale, materials possess clear physical, chemical, and biological attributes which are significantly different from the bulk materials. For instance, nanomaterials' increased surface area-to-volume ratio enhances their reactivity, thus making them extremely useful for a variety of applications in fields like chemical catalysis and drug delivery. These size-dependent properties are particularly critical in nanotechnology, where material control at the nanoscale can lead to new solutions for complex issues. The growth of research interest in nanomaterials has been driven by their potential to revolutionize a number of industries, such as medicine, electronics, and environmental sciencc, such that an understanding of their inherent properties becomes even more critical as we speculate about future applications. The nanomaterial category can broadly be categorized into zero-dimensional, one-dimensional, two-dimensional, and three-dimensional structures. Zero-dimensional nanomaterials, such as nanoparticles, are categorized according to their isotropic nature with diameters in nanoscale. One-dimensional nanomaterials like nanowires and nanotubes are identified by their fibrous nature and have unique electronic and mechanical properties. Two-dimensional materials, including graphene and transition metal dichalcogenides, have been of specific concern due to their strength and conductivity. Finally, three-dimensional nanostructures are generally made up of complex assemblies of the aforementioned and figure prominently in biomimetic applications. As global research focuses on specific areas of nanotechnology, namely chemistry and medicine, it is critical to understand these classifications as crucial to successfully taking advantage of their capabilities (Margon et al., 2025). Furthermore, the fast-developing nanomaterials industry falls under even more scrutiny regarding safety, environmental aspects, and regulation. As nanotechnology develops, it is imperative to conduct research on the potential dangers to health resulting from exposure to nanomaterials and their effects on the environment. Researchers advocate for the formulation of intensive control mechanisms to ensure public health and ecological security. Current regulatory structures have to be sensitive to the unique challenges that nanoscale materials pose, with a call for additional funding of research into risk and concerted regulation. Public engagement in discussing nanotechnology's social implications also underscores the call for transparency and informed choice (Davies, 2008). With humanity on the path to a nanotechnology future, a balanced approach with equal focus on innovation and safety will be crucial in securing sustainable development in this groundbreaking field.

Importance of Nanomaterials in Modern Science

The contribution of nanomaterials to modern science cannot be overstated since they are breakthrough factors in medicine, electronics, and sustainability. In medicine, for instance, the unique behavior of nanomaterials promotes targeted drug delivery and improved imaging techniques, basically optimizing the effectiveness of treatment. With the application of nanostructures, researchers are designing therapies to be delivered to disease-affected cells specifically with fewer side effects on healthy tissues. The new approach promises to revolutionize oncological treatments, as a demonstration of how advances in nanotechnology can directly lead to improved patient outcomes. The recent investigation into the scalability of such software is directed towards an imminent future direction in which the utilization of nanomaterials would impact the healthcare field in ways unimaginable before, thus marking their position in today's scientific endeavor. Furthermore, the creation of nanomaterials extends to environmental studies, where their utilization gives rise to sustainable methods and means. Nanotechnology enables the development of advanced materials that can effectively purify water, restore polluted land, and reduce greenhouse gas emissions. Nanomaterials, for example, may be employed in water filtration systems to rid water of contaminants at the molecular level, yielding clean drinking water in areas where scarcity prevails. Utilizations of the technologies are in line with global sustainability goals and are vital to combat climate change. With cleaner industries being demanded, the role of nanomaterials as environmental drivers increases increasingly imperative. Such an interdisciplinary nature of nanomaterials highlights their multidisciplinary relevance and future potential for breakthroughs in fields, substantiating their roles as principal features of current science. Furthermore, the research landscape in nanomaterials portrays a competitive and dynamic arena, particularly in the fields of scientific production and innovation. Areas such as Europe and North America have historically dominated nanotechnology studies, but new economies, especially in Asia, are quickly rewriting the book. The increasing interest in nanoscience is reflected in the rate of exponential growth of research and development, with fields such as chemistry, biochemistry, and engineering recording dramatic leaps forward. As global collaborations intensify and diversify, the number of nations that contribute to the cumulative knowledge and usage of nanomaterials increases. These trends exhibit a global trend toward greater research capability and utilization among economic players, further cementing the case for nanomaterials research investment as an option of priority. Hence, as these tendencies continue to evolve, the destiny of nanomaterials will persist in being at the forefront of the modern scientific agenda, influencing a wide array of fields and applications, as evidenced by current academic studies into this pioneering technological horizon (Margon et al., 2025) (Figure 10.2).

Figure 10.2. Importance of nanomaterials in modern science.

Source: https://mindthegraph.com/blog/wp-content/uploads/2023/02/characterising-nanomaterials-in-food-by-fff-icp-ms.jpg.

10.1. CURRENT STATE OF NANOMATERIALS

The nanomaterials discipline has witnessed significant leaps in the past few years, and their influence is being experienced in all lines of business, exhibiting their revolutionary potential. Nanomaterials possess distinctive properties that vary much from their bulk counterparts based on their size and surface-to-volume ratio. The distinctive properties result in enhanced functionality in the electronics, medicine, and energy industries. For instance, in electronics, nanomaterials allow for smaller, faster, and more efficient products, while in healthcare, they allow for targeted drug delivery systems with less side effect. As more research goes into the intricacies of nanomaterials, not only better uses but also opportunities for innovation that would reframe existing technology and social norms await in the future. However, the spread of nanomaterials raises serious questions regarding the health and well-being of the public and environmental protection. With their increased entry into everyday consumer products, regulative bodies are pressed to establish effective systems of oversight. Current legislation must be scrutinized and potentially changed in order to take into consideration the unique risks of nanotechnology, as shown in arguments on requiring more funding for risk assessments and public inclusion in regulation. The intersection of nanomaterials with health and safety concerns highlights the importance of harmonized regulatory frameworks balancing innovation and protecting the common good. These challenges are important to overcome in order to foster responsible development in the area and ensure public interest is protected alongside technological progress (Davies, 2008). The new applications for nanomaterials extend into fields such as biocompatible electronics, in which researchers propose using electronic devices based on plants. Concepts of computing systems using plant morphology are a fascinating intersection point for biology and technology. Specifically, by modifying the electrical conductance of plants using functional nanoparticles, researchers are looking into if plants can be employed as living electronic devices to perform sophisticated computations. A biohybrid approach such as this could lead to the development of green, sustainable computing systems using natural processes. With cross-disciplinary activities flourishing among computer scientists, engineers, and biologists, the potential to create living electronic systems will redefine nanomaterials usage and understanding and lead to new opportunities for radical and sustainable applications in the future (Adamatzky et al., 2017).

10.1.1. Types of Nanomaterials

Nanomaterials are typically classified into different types, each being characterized by some properties which occur as a result of the nanoscale size. One of the most well-known classes are nanowires, nanosheets, and nanoparticles. Nanoparticles, usually 1–100 nanometers in diameter, possess distinct optical, electrical, and magnetic properties that are quite different from bulk counterparts. Applications include drug delivery systems, catalysis, to cleaning up the environment. In contrast, one-dimensional nanomaterials such as nanowires have a high aspect ratio and large length to diameter ratio that enhance their electronic and mechanical properties of great importance in the application of nanomedicine and biosensing. Nanosheets, typically made up of two-dimensional materials like graphene and transition metal dichalcogenides, offer excellent strength and conductivity, which promises new avenues of innovation in flexible electronics and energy storage. Functionalization of nanomaterials is a critical process for forming their properties for specific applications.

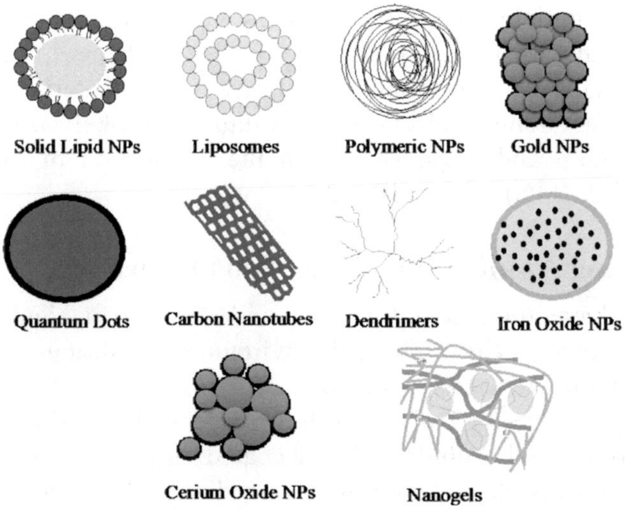

Figure 10.3. Types of nanomaterials.

Source: https://external-content.duckduckgo.com/iu/?u=https%3A%2F%2Ftse4.mm.bing.net%2Fth%2Fid%2FOIP.QI0RlTx809hV5ag3ogSCswHaF3%3Fpid%3DApi&f=1&ipt=fddfb4500c8b4704dd4814b9e29b5a8a566faca788f2c5a4753e2e0730c3da6e.

Through modulation of surface chemistry and structure, the solubility, stability, and cellular interaction can be enhanced, of particular importance for medical applications. For instance, amine-functionalized silicon nanowires offer a highly promising method of targeted drug delivery due to cell membrane penetrability by physical wrapping phenomena. This fundamental understanding of how nanomaterials interact with cells, and especially with one-dimensional materials, underscores rational design in nanomedicine, with potential for new clinical uses. Moreover, these alterations have the potential to address public health and environmental safety issues of paramount importance, thus emphasizing the need for harmonized regulation and financing towards risk research within the new area of nanotechnology (Davies, 2008). With the advancement of technology, the functionality and grouping of nanomaterials are likely to go even further, with the potential for advanced applications in various sectors. Sustainability and green nanomaterials are expected to be supported by future trends, potentially diminishing their influence on the environment but still maintaining efficiency. New areas such as nanobiotechnology and energy nanomaterials are expected to revolutionize industries from medicine to energy production. Increased research on composite nanomaterials, which combine properties from multiple categories, promises a shift towards multifunctional structures that can perhaps better address complex challenges. As researchers learn more about the interaction and mechanisms of nanomaterials, particularly their health and environmental impacts, the future will likely witness balance between innovation and responsibility in the application of the advanced materials (Davies, 2008) (Figure 10.3).

10.1.2. Existing Applications in Various Industries

Nanomaterials have transformed various industries to a significant level, especially in water purification and environmental cleanup. Particularly, the textile and oil industries generate enormous quantities of wastewater with toxic chemicals, typically carrying suspended oil droplets causing immense environmental pollution. Sophisticated current treatment processes such as photocatalytic degradation are being designed to eliminate such an issue effectively. By hybridizing catalysts such as titania (TiO_2) with carbon nanotubes (CNTs) and magnetic iron oxide (Fe_3O_4), researchers are paving the way for new solutions to degrade organic pollutants in wastewater. By bringing together these nanomaterials, one can stabilize Pickering emulsions, which has potential to be an effective and recoverable system in wastewater treatment systems, thereby harmonizing environmental sustainability and industry use (Mason et al., 2018). In medicine, applications of nanomaterials are extensive, especially for drug delivery and diagnostics. Nanoparticles can be engineered to encapsulate therapeutic agents for targeted delivery directly to the sick cells with decreased systemic toxicity. The process enhances the effectiveness of treatments, especially in cancer treatment, where conventional treatment regimens have adverse side effects. Furthermore, the development of nanosensors has revolutionized diagnostic procedures through the improvement of the sensitivity and specificity of disease diagnosis. With the use of the unique properties of nanomaterials, the devices allow for rapid and accurate analysis, making possible earlier intervention and better patient outcomes. Such advancements in nanotechnology not just enhance the quality of care in healthcare outcomes but also reflect an emerging trend towards personalized medicine in therapeutic approaches (Lakhtakia et al., 2005). The use of nanomaterials in the electronics sector also reflects the versatility and promise of these materials across different sectors. Nanotechnology is crucial in the manufacturing of smaller and efficient electronic devices, including transistors and sensors, thereby enabling miniaturization. With the application of materials such as quantum dots and nanowires, organizations are able to improve electrical conductivity and minimize power consumption by electronic devices. Nanoelectronics has also opened up possibilities for flexible and wearable technologies, whereby users can be able to engage with devices more easily. These advances are not only the core of consumer electronics but have relevance for more integrated applications such as IoT devices and smart technologies. As industries realize further the benefits of nanomaterials, further research will be anticipated to optimize their utilization and unlock additional functionality in electronics (Lakhtakia et al., 2005).

10.1.3. Recent Advancements in Nanotechnology

The fast-paced evolution in nanotechnology has unveiled unprecedented opportunities across fields, particularly biomedical applications. Recent developments have highlighted the shift of nanomaterials into the medical diagnostic and biosensing arena. Through the investigation of more than 290 research publications since 2015, scientists have revealed how nanomaterials can be used as multifunctional platforms for the detection of a wide variety of clinically relevant molecules. Such molecules include nucleic acids, viruses, and cancer antigens, which are of vital importance in early disease diagnosis. With the enhancement of diagnostic technology sensitivity and specificity, the entry of nanomaterials has created opportunities for more efficient point-of-care diagnostics and therefore the possibility of remodeling patient management and treatment plans. This intersection of nanotechnology and medicine illustrates the promising trajectory of nanomaterials in sketching the future direction of medical diagnostics and therapeutic interventions (Altintas et al., 2019). In addition to biotechnology, progress in nanotechnology also extends into agriculture, particularly in the Third World where traditional methods have a propensity to yield marginal gains. Nanomaterials application towards improved crop efficiency and sustainability has become a research and development priority as a way of addressing food security concerns. Nanoparticles, for instance, are being utilized to realize targeted delivery of pesticides and nutrients with significantly lower volumes at minimal environmental expense. In addition, nanotechnology allows the creation of smart sensors to monitor plant health, thereby enhancing irrigation and nutrient application. Such innovations not only optimize crop yields but also ensure sustainable agriculture. Nanotechnology's crucial role in agriculture speaks to its potential to solve key global issues so that food supply systems can keep pace with rising population pressures in developing nations (Abbott et al., 2025). Looking ahead, the coming together of nanotechnology with a range of disciplines should lead to new solutions for further improving quality of life and environmental care. As researchers find out more about the outstanding properties of nanomaterials, the potential for new applications is nearly unlimited. From new materials offering greater strength-to-weight advantage to drug delivery systems that target cells selectively without harming healthy tissue, the opportunities are vast. Moreover, the merger of nanotechnology with artificial intelligence and machine learning would open doors to innovations in personalized medicine that could deliver tailored treatments based on unique genetic predispositions. The advances in nanomaterials thus signify not just technological progress but also mirror a revolution in the way industries will change to address future challenges, boding well for a dynamism to research and application in the years ahead (Figure 10.4).

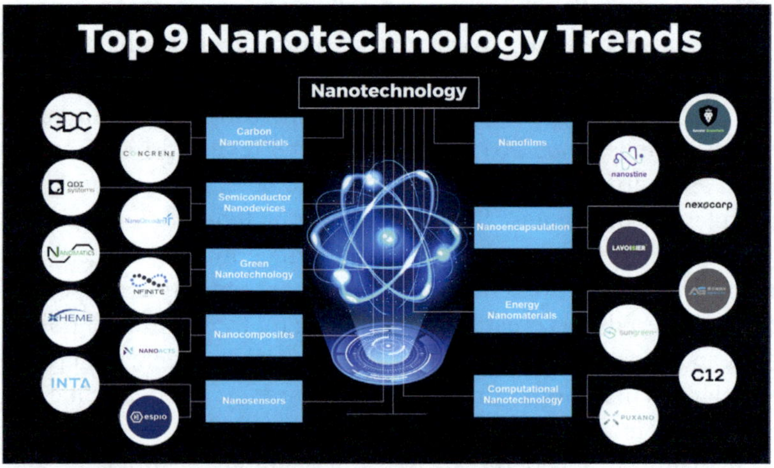

Figure 10.4. Recent advancements in nanotechnology.

Source: https://external-content.duckduckgo.com/iu/?u=https%3A%2F%2Fwww.startus-insights.com%2Fwp-content%2Fuploads%2F2022%2F11%2FNanotechnology-trends-InnovationMap-StartUs-Insights-noresize-2-900x644.webp&f=1&nofb=1&ipt=903b232ad0e952083301558ef5053ba85cc9c858d8da4094add8c4166c141ddc.

10.2. NANOMATERIALS IN MEDICINE

The use of nanomaterials in the biomedical field is leading-edge scientific research, a revolution in the creation of therapeutic intervention and diagnosis. As nanotechnology advances, researchers are taking advantage of the unique properties of nanomaterials—increased surface area, increased reactivity, and tunable size—to create novel solutions to longstanding medical issues. Above all, their versatility facilitates multipurpose uses ranging from drug delivery systems to biosensing and imaging platforms. The growing dominance of nanomaterials over chemical engineering and biochemistry illustrates this reality, demonstrating the paramount value of interdisciplinary research in creating medical technology. Besides, it has been discovered that Europe and North America are leading the way in this field of study, with increasing contributions from countries like China being a reflection of potential re-shaping of global research trends in nanotechnology (Margon et al., 2025). Of key interest is the contact between one-dimensional (1D) nanomaterials and living tissues, which has generated considerable interest among scientists in the field of nanomedicine. These interactions must be interpreted for the rational design of nanomaterials to effectively engage with cellular structure. For instance, studies on amine-functionalized silicon nanowires ($SiNW-NH_2$) have demonstrated their uptake pathways within cells. Utilizing advanced imaging technologies such as optical and transmission electron microscopy, scientists have documented cellular responses to the nanostructures, describing the underlying concept of membrane mechanics and cellular internalization.

This research reflects not only the versatility of 1D nanomaterials but also their potential to transform drug delivery systems and cellular engineering applications, thereby accelerating the development process from the laboratory to the clinic (Huang et al., 2017). Nanomaterials in medicine hold a promising future with a range of innovations that can genuinely enhance patient recovery and streamline medical procedures. Existing studies seek to create nanocarriers that are multifunctional and can carry out imaging, diagnosis, and targeted therapy simultaneously, a procedure that holds the promise of significantly improving medical therapy accuracy. Nanomaterials-based biosensing applications also hold the promise of leading to faster and more precise disease detection to allow timely interventions. As regulatory bodies begin to acclimatize to such developments, the test will be to ensure these technologies are safe for use by humans as well as effective. Overall, increased research and application of nanomaterials in medicine are an expanding frontier with the potential to overhaul therapeutic possibilities and set new benchmarks for medical practice (Margon et al., 2025) (Figure 10.5).

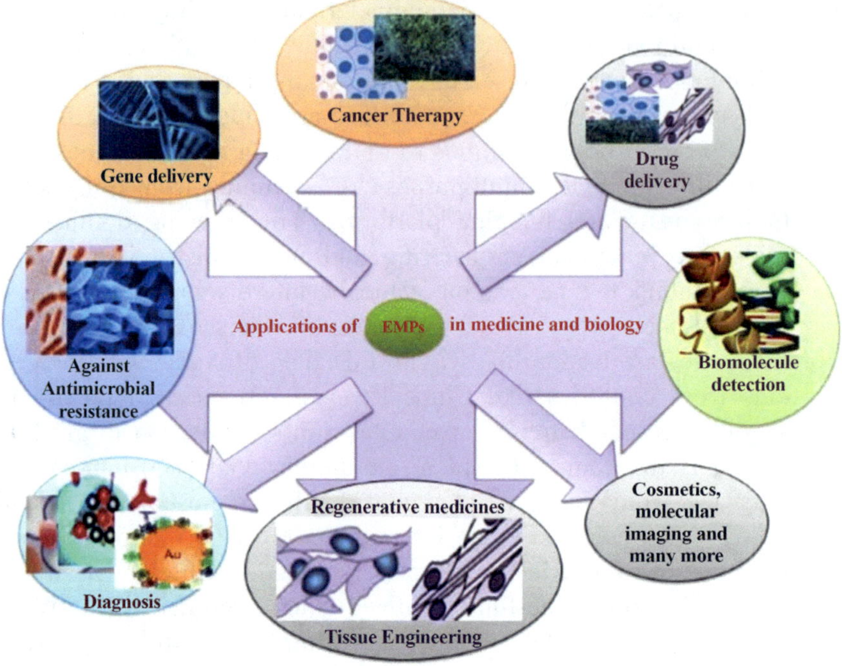

Figure 10.5. Nanomaterials in medicine.

Source: https://media.springernature.com/lw1200/springer-static/image/art%3A10.1007%2Fs00775-018-1600-6/MediaObjects/775_2018_1600_Fig1_HTML.png.

10.2.1. Drug Delivery Systems

The evolution of drug delivery systems has been immensely impacted by nanotechnology, particularly the improvement of the precision and efficiency of therapeutic drugs. Nanocarriers such as liposomes, dendrimers, and polymeric nanoparticles have been recognized to deliver drugs directly to the targeted cells or tissues while restricting exposure to the systemic circulation. This particular approach is particularly important in cancer, where traditional chemotherapy creates crippling side effects due to the fact that the drugs are delivered nonspecifically. By engineering nanocarriers to react to specific stimuli, such as pH or temperature, researchers are hoping to create systems that react in an adaptive way to the physiology and improve therapeutic effectiveness. As Grenha et al. (2020) have indicated, the strategic design of such drug delivery systems prioritizes intrinsic benefits as well as barrier-evading properties inherent in conventional drug delivery systems, leading to more effective treatment that can significantly enhance patient quality of life. New drug delivery systems designed using nanotechnology have created new avenues for the management of pulmonary disease in respiratory medicine. Nanocarriers have been increasingly proposed for lung drug delivery, combining unique strengths of nanostructures and therapeutic targeting. Targeted therapy comprises the delivery of anticancer drugs, antibiotics, and proteins directly into the lungs, improving efficiency and reducing side effects associated with systemic administration. For instance, surface engineering techniques offer the option of modulating nanocarrier properties to adhere to specific biological targets or release drug agents based on environmental inputs. Such progress illustrates the versatility of nanocarriers, as mentioned in Grenha et al. (2020), and inspire new research activities towards multifunctional delivery systems to treat complicated therapeutic problems in respiratory medicine. The development of PEGylated nanomaterials is an excellent illustration of chemical modifications that can significantly enhance drug delivery systems. The process of PEGylation, where polyethylene glycol is linked to nanomaterials, improves their biocompatibility, stability, and the length of time they circulate in blood. This modification has been shown to alter the physicochemical characteristics of nanocarriers, such as pH values, which can consequently enhance their efficacy as drug delivery systems. Experiments have established that derivatives of graphene oxide with PEGylated ones exhibit antifungal properties, which is a litmus test for their potential beyond conventional use. The inclusion of metals in nanocarriers enables synergistic effects, as evidenced by studies presented in. That these systems exhibit multifunctional properties is the pointer to the colossal potential for nanomaterials to revolutionize drug delivery strategies and provide the entry points for new treatment modalities across all medical branches.

10.2.2. Diagnostic Imaging Enhancements

The use of nanomaterials in diagnostic imaging has begun to significantly increase medical imaging techniques, making them more accurate and efficient. By improving nanotechnology, nanoparticles can be engineered to act as targeted contrast agents and allow more accurate imaging of tumors and other pathological conditions. For instance, the nanoparticles may be engineered to attack cancer cells so that images are taken better to assist doctors in treatment and diagnosis planning. This precision not only enhances the early diagnosis but also minimizes the risk of misdiagnosis, which in most situations is a determining factor in patient care. With more researchers exploring various nanomaterials, the potential for obtaining high selectivity imaging agents grows, leading to dramatic improvements in the general efficiency of the diagnostic methods when applied to clinics (Anurogo et al., 2017). In addition, nanomaterials' well-established properties, such as a high surface area-to-volume ratio and ability to encapsulate therapeutic compounds, present a significant potential for theranostic use for imaging. Nanoparticles have the ability to provide both imaging and therapeutic functionality, enabling physicians to understand disease states while also delivering targeted therapy. This combined ability has the potential to transform the treatment of chronic diseases, including cancer, where rapid and precise measurement of tumor response to therapy is vital. Recent developments indicate that nanomaterial technology has the ability to produce multifunctional probes not just to enhance the contrast of imaging but also to monitor therapeutic response in real time, effectively linking therapy to diagnosis (The Boston University Photonics Center, 2016). The future of diagnostic imaging will be significantly benefited by multidisciplinary research across disciplines. By uniting researchers in nanomedicine, bioinformatics, and materials science, they are setting the stage for unprecedented innovation to enhance diagnostic accuracy with nanoparticles and nanoscale devices. The coming together of such distinctive areas of research facilitates rapid translation of lab findings to real-world applications, which results in the development of next-generation imaging technologies. Such a collaboration not only accelerates the cycle of innovation but also addresses complex healthcare issues. Lastly, constant research with nanomaterials in diagnostic imaging promises to revolutionize the treatment of patients by making it more personalized and efficient in the quest for improved health outcomes (Anurogo et al., 2017).

10.2.3. Cancer Treatment Innovations

The application of nanomaterials in cancer therapy is a revolutionary shift in treatment modalities, showing constant innovation in medical science. Nanoparticles are engineered to enhance drug delivery systems so that therapy is targeted with lower side effects and improved therapeutic efficacy. By encapsulating cytotoxic drugs in nanocarriers, these materials allow for direct delivery of chemotherapeutics to target cancer sites, with significant reductions in collateral harm to normal tissues. Not only does this precision medicine increase drug efficiency, but it also enhances patient benefits by reducing treatment protocols. The research of these applications shows that with the advancement of nanotechnology, it is more and more going to become a part of oncological care, redefining the context of cancer therapy and patient care. Moreover, the advancement of diagnostics has also been enhanced by using nanomaterials, which play a significant role in the diagnosis of cancer at the early stage. Nanosensors are being developed to identify biomarkers for various cancers so that physicians can diagnose and monitor disease progression at unprecedented sensitivity levels. The ability to detect cancer during its early stages can revolutionize treatment success rates because, as a rule, early treatment is associated with good prognoses. The combination of nanotechnology with imaging diagnostics also enables higher dynamic understanding of tumor activity and treatment response.

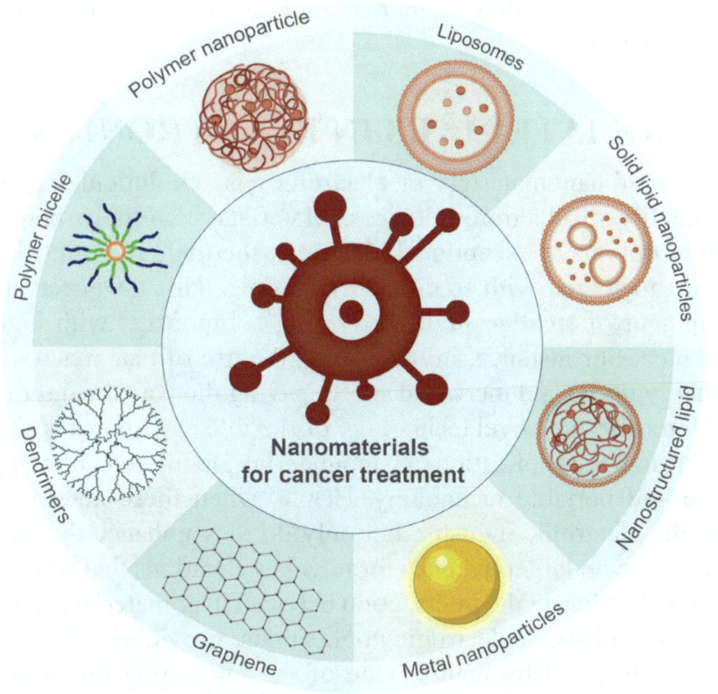

Figure 10.6. Nanomaterials in cancer treatment innovations.

Source: https://media.springernature.com/lw1200/springer-static/image/art%3A10.1007%2Fs10565-024-09943-9/MediaObjects/10565_2024_9943_Fig1_HTML.jpg.

It is crucial, as emphasized in FORTINA et al. (2015), that physicians are always aware of what is happening with technology and how it might impact patient treatment. Moreover, the capacity of nanomaterials to revolutionize cancer treatment advancements carries over to individualized treatment, which is directed at the genetic profile of the tumor in a patient. This personalized approach is essential in light of cancer heterogeneity and variability in response to standard therapies. With the help of nanotechnology, researchers are trying to read the unique molecular signature of the tumor and develop personalized therapy that addresses specific pathways exploited by cancer for its growth. Not only does this improve the likelihood of treatment, but it also minimizes the likelihood of resistance that typically plagues conventional therapeutic strategies. As nanomaterial advancements yet continue to uncover, they mark a new beginning for paradigms in oncological treatment, attested to by their ability to amalgamate multi-faceted strategies towards combat against this versatile disease, as in healthcare innovation future discourse (Debackere et al., 2025) (Figure 10.6).

10.3. NANOMATERIALS IN ELECTRONICS

Increased use of nanomaterials in electronics is revolutionizing the design and functionality of electronic devices. Materials such as carbon nanotubes and quantum dots have exceptional electrical, thermal, and optical properties that are hard to match with traditional materials. This advancement enables the development of smaller, high-performing components with higher levels of performance. For instance, in transistors, the use of nanomaterials leads to reduced energy usage and increased speed, paving the way for future computer systems. Moreover, the novel technology of flexible electronics is significantly improved through the application of nanomaterials, facilitating new applications in wearable and portable technology. Hence, when these new materials are utilized by the electronic industry, not only do they enhance the performance of devices but create larger avenues for novel applications that were deemed to be impossible. Exploring the intersection between nanomaterials and biological systems further enhances electronic applications, providing glimpses towards future designs inspired by nature. One of the interesting possibilities is the use of plant-derived components with electronic systems. Research highlights the potential of electrically modified plants to create functional electronic devices, such as resistors and capacitors, by incorporating nanoparticles or conducting polymers. Root morphological plant development and root interaction present a new scaffold for complex computation like that required in conventional electronics. This innovative method may lead to the creation of bio-hybrid systems with the living organisms serving as aid to processing and passing on information, thus creating an inter-disciplinary biologists-engineers-computer scientists network engaged in the development of living electronic devices with a reflection of syntheses of biological and technology paradigms (Adamatzky et al., 2017). The corresponding accelerating growth in nanomaterials not only enhances the performance of isolated electronic devices but also signals a direction towards greener and more sustainable technologies. Using nanotechnology, companies are able to create low-energy consumption, lightweight devices with minimal environmental impact. In addition, the ability to create the devices out of renewable materials, such as plant materials, signifies an overwhelming tendency towards waste minimization and environmental sustainability of electronics. With the escalating need for cleaner technology, integration of nanomaterials is an attractive option for electronics firms to lead the way in green technology while enhancing the performance of products. This double imperative for sustainability and innovation will likely set the course for the future of nanomaterials in electronics, ushering in an era both of technological advancement and of ecological consciousness (Figure 10.7).

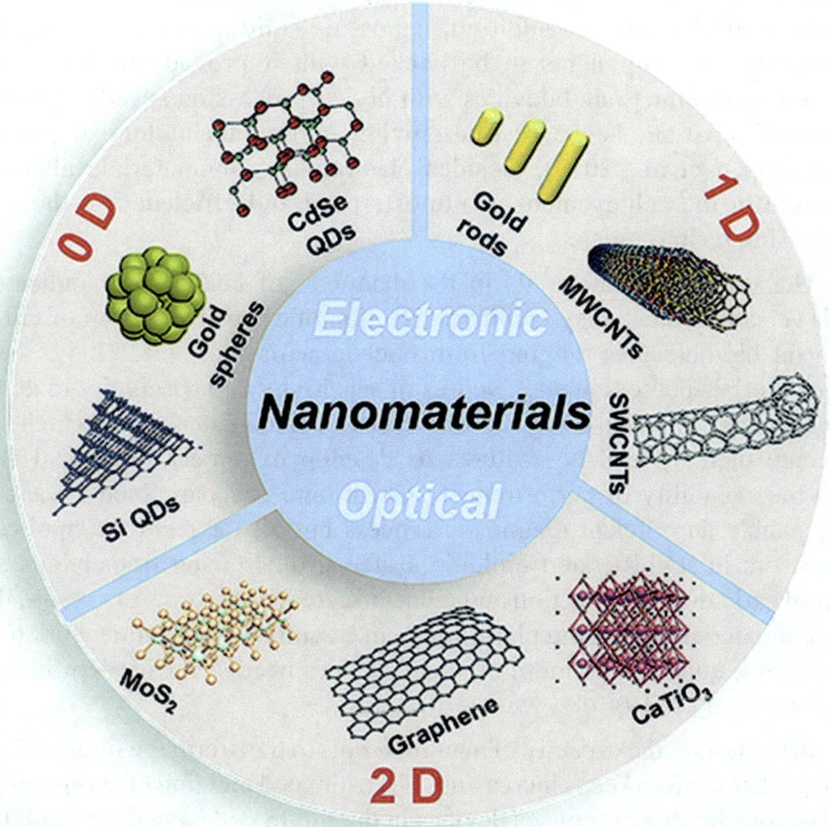

Figure 10.7. Nanomaterials in electronics.

Source: https://pubs.rsc.org/image/article/2020/NR/d0nr04156d/d0nr04156d-f1_hi-res.gif.

10.3.1. Development of Smaller and Faster Devices

The relentless race for smaller and faster devices has spurred revolutionary advances in nanotechnology, primarily through the inventive manipulation of nanomaterials. These materials, whose physical and chemical attributes at the nanoscale differ from those at the bulk scale, enable the device to be miniaturized as well as optimized. For instance, semiconducting quantum dots and nanowires offer superior functionality essential for the development of future electronic devices. The emergent properties at lower dimensions result allow designers to control matter and energy with unprecedented precision, giving rise to innovations based on heterogeneous composition and structure of elements. In addition, nanoscale solid systems' self-organizing nature allows components to be packed with improved efficiency, finally enabling more integrated devices with higher processing speeds and reduced power dissipation, hence revolutionizing current technological paradigms (Meyyappan et al., 2013). Besides electronics, nanomaterials also play a key role in the achievement of utmost speed and efficiency in the case of biosensing technologies.

The use of nanomaterials in the diagnosis of health is an indication of their versatility since they facilitate detection of a broad spectrum of clinically relevant biomolecules ranging from nucleic acids to toxins. This is because nanomaterials possess innate features of sensitivity and specificity in detection mechanisms. By combining various types of nanomaterials, researchers can leverage their synergistic features to develop extremely functional sensors with the capability of performing fast, real-time analysis. Such advances not only enable an efficient diagnostic process but also expand the applicability of sensors in sophisticated biologic systems, thus taking tremendous strides toward early disease detection and point-of-care diagnostics. As a result, the use of nanomaterials in sensing platforms is an essential line of future work towards developing quicker and more efficient devices needed for modern healthcare (Altintas et al., 2019).

In the future, the capacity of nanomaterials to transform the future of smaller and quicker devices keeps increasing. The estimated melding of computer power and nanotechnology promises devices in the future will have the potential to be interfaced with artificial intelligence, enabling more functionality in smaller form factors. New technologies such as flexible electronics and miniature sensors will enable technology to be molded into specific applications from wearable medical monitoring to more advanced communication devices. In addition, ongoing studies of nanomaterials' scalability and self-assembly pinpoint the promise of future large-scale production of sophisticated nanoscale devices. Finally, the trajectory of development for the materials not only points to their vital role in new technologies but also raises significant questions regarding sustainability, ethics, and the possibility of an ultra-tiny, high-speed device-driven future (Meyyappan et al., 2013).

10.3.2. Role in Renewable Energy Technologies

New nanomaterials play an important role in developing renewable energy technology, particularly enhancing the performance and efficiency of energy storage and conversion devices. Nanostructured materials, specifically, have enhanced characteristics that improve photovoltaic cell performances. Their increased conductivity and surface area allow for greater charge carrier mobility, which in turn provides higher energy conversion efficiencies. With rising demand for renewable energy sources, the use of such advanced materials in solar technologies is an emerging trend.

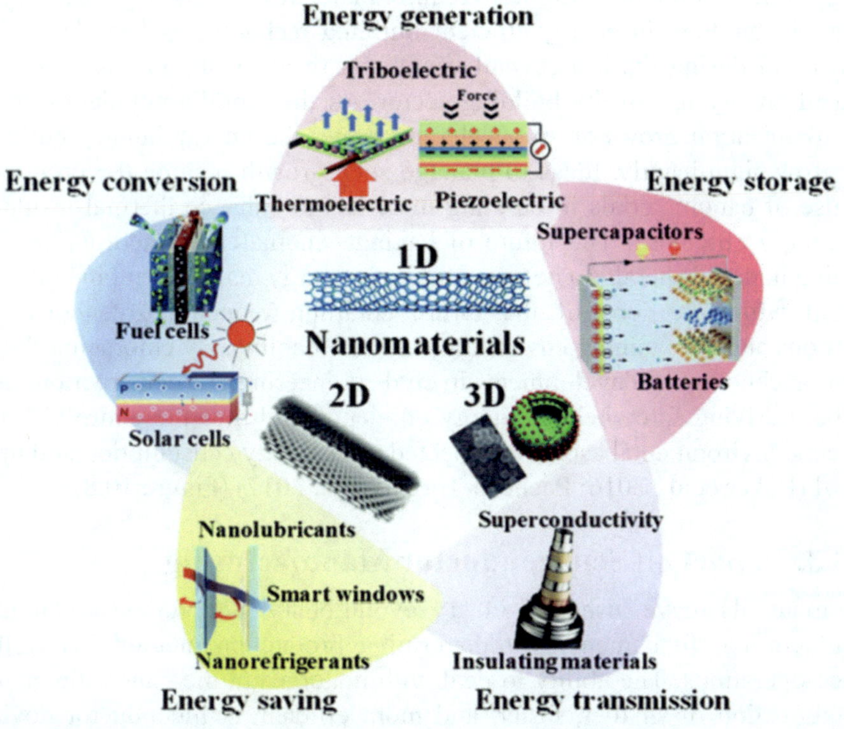

Figure 10.8. Nanomaterials role in renewable energy technologies.

Source: https://media.springernature.com/lw685/springer-static/image/chp%3A10.1007%2F978-981-16-1888-8_7/MediaObjects/511627_1_En_7_Fig3_HTML.png.

The quality of nanomaterials to improve energy harvesting and storage systems is equally essential to address global energy challenges, broadening the boundaries of what can be achieved with renewable sources and paving the way for more resilient energy grids. Apart from solar energy use, nanomaterials are also a crucial contribution to the efficiency of energy storage devices such as supercapacitors and batteries. This is due to the utilization of nanostructured electrodes, which contribute to better electron and ion transport. Enhanced properties include higher energy density, minimized charging time, and longer life cycle, which are significant parameters for addressing the intermittency associated with renewable energy resources such as wind and sunlight. With rising global populations, the energy demand explosion makes essential the need for energy storage technology to offset supply. Thus, nanomaterials development puts the renewable energy sector in a position to overcome existing energy storage limitations to future requirements in sustainability. Besides, use of nanotechnology in energy-efficient building technologies has also gained momentum during the last several years as a result of the pressing need for reduced energy use in the building sector. As the world's population grows and urbanization grows at an accelerating rate, the energy-hungry buildings will grow significantly. Ideas to manage such growth include the production and use of nanomaterials in building materials to enhance thermal insulation and energy efficiency. The future of having nanomaterials incorporated into building materials heralds a new revolution in energy management in buildings, aligned with directions of the future entailing reducing greenhouse gas emissions and maintaining environmental sustainability. By embracing the use of nanotechnological developments in civil engineering, the construction sector can be a driving force behind energy efficiency and thus the entire vision of reducing environmental aspects connected with energy consumption and urban sprawl (Baker et al., 2016; Pacheco-Torgal et al., 2017) (Figure 10.8).

10.3.3. Impact on Semiconductor Manufacturing

Nanomaterials have ushered in a revolutionary era for semiconductor manufacturing, fundamentally transforming production methods as well as device operations. The ability to deal with materials at the nanoscale enables the fabrication of smaller, faster, and more efficient semiconductor devices. Miniaturization is imperative because it forms the foundation upon which the relentless pursuit of heightened levels of integration in electronic devices eventually results in higher computation capacity and energy efficiency. Also, nanotechnology innovations allow for the development of high-quality semiconductor materials with superior electronic properties required in the development of next-generation technologies. As the industry is in a quest to uncover these nanoscale innovations, innovation arising from the same is poised to transform the microelectronics arena, and nanomaterials find themselves at the epicenter of semiconductor innovation. The coordination between researchers, engineers, and industry players is central to the proper utilization of these innovations and transforming them into viable applications. As semiconductor manufacturing evolves, the use of nanomaterials not only affects the design of the products but also the supply chain as a whole. Production of nanostructured materials tends to be facilitated by specialized equipment and procedures, prompting producers to invest in cutting-edge technologies and procedures. For instance, bottom-up methods like CVD and ALD enable control of material properties with precision, which is increasingly critical in the fabrication of devices with strict specifications. Scaling up these processes and maintaining reliability and economic feasibility is posed by the nanomaterials' complex nature.

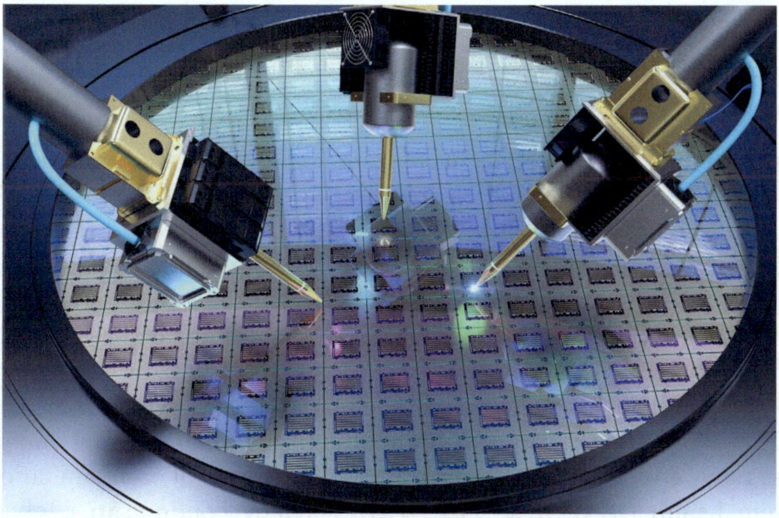

Figure 10.9. Nanomaterials Impact on semiconductor manufacturing.

Source: https://www.technology-innovators.com/wp-content/uploads/2023/05/Advanced-Materials-and-Nanotechnology-in-Manufacturing-min.jpg.

Consequently, business managers must balance innovation and pragmatism, such as the levels of yield and the production cost, as they try to introduce nanomaterials in various scales within the firm. These innovations will ultimately shape the competitiveness of semiconductor companies within a dynamic market environment that requires advanced materials (Lakhtakia et al., 2005). Beyond technological advancements, the application of nanomaterials in semiconductor manufacturing also has far-reaching sustainability and ethical practices implications. With growing environmental pressures alongside fast-emerging technologies, industry stakeholders must account for the lifecycle impacts of nanomaterials, such as extraction and post-production waste disposal. Green production models and materials play central roles in balancing semiconductor manufacturing with overarching sustainability goals. Managing ethical concerns also involves managing public attitudes toward and regulatory frameworks for nanotechnology, which can have far-reaching impacts on market acceptability. Furthermore, interdisciplinary collaboration among scientists, engineers, policymakers, and ethicists plays a key role in defining guidelines for responsible research and development. As the semiconductor industry advances toward a future enriched by nanomaterials, so too must it adopt practices that defend environmental integrity and foster public trust in technological innovation (Figure 10.9).

10.4. ENVIRONMENTAL APPLICATIONS OF NANOMATERIALS

The application of nanomaterials in environmental sustainability is increasingly perceived to have the potential to solve some of the most serious ecological problems of the world. A new trend is the development of nanosensors and bionanosensors, which enable advanced environmental sensing and characterization. These technologies enable real-time monitoring of toxic substances and pollutants with high effectiveness in enhancing the ability to control and prevent environmental risks. Research from the NINE congress focuses on heightened attention to such applications, suggesting that the integration of nanotechnology will revolutionize environmental monitoring processes by providing in-depth insights on levels of contamination and ecosystem health (Bavasso et al., 2016). As they continue to evolve, they may encourage more anticipatory strategies in pollution management, hence fostering a healthier environment while also endorsing global efforts towards sustainability. Nanomaterials are also essential in cutting-edge remediation technologies for cleaning up contaminated environments. Emerging nanostructured materials, such as engineered nanoparticles, have been found to be very efficient in the remediation of toxic pollutants and heavy metals in terrestrial and aquatic ecosystems. These materials can be engineered at the molecular level to selectively target selected pollutants, thereby making their remediation more effective than traditional remediation technologies. This ability is promising for the future where nanotechnology-based remedies can accelerate clean-up activities and reduce the overall environmental footprint of industrial activities (Bavasso et al., 2016). The ongoing studies on these materials hold promise for a paradigm shift towards merging nanotechnology with environmental remediation efforts, keeping with global efforts at sustainability and environmental health. However, nanomaterial use for environmental clean-up comes with its set of issues. Regulatory approaches must adapt in order to handle the unique characteristics and potential hazards of nanoscale materials, particularly in the areas of public health and environmental protection. Current literature highlights greater risk assessment methods and combined regulatory oversight to guarantee that the benefits of nanotechnology are realized without sacrificing safety. Public participation in the regulation process can also facilitate improved transparency and acceptance for the technologies. As science accelerates, it will be necessary to balance the promise of nanotechnology with strong safeguards in forging environmental uses while preventing possible adverse effects. Through ongoing collaboration between scientists, policymakers, and the public, an anticipatory approach can be forged so as to achieve the full potential of nanomaterials for environmental sustainability.

10.4.1. Water Purification Technologies

The quest for efficient water purification technologies has been brought into the spotlight owing to increased worldwide water scarcity and pollution. Conventional treatments, including sedimentation and chemical disinfection, are unable to handle complicated contaminants, and hence alternative innovative solutions must be explored. The nanomaterial-based emerging technologies, specifically, have been shown to offer remarkable capacity in the improvement of purification activities. Nanomaterials possess an exceedingly high surface area-to-volume ratio and specially enhanced reactivity, which make them highly effective in the removal of contaminants, including heavy metals, pathogens, and organic compounds. Experiments have determined that nanotechnology can be tailored to target specifically desired impurities, not just making water treatment systems more effective but also more efficient. This shift toward the introduction of new methods of purification is less a trend and more a natural evolution to ensure clean drinking water in an increasingly contaminated world. Specifically, water stabilization ponds (WSPs) have been updated through the addition of nanomaterials, addressing some of the deficiencies that have long plagued their effectiveness. WSPs are intriguing advantages such as low operating costs and simplicity but are confronted with meeting stringent discharge constraints and treating new pollutants. The incorporation of nanotechnology in WSPs can enhance pollutant removal efficiency and reduce greenhouse gas emissions—two pillars in sustainable wastewater management. As recent research has pointed out, new treatment technologies that can cope with dynamic changes in wastewater properties and environmental conditions are much needed. With the incorporation of advanced models founded on nanomaterial behavior, WSPs can continue to serve as a viable alternative in modern wastewater treatment, ultimately ensuring improved public health outcomes and environmental responsibility (Goethals et al., 2020). Moreover, the role of nanotechnology extends beyond treatment to regulating agricultural activities impacting water quality. The population growth and ensuing agricultural demands have led to increased use of pesticides and fertilizers, which in turn have some tendency to infiltrate groundwater systems. Remediation is thus essential in order to counteract this type of contamination. Nanomaterials have been promising in halting the progress of pollutants as well as tracking contaminants at trace levels. Recent studies have shown that specific nanomaterials can be employed to create targeted interventions, enhancing soil and water conditions. Notably, even though there has been controversy regarding the toxic impacts of certain nanomaterials, overall advantage in environmental management and purification techniques emphasizes their growing significance in next-generation water purification technologies (Bacchini et al., 2013) (Figure 10.10).

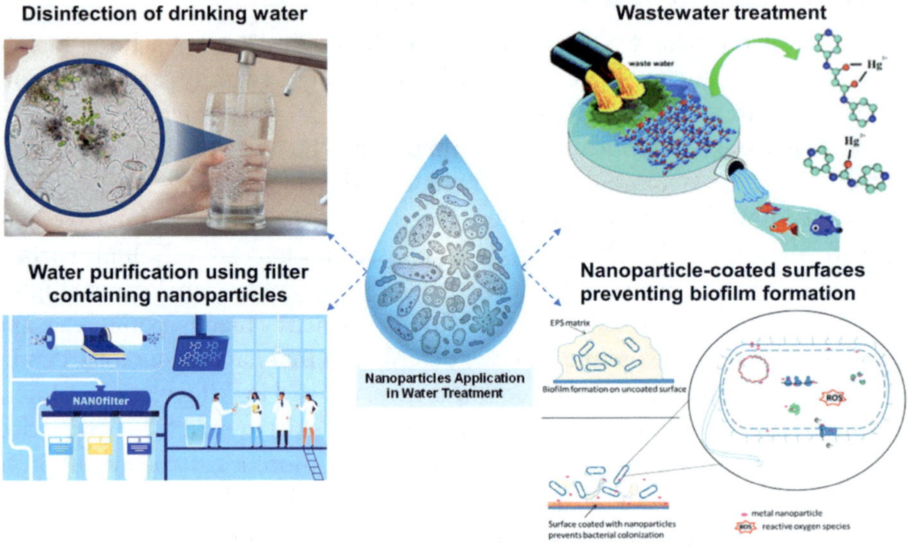

Figure 10.10. Nanomaterials in water purification technologies.

Source: https://www.frontiersin.org/files/Articles/1427843/fnano-06-1427843-HTML/image_m/fnano-06-1427843-g002.jpg.

10.4.2. Air Quality Improvement Methods

Subduing air quality deterioration has become imperative in today's environmental management, which requires new technologies to regulate pollutants suitably. Out of the various methodologies, the application of nanomaterials provides a revolutionary approach. Nanomaterials, due to their unique physicochemical properties, can be designed to adsorb and detoxify harmful airborne contaminants. Nanoscale adsorbent technology, such as carbon-based materials and metal oxides, has proven to be highly effective in adsorbing volatile organic compounds (VOCs) and particulate matter (PM). As urban areas struggle with rising levels of pollution, use of such advanced materials on a citywide scale can not only enhance air quality but also lead to public health improvements, ushering in a new era of environmental security and sustainability. Another possible avenue towards improving air quality is the inclusion of nanotechnology in filters. Traditional air purifying technology lacks the ability to eliminate ultrafine particles and microorganisms. But recent developments in nanofiber technology have given rise to filters with high filtration efficacy. The nanofiber membranes can be designed with high surface areas and pore size tunability, enabling them to target specific contaminants, like viruses and bacteria. Furthermore, functionalized nanomaterials can be incorporated to enhance the antimicrobial performance of these filter systems, delivering healthier indoor air. As urban areas transition to greener living, such systems are a two-pronged advantage of purer air and reduced health hazard because of airborne transmission of disease. Besides the above practices, the development of smart nanomaterials for real-time air quality monitoring is an elementary trend that will revolutionize environmental health. Such materials are very sensitive in the detection of some air pollutants and may provide real-time results regarding conditions of air quality. For instance, nanoconductors and nanosensors may be used in wearable devices or monitoring sites for the purpose of information provision to citizens and policymakers to make decisions regarding management of air quality based on knowledge. The interconnection of this technology can facilitate rapid identification of pollution sources and guide targeted interventions. Together, these advances not only aim to minimize air pollution but also emphasize the critical role of nanomaterials towards long-term environmental improvement and public health interventions in urban life (Bacon et al., 2011; Grenha et al., 2020).

10.4.3. Soil Remediation Techniques

Soil contamination has become a serious environmental issue due to industry, agriculture, and urbanization. Traditional remediation processes, such as excavation and phytoremediation, usually do not work well in extremely contaminated or complicated sites. With researchers looking for an alternative, nanotechnology seems to be a good option for contaminated soil remediation. Nanoscale material use has the potential to enhance traditional remediation by increasing availability of contaminants and facilitating their degradation. Specifically, these materials will be capable of interacting at the molecular level with pollutants and soil, thus creating more effective treatments which not only provide remediation but sustainable soil management practices. This integration of nanotechnology with existing remediation technologies brings with it promising prospects for the future, marking a new era for environmental remediation. Nanomaterials show great potential to enhance the effectiveness of soil remediation owing to their unique characteristics of high surface area and reactivity. As an instance, nano-zero-valent iron (nZVI) has shown great potential as a material owing to its ability to reduce a wide range of pollutants, such as heavy metals and organic pollutants, through chemical reduction to non-toxic forms. The present overview of the applications of nanotechnologies in agro-environmental science highlights the advantages of these materials, notably in improving soil quality and the detection of contaminants. These developments have the potential to directly impact higher agricultural productivity and sustainable land use. Not only does the application of nanomaterials detoxify polluted areas, but it also lowers the toxicological effects that are normally associated with conventional remediation techniques, hence paving the way for a safer agricultural practice and a cleaner environment (Bacchini et al., 2013). The application of nanomaterials in soil remediation also offers potential for enhanced monitoring and assessment of soil quality. Advanced nanosensors can provide real-time estimations of concentrations of contaminants and soil condition, enabling timely interventions to mitigate the hazards of pollution. As emphasized by the study, residual and rising soil contaminants have deep food security and public health concerns, which necessitate the institution of robust monitoring technologies. Through integrating the potential of nanotechnology with traditional remediation technologies, stakeholders can more effectively manage contaminated sites and meet both environmental and health demands. Thus, future trends in soil remediation are still moving towards a paradigm shift through ongoing improvements in nanomaterials, which hold promise for more effective and more sustainable solutions to environmental management.

10.5. CHALLENGES IN NANOMATERIALS RESEARCH

Research work in the field of nanomaterials entails a collection of unique challenges, such as those experienced in environmental applications. Although increasing evidence indicates that nanotechnology can provide solutions for environmental issues, scientists face regulatory regimes and safety assessments difficulties. The development of nanosensors and bionanosensors to detect the environment is an example, with a wide need for testing and verification prior to use on a wide scope. These obstacles are not only technological but are also a consequence of the complexities involved in implementing nanotechnology in existing environmental policy. The challenge of ensuring safe application without losing benefits represents not only a science challenge but a socio-political challenge, as regulatory agencies and the public wonder about the innovations that arise from this research arena. These concerns put the topic of sustainable innovations in nanotechnology into perspective, emphasizing the necessity of adequate evaluation as evidenced in Bavasso et al. (2016). The production technologies and processes of nanomaterials also raise significant concerns. As scientists drive the mass production of nanostructured materials, ensuring quality and consistency and minimizing environmental impact becomes a delicate balancing act.

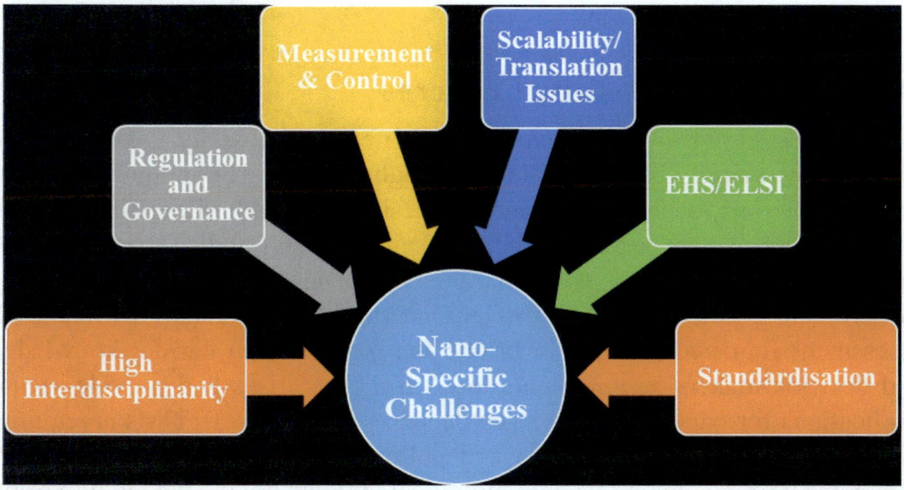

Figure 10.11. Challenges in nanomaterials research.

Source: https://www.researchgate.net/publication/349484950/figure/fig2/AS:993724469354499@1613933784826/Nanotechnology-Specific-Challenges.ppm.

The complexity in synthesizing nanomaterials, both chemically and physically, is aggravated in transferring from laboratory to industrial level. Over the innate issue of reproducibility, the new science has to deal with the economic matters of production technology. This demands fresh thinking that not only optimizes yield but makes sustainable methods possible. With increasing industries venturing into nanotechnology, there will be an emphasis on cleaner production processes as the solution in moving past the present shortcomings and paving the way for future success in this field, as seen through analysis as put forth in Bavasso et al. (2016). Moreover, the multimodal behavior of nanocarriers, especially in biomedicine such as pulmonary delivery of drugs, says a lot about how difficult it is to engineer effective nanomaterials. As new approaches are being employed to optimize therapeutic efficacy, researchers must overcome obstacles of biodistribution, toxicity, and long-term stability. For instance, precision needed in surface engineering—material structure manipulation or addition of targeted ligands—requires understanding of the interaction on a nanoscale. This level of complexity poses challenges to regulatory assessment and compound the delay in advancing clinically impactful applications. Further, appending these multifunctional nanocarriers to specific therapeutic targets requires sustained multidisciplinary research efforts. The constraints of incorporating these nanomaterials are at last imposing rigorous research frameworks, drawing lessons from other sciences to ensure safe and effective utilization under real-world conditions bounded by future demands considered in (Figure 10.11).

10.5.1. Safety and Toxicity Concerns

With the development of nanomaterials, safety and toxicity concerns have also become priority issues needing thorough scrutiny. The unique properties of nanomaterials, such as their increased surface area and reactivity, could potentially pose unexpected hazards to human health and the environment. Environmental impact assessments are increasingly becoming key tools for evaluating potential hazards of engineered nanomaterials (ENMs). Risk assessment frameworks need to bridge the gap between current knowledge and decision-makers' requirements to control these risks and provide secure applications for nanotechnology. The deeper exploration of ENMs is required in order to inform best practices and policy regulations since the complex engagement of nanomaterials with biological systems has the potential to bring about varied effects, which are not fully understood (Bilal et al., 2017). Thus, addressing safety concerns with the aid of robust regulatory frameworks is imperative for the safe development of nanotechnology. The potential toxicity of nanomaterials, particularly in biomedicine, is posing significant challenges with the expansion of the nano-medicine field. For instance, although gold nanoparticles have high potential for the targeted delivery of drugs, their safety profile has been controversial. Their effect on human cells has provided conflicting evidence, demonstrating that there is still no agreement among the scientific community regarding their potential risk. Parameters like particle size, shape, and surface chemistry are key factors influencing interactions between nanoparticles and biological systems and therefore toxicity outcomes (Cametti et al., 2015). Such inconsistency calls for the need for standardized experimental models to be employed in assessing the safety of nanomaterials effectively. It is important to know these variables in detail to ensure the secure integration of nanomaterials into therapeutic applications and reduce any potential risk due to their utilization. Addressing the safety and toxicity concerns associated with nanomaterials is vital to achieve public confidence and acceptability of technological advancements. The concern of potential health risks may hinder the application of nanotechnology in many areas, from medicine, electronics, to environmental science. To rectify such challenges, effective communication of risks and benefits to stakeholders is crucial. Encouraging the public, regulators, and industry to discuss safety protocols can not only enhance transparency but also allow the development of well-informed policies that guide the appropriate use of nanomaterials. Continued research into toxicological profiles of nanoparticles remains essential in developing comprehensive safety protocols. With the future trends of nanotechnology unfolding, safety will be at the forefront to ensure that the technological breakthroughs are safe and useful for use by society.

10.5.2. Regulatory and Ethical Considerations

The regulatory context of nanomaterials is among the grand challenges encountered in handling such materials, particularly given that they are being used in a variety of applications, from medicine to environmental clean-up. The fast-paced development of nanotechnology tends to leave regulatory mechanisms in the lurch with their development, ensuring safety and effectiveness. Therefore, the regulatory authorities have to take a proactive approach, retaining aspects of the precautionary principle. This principle is meant to protect public health and the environment from future hazards arising from the lack of full scientific certainty. Application of this principle needs to be addressed with clear observation of the seriousness of potential harm, the level of available evidence, and the nature of actions employed in regulation. Nevertheless, as observed in various legislative climates, application of this principle may become uncoordinated and generate regulatory uncertainty that renders crisp decision-making challenging and in the end affect public confidence in nanomaterial technologies (Ahteensuu et al., 2016). Ethical concerns also play a crucial role in the regulation of nanomaterials, particularly in relation to how they intersect with public health and environmental sustainability. The ethics of applying nanotechnology involve close consideration of the potential effects on society as a whole, including concerns about equity and access. For instance, healthcare applications of nanobiotechnology hold out promises for revolutionary health solutions but may exacerbate health inequalities if they are not able to be delivered to all groups. Secondly, the employment of nanomaterials must be complemented by the rigorous risk assessments that weigh likely benefits against ethics issues to do with human and environmental well-being. As regulatory bodies and institutions shift towards fully integrating these ethics matters, they must also ensure there is public debate that is informed about the risks and applications of nanotechnology (Anne-Katrin et al., 2008). Lastly, international cooperation must be employed to develop regulatory frameworks that accommodate the growing complexity of nanomaterials and their applications. Since nanotechnology research and development are worldwide efforts, international harmonization of regulation is essential to ensuring safety and fostering innovation. Collective action can create the sharing of best practices, collaboration on mutual ethical issues, and a unified regulatory system capturing the particularities of different national environments. Such a partnership must also involve engagement with stakeholders like scientists, policymakers, and the public in order to guide regulatory decisions to consider local issues and values. Lastly, a robust nanomaterials regulatory framework must balance innovation with health and ethical protection and with setting a foundation for sustainable development in the industry (Anne-Katrin et al., 2008).

10.5.3. Technical Limitations in Production

Production of nanomaterials is the frontier in materials science, with promises of enhanced performance in a wide range of industries. However, significant technical limitations inhibit widespread application. Chief among these is that techniques employed to produce nanomaterials often have scalability issues. Techniques effective on a laboratory scale, like sol-gel synthesis or chemical vapor deposition, often struggle to be scaled up to industrial scales. This inconsistency generates inconsistent material properties, which can adversely impact performance in practical applications. In addition, the complexity of the synthesis processes typically leads to increased production costs, additional limiting economic feasibility in everyday use in most industries. Hence, the development of simplified manufacturing processes that ensure efficiency at the same time ensuring scalability forms an essential challenge for scientists and industries as a whole within the nanomaterials context. Yet another pressing technical constraint is the characterization and quality control of nanomaterials. Due to their small size and unique properties, traditional analytical techniques are found to be ineffective in adequately assessing the quality of the materials. For instance, procedures such as scanning electron microscopy (SEM) or atomic force microscopy (AFM) may not provide the comprehensive information needed for valid material verification. This lack can lead to differences between real and expected performance in products ranging from electronics to biomedicine. Besides, the information on nanomaterials long-term stability and compatibility with other materials under complicated conditions is still evolving. Therefore, the establishment of standardized procedures for the characterization of nanomaterials is vital in the role of delivering constant quality and reliability during production and application, an exercise that is still one of the problematic areas of this subject. Energy efficiency during nanomaterial production also represents a significant challenge, echoing the challenge inherent within related industries. Conventional energy technologies used in the production of nanomaterials tend to be wasteful, with the consequence of resources going to waste. As can be seen in recent studies, the desalination industry, for instance, has been plagued by energy inefficiencies arising from compromised assessment processes to ascertain energy input grades (Burhan et al., 2019). These problems resonate through the manufacturing process for nanomaterials, where more accurate calculations in energy become apparent. If such inefficiencies are not corrected, the production process becomes economically unviable, particularly in the light of sustainable processes becoming more in demand by consumers as well as regulative bodies. Overall, enhancing the energy efficiency of nanomaterials production is imperative in order to align with the future agenda of sustainability, finding harmony with the need for innovation in the entire production process (Burhan et al., 2019).

10.6. FUTURE TRENDS IN NANOMATERIALS

Future trends for nanomaterials are more influenced by the necessity for sustainable production processes and high-efficiency pursuit. Conventional wet chemical synthesis processes, though they yield greater amounts, usually consist of dangerous chemicals with environmental issues. Alternatively, newer approaches such as Laser Ablation Synthesis in Solution (LASiS) have come into view and, with their capability to reduce toxic byproducts, show promise as a green process. Recent advances in LASiS technology have focused on scaling up the production rate through a move away from batch-processing to continuous production systems, thus avoiding the drawbacks of low yields that dominate traditional processes. For example, the recent LASiS system that used a low power Nd:YAG laser was implemented in a bid to capture production rates as high as 17± 0.7 mg/hr, which is a serious step above earlier technologies and marking the move towards sustainable and efficient production of nanomaterials (Brabazon et al., 2020). Nanomaterials research also heavily intersects with the public awareness of the respective technologies. Recent polls and focus groups raise to the surface that while there is growing curiosity about nanotechnology, its use and impact are still unfamiliar to most people. As society becomes more familiar with nanotechnology and synthetic biology, there is a need to identify both perceived benefits and corresponding risks that the public perceives. These beliefs not only direct future research and development strategies but also assist policymakers in the design of regulations and public outreach programs to establish public confidence. Public debate about the advantages of nanomaterials, compared to their possible risks, will definitely inform future innovation in this rapidly expanding sector (N/A, 2008). Looking ahead, nanomaterials could transform industries from medicine to energy to electronics, and researchers already are exploring new uses capitalizing on their unique properties. In biomedicine, for instance, drug-delivery systems based on nanoparticles have shown enormous potential in optimizing therapeutic effects while minimizing side reactions. In addition, in the arena of energy, nanomaterials are being designed to make solar cells more efficient and allow for storage choices for energy, all towards obtaining clean energy sources. Also, accompanying those advancements, research into engineering nanomaterials will continue to allow them to be designed to their precise properties, further increasing their capabilities across industries. With the increasing incorporation of nanomaterials, the innovations brought about by the latter will not only address current challenges but also pave the way for future technological advancements that are harmonized with societal needs and environmental requirements.

10.6.1. Emerging Research Areas

One of the most significant future research topics in nanomaterials is their application in the medical field, namely drug delivery and diagnostics. The new properties of nanomaterials, such as their nanometer size and high surface area, can facilitate targeted delivery mechanisms that ensure the optimal effectiveness of treatments with reduced side effects. New research indicates the direction towards the application of nanoparticles to facilitate the controlled delivery of chemotherapy drugs that can dramatically improve patient outcomes. Furthermore, the progress in nano-diagnostics based on nanosensors and imaging modalities has opened up new avenues for the early diagnosis of disease. With these applications growing, they necessitate a reassessment of existing regulatory models to supply public health and environmental security, thereby highlighting the need for end-to-end notions of regulation of nanotechnology (Davies, 2008). This intersection of medicine and nanotechnology is a frontier replete with promise but plagued with issues that require careful navigating. Another research direction that is fascinating in nanomaterials involves environmental applications, particularly cleaning up pollution and clean energy. Nanomaterials possess the ability to adsorb and destroy contaminants, thereby providing effective solutions for cleaning water and air pollution. Recent research indicates that nanomaterials such as titanium dioxide and carbon nanotubes are being utilized in photocatalytic reactions to degrade poisonous chemicals, marking a significant development in environmental engineering. Similarly, the role of nanotechnology in energy production through the generation of efficient solar cells and batteries also triggers the pursuit of renewable energy sources. Its growing significance triggers investment and policy towards sustainable business and research use (Margon et al., 2025). Ultimately, the application of nanotechnology to environmental purposes is a double opportunity to respond to pressing environmental issues while guaranteeing sustainable development. In addition to medical and environmental strategies, future nanomaterials offer revolutionary potential in electronics and materials science. The constant downsizing of devices has created a demand for nanoscale components that enhance performance and efficiency. Research focuses on the development of nanoelectronics, which means the creation of smaller, faster, and better transistors, therefore driving computer technology. Furthermore, the use of nanomaterials to manufacture lightweight, yet stronger composites can create better-performing products, ranging from aerospace to consumer electronics. Since the competitive landscape of this sector continues to shift, the international research community is an area that demonstrates an impressive focus on effort in regions like Europe and North America with growing inputs from Asia, notably China and South Korea, reflecting the significance of cooperation on an international scale (Margon et al., 2025). The change in these new directions will definitely leave a lasting influence on subsequent innovations in other sectors.

10.6.2. Potential for Commercialization

The commercialization potential of the nanomaterials is growing into the limelight due to their tailored properties, which can be tuned for specific applications in specific industries. As the need for more excellent materials that improve product performance arises, significant advances in synthesis approaches have been seen. Most specifically, methods like Laser Ablation Synthesis in Solution (LASiS) present a green way of producing nanomaterials, which aligns with growing environmental needs. While traditional processes are generally more productive, they carry serious environmental drawbacks of toxic solvents and reagents. Therefore, advances driven by green manufacturing, such as the integration of low-cost and low-power laser systems with solvent flow systems, are a key step toward productivity improvement while concurrently attacking problems of environmental sustainability. This milestone is a pivotal step towards commercialization, which has the capability to replace current bottlenecks in the efficiency of production in the current methods, as demonstrated in new studies that have exhibited enhanced yield (Brabazon et al., 2020). In addition to enhanced modes of synthesis, the regulatory framework plays an important role in the commercialization of nanomaterials.

The evolution of regulatory frameworks, particularly in the European Union, indicates the necessity of resilient and flexible governance to maintain the rhythm of technological progress. The two observed trends—cooperative regulation during the 2000s and sector-based lawmaking in the example of cosmetics and biocides—indicate the complex regulatory environment for nanomaterials. The hybrid regime comes with a variety of challenges, particularly the possibility of incoherence and reactive governance as emerging threats present themselves. However, strict regulation can also facilitate innovation by offering a secure environment that encourages investment with the adherence to safety and ethical production standards. Hence, there is a need to maneuver such regulatory balances to realize the potential for commercialization of nanomaterials, so that they can be responsibly developed in accordance with the needs of the market (Ruggiu et al., 2015). Overall, commercialization viability of nanomaterials will be contingent upon a combination of new manufacturing methods, regulatory adaptability, and market feasibility. While the industry strives for increased efficiency and scalability, research must focus not only on improving manufacturing processes but also on collaborating with regulation agencies towards an integrated and supportive governance policy. Companies that have already invested in nanotechnology need to concentrate on these sectors so that their materials science research will find commercial payoffs. Apart from this, constant communication among researchers, manufacturers, and regulators will be needed to address safety concerns and foster consumer confidence. As this environment expands and evolves, the future of nanomaterials remains promising, with the potential to transform various industries, from healthcare to electronics, thereby solidifying its place in the market landscape and nurturing sustainable economic growth (Brabazon et al., 2020; Ruggiu et al., 2015).

10.6.3. Predictions for Market Growth

While the horizon of nanomaterials continues to evolve, market growth projections indicate a robust expansion driven by emerging developments across most sectors. The global market for nanomaterials is forecast to see rapid growth with advances in technology and increasing demand for high-performance materials, as per different industry reports. The applications of nanomaterials across different industries like medicine, electronics, and energy suggest a diversified approach in their utilization. As researchers identify new ways to leverage the unique properties of materials at the nanoscale, the marriage of scientific breakthrough and business viability becomes increasingly clear. Furthermore, key drivers such as environmental sustainability and the need for smarter resource management will spur quicker adoption, toward an energetic market poised to deliver in meeting challenges to come. The intersection of innovation and market forces will be the key driver in shaping this disruptive industry. As companies evolve to find room for new technologies, the potential for new applications of nanomaterials is greater than ever. Forecasted growth trends show major investment and growth will take place in key markets like healthcare and nanocomposites. The healthcare sector alone holds out the promise of transforming patient treatment by means of innovation in drug delivery systems and diagnostic devices that leverage nanoscale innovation. For instance, the use of nanomaterials in imaging devices has the potential for increased resolution and sensitivity, again driving market demand and facilitating rapid innovation in medical therapies. The construction sector also is considering the benefits of more resilient and lighter nanocomposites. As markets respond to this shift in technology, players must remain agile and informed in order to capitalize on new opportunities while also keeping away from potential problems. But there are no problems with the progress of the market to growth. Prospects are good, yet there is still a degree of uncertainty regarding regulatory policies, people's opinion, and ethical considerations for the use of nanomaterials. These must be managed carefully in order to allow for sustainable growth of the market. Additionally, the history of past predictions of technology also provides valuable lessons, for example, as pointed out in the reported differences in past weather forecasts in relation to laboratory medicine and its evolution (FORTINA et al., 2015). The above facts serve as a timely reminder that with future potential for market expansion in research projections, individuals should have a critical mindset. Stakeholders must factor in possible risks and remain adaptable in their strategies, understanding that the regulatory environment and scientific developments will ultimately decide market trends. As the industry of nanomaterials evolves, caution and foresight will be essential to tapping its full potential with solutions to any inherent risks.

SUMMARY

The pursuit of nanomaterials has brought amazing directions for innovation in various fields of study, including medicine, electronics, and environmental research. As has been demonstrated through this essay, the potential applications of nanomaterials are numerous and continue to expand as techniques of synthesis and product engineering advance. One of the primary trends is sustainable process development, such as Laser Ablation Synthesis in Solution (LASiS) use. The process is a significant contribution to nanomaterials production in an environmentally friendly manner by reducing toxic byproducts of standard chemical synthesis. Challenges in low yields nonetheless require ongoing work to counteract gains in efficiency and scalability of production, seen in recent advances in the integration of low-cost laser and solvent flow systems that have a potential productivity boost (Brabazon et al., 2020). Ultimately, to capture the full transformative potential of nanomaterials, the future relies on achieving a balance of innovation and sustainability. Parallel with technological advance, of equal or even greater significance is the regulatory context of nanomaterials and applications. When new nanotechnology evolves, use of the precautionary principle becomes increasingly significant. The precautionary principle is an approach to safeguarding public health and the environment against risks from novel substances, particularly where there are scientific uncertainties. A scrutiny of legislations implemented is essential in order to demonstrate contradiction in the implementation of this principle, and improved rules are stressed that ensure safe use of nanomaterials in all industries (Ahteensuu et al., 2016). Through demanding a stronger policy, stakeholders are able to catalyze innovation while ensuring risks are not taken, hence securing public trust in the use of nanotechnology. The emergent union of new synthesis methods and the creation of the requisite regulatory environments can propel the nanomaterials industry toward a brighter future. Greater coordination between scientists, policymakers, and business executives is required to map the path for this field that is undergoing rapid transformation. As the uses of the nanotechnologies are poised to increase manifold, the common aim should be to harvest the gains without diluting the safety controls. Informing stakeholders and the public about the risks and benefits of nanomaterials is likely to boost acceptance and promote research investment. By achieving both the thrilling promise of nanotechnology and a prudent path to regulation, society can ensure that future trends in nanomaterials not only propel technological innovations but also promote sustainable development and safety.

Summary of Key Points

The nanomaterial technology is still growing extremely rapidly with broad implications across numerous fields, particularly medicine and environmental science. The most interesting one among them is the interaction between cells and one-dimensional (1D) nanomaterials, e.g., silicon nanowires. The rational design of nanomaterials for nanomedicine relies on an understanding of this interaction. Recent studies successfully visually observed the invasion of amine-functionalized silicon nanowires into CHO-β cells through a physical membrane wrapping mechanism, which gave elementary insights into membrane mechanics, along with applications of the nanomaterials in targeted drug delivery systems. The study emphasizes the significance of sophisticated imaging technologies to investigate cellular reactions to nanostructures, paving the way for future advancements in clinical usage and therapeutics (Huang et al., 2017). Amid these developments, regulation of nanotechnology is a major priority area that is still in development. The rapid development of nanomaterials raises unique challenges to environmental and public health protection, necessitating reconsideration of existing legislation and regulations. Specialists suggest the adaptation of existing frameworks to ensure adequate control over nanotechnology's applications, advocating greater funding for risk research. Regulatory systems that include public participation are required to counterbalance sophistication resultant from nanomaterials and avoid being safe at the expense of stifling innovation. Putting emphasis on careful risk evaluation and transparent growth procedures, stakeholders stand in good stead to deal with nanotechnology's ethical concerns and societal challenges (Davies, 2008). Future growth trends for nanomaterials signal a transition away from traditional applications, and interdisciplinary research developing innovative solutions to global challenges. With the advent of nanotechnology, new technology is being discovered in uses such as renewable energy, water purification, and intricate manufacturing processes. The technologies are not only striving to maximize efficiency but also labor on reducing the impact on the environment, highlighting the complementary strengths of technological progress and sustainability. As science continues to uncover the promise of nanomaterials, a collaborative initiative will be needed to access these capabilities responsibly, aligning scientific breakthrough with ethical reflection and public good. Ongoing progress in the field holds the promise of breakthroughs of transformative potential that can help resolve essential problems in health, environmental sustainability, and beyond.

REFERENCES

1. Abbott, L., Gruère, G., & Narrod, C. (2025). Agricultural, food, and water nanotechnologies for the poor: Opportunities, constraints, and role of the Consultative Group on International Agricultural Research. https://core.ac.uk/download/pdf/6237698.pdf (accessed on 25th September 2025).

2. Adamatzky, A., Trewavas, A., Volkov, A. G., et al. (2017). Computers from plants we never made. Speculations. http://arxiv.org/abs/1702.08889 (accessed on 25th September 2025).

3. Ahteensuu, A., Belvèze, C., Cameron, D., Crawford-Brown, D., Eisenhardt, S., et al. (2016). Multi-case review of the application of the precautionary principle in European Union law and case law. https://core.ac.uk/download/42144488.pdf (accessed on 25th September 2025).

4. Altintas, Z., & Pirzada, M. (2019). Nanomaterials for healthcare biosensing applications. https://core.ac.uk/download/287885382.pdf (accessed on 25th September 2025).

5. Anurogo, D., Ikrar, T., & Parikesit, A. A. (2017). Bio nanomedicine: A "panacea" in medicine? https://media.neliti.com/media/publications/179998-EN-none.pdf (accessed on 25th September 2025).

6. Arce, M., Benito, A., Cerpa-Naranjo, A., et al. (2020). Synergistic antifungal study of PEGylated graphene oxides and copper nanoparticles against Candida albicans. https://core.ac.uk/download/323056246.pdf (accessed on 25th September 2025).

7. Bacchini, F., Greppi, G., Mura, S., et al. (2013). Advances of nanotechnology in agro-environmental studies. https://core.ac.uk/download/16749559.pdf (accessed on 25th September 2025).

8. Bacon, J., Cundill, A., Dale, P., et al. (2011). Contamination. https://core.ac.uk/download/384902.pdf (accessed on 25th September 2025).

9. Baker, L. A., Boyer, T. H., Drechsel, P., et al. (2016). Total value of phosphorus recovery. https://core.ac.uk/download/213082316.pdf (accessed on 25th September 2025).

10. Bavasso, I., Chianese, A., Di Palma, L., et al. (2016). Perspectives in nanotechnology-based innovative applications for the environment. https://core.ac.uk/download/54530773.pdf (accessed on 25th September 2025).

11. Bilal, M., Cohen, Y., Godwin, H. A., et al. (2017). Needs and challenges for assessing the environmental impacts of engineered nanomaterials (ENMs). https://core.ac.uk/download/323076183.pdf (accessed on 25th September 2025).

12. Bock, A. K., Gaisser, S., Hüsing, B., & Wagner, V. (2008). Nanomedicine–Drivers for development and possible impacts. https://core.ac.uk/download/38611572.pdf (accessed on 25th September 2025).
13. Brabazon, D., Foley, G., Freeland, B., et al. (2020). High-efficiency generation of nanomaterials via laser ablation synthesis in solution with *in-situ* diagnostics for closed-loop control. https://core.ac.uk/download/288473313.pdf (accessed on 25th September 2025).
14. Burhan, M., Ng, K., Shahzad, M. W., et al. (2019). Desalination processes' efficiency and future roadmap. https://core.ac.uk/download/287613538.pdf (accessed on 25th September 2025).
15. Cametti, C., Fratoddi, I., Russo, M. V., et al. (2015). How toxic are gold nanoparticles? The state-of-the-art. https://core.ac.uk/download/54248998.pdf (accessed on 25th September 2025).
16. Davies, J. C. (2008). Nanotechnology oversight: An agenda for the new administration. https://core.ac.uk/download/71353341.pdf (accessed on 25th September 2025).
17. Debackere, K., Monard, E., Raspoet, D., et al. (2025). Technologie en innovatie in Vlaanderen: Prioriteiten. https://core.ac.uk/download/pdf/6468460.pdf (accessed on 25th September 2025).
18. Fortina, P., Kricka, L. J., Park, J. Y., & Polsky, T. G. (2015). The future of laboratory medicine–A 2014 perspective. https://core.ac.uk/download/54509579.pdf (accessed on 25th September 2025).
19. Goethals, P., & Ho, L. T. (2020). Municipal wastewater treatment with pond technology: Historical review and future outlook.
20. Grenha, A., & Pontes, J. F. (2020). Multifunctional nanocarriers for lung drug delivery. https://core.ac.uk/download/304375181.pdf (accessed on 25th September 2025).
21. Huang, Y., McNear, K., & Yang, C. (2017). Understanding cellular internalization pathways of silicon nanowires. https://core.ac.uk/download/215928782.pdf (accessed on 25th September 2025).
22. Lakhtakia, A., Altmann, J., Arnall, A., Bagchi, B., Bainbridge, W. S., Bartlett, A., Baudrillard, J., et al. (2005). A map of the nanoworld: Sizing up the science, politics, and business of the infinitesimal. http://arxiv.org/abs/physics/0510209 (accessed on 25th September 2025).
23. Margon, D., Coccia, M., & Finardi, U. (2025). Research trends in nanotechnology studies across geo-economic areas. https://core.ac.uk/download/pdf/6438584.pdf (accessed on 25th September 2025).
24. Mason, E. (2018). Synthesis and photocatalytic performance of TiO_2-CNT and magnetized Fe_3O_4-TiO_2-CNT multifunctional hybrids: A

Pickering emulsion platform for organic degradation. https://core.ac.uk/download/211346833.pdf (accessed on 25th September 2025).

25. Meyyappan, M., Neyts, E. C., & Ostrikov, K. (2013). Plasma nanoscience: From nano-solids in plasmas to nano-plasmas in solids. http://arxiv.org/abs/1306.6711 (accessed on 25th September 2025).
26. N/A. (2008). Awareness of and attitudes toward nanotechnology and synthetic biology. https://core.ac.uk/download/71349451.pdf (accessed on 25th September 2025).
27. N/A. (2016). The Boston University Photonics Center Annual Report 2015–2016. https://open.bu.edu/bitstream/2144/29754/1/Photonics-Center-Annual-Report-2016-final.pdf (accessed on 25th September 2025).
28. Pacheco-Torgal, F. (2017). High tech startup creation for energy efficient built environment. https://core.ac.uk/download/76177771.pdf (accessed on 25th September 2025).
29. Ruggiu, D. (2015). The consolidation process of the EU regulatory framework on nanotechnologies: Within and beyond the EU case-by-case approach. https://core.ac.uk/download/53525097.pdf (accessed on 25th September 2025).

INDEX

C

F

G

H

N

O

P

Q

R